THE DERIVATION OF MATHEMATICS

Mastering Secondary School Mathematics

PAUL McNAMARA

Published in Australia in 2018 by Paul McNamara

Website: www.wcmpt.com

Email: dr.paul.mcnamara@gmail.com

National Library of Australia Cataloguing-in-Publication entry:

Author:	McNamara, Paul, 1952–
Title:	The Derivation of Mathematics : Mastering secondary school mathematics / Paul McNamara
ISBN:	9780995449602 (paperback)
Target Audience:	For secondary school age
Subjects:	Mathematics – Study and teaching (Secondary).
	Mathematics – Problems, exercises, etc.

Disclaimer

The author has made every effort to ensure the accuracy of the information within this book was correct at the time of publication. The author does not assume and hereby disclaims any liability to any party for any loss, damage or disruption caused by errors or omissions, whether such errors or omissions result from accident, negligence, or any other cause.

CONTENTS

ACKNOWLEDGEMENTS

I'm sure any author would agree that publishing a book involves many hours of hard work, and often a lot of assistance from others. In my case, I am grateful to the following people, who all contributed in some way to the successful publication of this book.

Firstly, I am indebted to my parents for their love and support. In particular, to my mother Iris for making things appear to be easy while instilling in me the belief that anything is possible if you are willing to put in time and effort. Also, my father Ben (who was a Mathematics teacher for more than 50 years) for conveying to me the importance of integrity, commitment and the benefits of developing an in-depth understanding of Mathematics, as well as his generosity in spending countless hours discussing mathematical concepts with me.

Likewise, I thank all my siblings who have enriched my life immensely and been very supportive throughout the writing of this book.

My heartfelt appreciation to my daughter Sally and son David for their understanding and support over the many years involved in writing this book (which has spanned a significant portion of their lives).

Thanks to my many friends and, in particular, my long-term friend and colleague Lloyd Williams. I am very appreciative of Lloyd's generous undertaking in reviewing the book from cover to cover and making suggestions to improve the logic, accessibility and readability of the mathematical concepts. Also, his willingness to translate the mathematical content into its final published form has been an invaluable contribution.

My sincere thanks to Kirsty Ogden from Epiphany Editing & Publishing who was instrumental in bringing this book to life. Her contribution and commitment to project managing my publishing journey has been inspirational. Kirsty's persistence, sensitivity and determination to make this book the best it could be while, at the same time, remaining calm, creative and supportive helped to ensure that my journey to final publication was as painless as it could possibly be.

Finally, this book would never have been completed without the unwavering support of my wife Helen over more than a five-year period. By any measure, Helen is a brilliant analyst and her skills were invaluable in ensuring the consistency and readability of the English and Mathematical content of this book. This book is a testament to her kindness, commitment, integrity and compassion for secondary school students everywhere, who are currently struggling to master the challenges of Mathematics.

PREFACE

When I first set out to write this book, my goal was not to make the many excellent secondary school Mathematics textbooks that are currently available, redundant. Rather, my aim has always been to create clarity and ignite a passion for Mathematics in school students, thereby bringing these books to life for them. My sincere desire is to demonstrate the undeniable beauty and simplicity of Mathematics and its applications to the world we experience around us.

So what originally inspired me to write this book? During my formative secondary school years, I developed an unshakeable belief that if Mathematics (and its practical applications to Physics) was approached the right way, it could be simple, logical and very easy to learn. My initial understanding of Mathematics developed at a moderate pace and was greatly assisted by many conversations I had with my father, who was a Mathematics teacher with a genuine love of his field. By the end of Year 10, I had conquered the basics of Newtonian Physics and Special Relativity. This achievement motivated me to focus on my studies during the remainder of secondary school.

At university, I pursued my growing interest in Pure and Applied Mathematics, as well as Physics, by enrolling in a science degree with the goal of developing the deepest possible understanding of these fields. Although I was only studying standard-level subjects, by the third year of my degree I was gaining good results and so was promoted to the Honours program in Physics. This led to me obtaining a BSc with a double major in Mathematics and Physics, and then, later, an Honours degree in Physics.

After my graduation, I decided to embark on a career in the emerging field of computing. I became a computer technician and, within my job, I applied basic logic and Physics to the manufacture and repair of computer hardware. After two years in manufacturing it became clear to me that, in order to create a successful career in computing, I would have to study for an Electrical Engineering degree (or equivalent). At this point, I returned to university as a mature-age student to study for a qualification that would support my fledgling computing career. When I applied for the Engineering course, the only job at the university I was eligible for was a full-time tutoring position in Mathematics which meant that I had to obtain a fourth year qualification in Pure Mathematics. Over the following two years, I studied this course part-time and was then accepted into a Masters of Pure Mathematics program.

I then decided to follow my love of Relativity and switched faculties to enrol in a Masters of Physics degree. After six months, the Physics faculty decided to upgrade my course to a PhD program which led to five years of tutoring work in Physics and my PhD degree (with a thesis in General Relativity). After this seven year detour, I returned to the corporate world and my former career in computing.

For the past 22 years, I have worked as a project manager, a program manager and a consultant. Throughout this period, my love of Mathematics and Physics has never diminished, and my hobby has been to continue to improve my understanding of the foundations of Mathematics and Physics.

Around 2001, I started to write a book on basic Physics for secondary school students. After four years, I had completed the book (although I didn't publish it). However, I still felt that I had left one major question unanswered: 'If Physics is the subject in which Mathematics is used to model simplified physical systems (e.g. describing an object falling due to gravity), then what do the basic mathematical symbols of Physics (i.e. 'm' for mass, 'l' for length and 't' for time) really mean?'

For the past 10 years, my spare time has been consumed with tapping into my in-depth understanding of the foundations of Mathematics and Arithmetic so I could write this textbook for secondary school students. I have designed this book to be a simple, logical and very easy-to-learn approach to Mathematics. My hope is that you will also find this 'missing' information about Arithmetic which I have explained in a logical learning sequence beneficial to mastering your secondary school Mathematics course; the absence of which was the source of my frustration and disappointment when I studied Mathematics as a teenager.

Paul McNamara

INTRODUCTION

For many years I had been disappointed and frustrated with what I considered to be an inefficient and ineffective learning system that traditional secondary school Mathematics teaching and textbooks provided to students. Although I'd attended four different high schools during my youth and was fortunate to have good Mathematics teachers at all of these schools (and I'd always achieved results beyond my own and others' expectations), I'd always felt that: 'there has to be a better way of learning Mathematics'.

Even earlier in my schooling, during my last two years of primary school, I was fortunate to study tennis with Charlie Hollis, who is one of the greatest Australian tennis coaches of all time. His fundamental premise for learning tennis was: 'if you can't learn to play tennis perfectly without a tennis ball, you will never be able to master tennis with a tennis ball'. With this strong focus on technique, Charlie claimed that anyone could master tennis within five years. My subsequent years of tennis coaching during high school left me with an unshakable conviction that his claim was well justified.

This practical experience with using the 'best approach' to achieve success in tennis set up an expectation in me that if I could only find the right approach to learning Mathematics at high school I would enjoy the same level of success. Although I tried several different approaches and adapted my study strategy to match the subject matter, real success eluded me. It wasn't until a decade later that I discovered a generic learning approach based on the premise that you needed to: 'start with the simplest area of a subject', 'attribute as much meaning in this simplest area as possible' and only then 'extend your learning to a broader area of that subject'. Over the past ten years, this approach applied to learning Mathematics became the driving force behind me writing this book and developing the Mastering Secondary School Mathematics (MSSM) Program.

The Mastering Secondary School Mathematics (MSSM) Program

This book explores and explains in detail the most basic and simple branch of Mathematics known as 'Arithmetic'. Based on the elementary number systems, Arithmetic is the process of counting, adding and multiplying numbers. Within this book, I will be introducing the concepts of basic Algebra and the first steps to Geometry. However, other branches of Mathematics such as Trigonometry, Coordinate Geometry and Calculus will not be addressed.

The aim of this book is to provide a program that will lead us to the most direct route to Mastering Secondary School Mathematics for students, parents and teachers. This mastering process involves developing an in-depth understanding of the foundations of secondary school Mathematics in the form of the elementary number systems of mathematics that we began learning in primary school. By studying the MSSM Program, students will achieve benefits not only in secondary school Mathematics, but in other secondary school subjects as well.

The MSSM Program for Arithmetic could also rightfully be called 'The Genesis of Mathematics' as it assumes readers recall some of their times tables and little else of what was taught to them in primary school. It provides the opportunity for secondary school students to start over again and learn Mathematics from the beginning, starting with the Natural Number System (the first of the four Elementary Number Systems).

The process presented in this book differs from a typical secondary school Mathematics syllabus that has the primary objective of covering a broad range of mathematical concepts and techniques with the aim of giving the student the necessary background to be able to apply Mathematics to whatever other discipline they may be interested in.

The MSSM Program leverages off existing aptitudes secondary school students have in terms of games, modelling, language, logical reasoning, systems thinking or a natural learning ability (the vast majority of students have at least one of these). In this way, it demonstrates that Mathematics is the starting point for developing a fundamental understanding that students can apply to the world around them.

Format of the MSSM Program

The various chapters of this book follow a standard layout structure that enables readers to progress through each chapter in a methodical way. This, in turn, ensures that continuity of learning occurs. Each chapter is a discrete module which provides mathematical content that is developed from the most basic number system (i.e. the Natural Number System) and that follows a logical progression to the final elementary number system (i.e. the Real Number System).

Each chapter is arranged into the following sections:

- Background and Context
- Approach

- Theory
- Application
- Summary.

Within each chapter, the MSSM Program embodies a natural learning sequence by starting with answers to the 'Why' questions, then moving on to the 'What', 'Which' and 'How' questions in that order. Traditional Mathematics teaching is mostly prescriptive and concentrates on the 'How' questions by using algorithms (or 'recipes') to get results that hopefully justify the effort involved in using them. Although answering the 'How' questions has become the standard, this approach lacks the motivation and the ability to engage students' critical and creative skills, thereby resulting in suboptimal learning. It is only when students master the full spectrum of questions that they become empowered and passionate learners, and accept accountability for their learning effectiveness, efficiency and overall progress.

Overview of the LEAD Process

Within the MSSM Program, the progressive development of the Arithmetic of each elementary number system is delineated using the **LEAD** process. This process is represented in the associated diagram (from top to bottom) for each elementary number system. This LEAD Process is explained in more detail below.

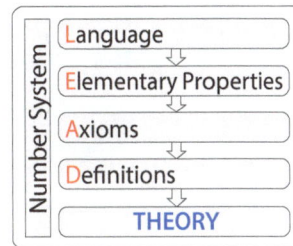

Language of the Elementary Number Systems

The first step in the LEAD Process is to incrementally develop a mathematical language that can be used in each of the elementary number systems. This language consists of its:

1. Alphabet
2. Terms
3. Equations.

This mathematical language commences with a basic 'alphabet' of symbols that we intuitively understand. The basic alphabet of this language begins with the following **seven basic symbols**: '0', '1', '+', '×', '(', ')' and '='.

The symbols '0' and '1' are called 'numbers'; the symbols '+' and '×' are called 'operators'; the symbols '(' and ')' are called grouping symbols; and finally the symbol '=' is called the 'equal' symbol. In particular, the operators '+' and '×' are called **binary** operators as they combine two numbers to generate a third number.

In order to be able to express properties (i.e. fundamental behaviours) of a number system, it is essential to extend the basic alphabet of this language to an alphabet that includes symbols that can be assigned any number within that alphabet. These symbols are called **variable symbols** or just **variables** to indicate that we can arbitrarily vary the number we choose to assign to each of these variable symbols of the alphabet. These variable symbols are chosen from the Latin and Greek alphabets and are selected from the sets $\{a, b, c, \dots, x, y, z\}$ and $\{\alpha, \beta, \gamma, \dots, \tau, \varphi, \omega\}$, respectively. Our extended (or new) alphabet now consists of the set of elements $\{0, 1, +, \times, (,), =, a, b, c, \dots, x, y, z, \alpha, \beta, \gamma, \dots, \tau, \varphi, \omega, \dots\}$.

This alphabet allows us to construct **finite valid** strings of symbols which make up a language used to describe an elementary number system. A finite valid string of symbols has a fixed number of symbols and does not go on indefinitely.

A finite valid string of symbols from the alphabet is called a '**term**' (provided that it does **not** contain an equal or not-equal symbol). A term is **valid** if it uses the symbols of the alphabet in the correct sequence. For example, if an operator occurs in a term, then there must be a number on either side of that operator. Examples of valid terms include: '7', 'a', '2×3', '$5 \times (a \times 9)$', '$2 \times (5 + 3)$', '$a + b$' and '$(a \times b \times c) + (d + 3)$'. However, the following terms are invalid: '$2 \times \times 3$', '$2 + \times 3$' and '$2)1 + 3($'. These terms are invalid because they are not using the operator and grouping symbols as they are specified in our language of Mathematics.

A finite valid string of symbols that consists of a term on either side of an equal symbol is called an '**equation**'. As with terms, an equation is **valid** if it uses the symbols of the alphabet in the correct sequence. Examples of valid equations include: '$0 = 0$', '$1 + 1 = 2$', '$2 \times 4 = 3 + 5$', '$a + 2 = 3$', '$a + b = b + a$', '$a \times b + 3 = 2 \times c$', '$a = 5$' and '$a + b = 0$'. However, '$2 = = 3$', '$2 = \times 3$' and '$2)1 = 3($' are invalid equations.

As terms and equations are constructed from symbols of the alphabet, it is convenient to classify terms and equations as follows:

1. **Arithmetic** terms and equations – terms and equations which **don't** contain a variable

2. **Algebraic** terms and equations – terms and equations which **must** contain at least one variable.

The symbols in the basic alphabet are said to be '**implicitly-defined**', that is, we **indirectly** give meaning to these symbols through equations called '**axioms**'. This meaning is created by the interrelationships between the symbols described by a specific set of axioms.

Once symbols of our alphabet have been given their meaning, we can create new symbols by using specific combinations of these existing symbols in our alphabet. These new symbols are used to model some characteristic of an elementary number system. The statement that gives the description of these new symbols is called a 'definition'. This is the last step in the modelling process before we can start deriving new properties from existing basic properties.

The four languages associated with the four elementary number systems are the primary focus on languages used in this book. Before starting to learn these number systems languages, it is helpful to be aware of an old Mathematical saying: 'The reason Mathematics is so difficult to learn is because it is so simple!' If this saying is going to be a useful guide, then we need a new approach to the standard learning process. This new approach will identify the easiest place to start and a method for keeping our work as simple as possible.

Elementary Properties of the Elementary Number Systems

The second step in the LEAD Process is to identify, within the appropriate chapters, the elementary properties of the number system being studied. A **property** of a number system is a fundamental **behaviour** of that number system expressed as an equation. These properties will tell us how the numbers in each of the number systems behave when they are manipulated to provide us with true answers.

Axioms of the Elementary Number Systems

The third step in the LEAD Process, within the appropriate chapters, will show how the elementary properties of each number system are expressed in 'true' equations. These equations are called **axioms**.

Axioms of a number system give meaning to the basic symbols of its alphabet: e.g. '$1 + 0 = 1$', '$1 \times 1 = 1$', '$a + b = b + a$' and '$a \times b = b \times a$'.

Definitions of the Elementary Number Systems

The fourth step in the LEAD Process is to define, in the Definition Section of each chapter, the new symbols and words that we will need to use in our language. The new symbols that are defined using symbols from the basic alphabet are said to be '**explicitly-defined**', that is, we **directly** give meaning to a new symbol in terms of other symbols. For example, the symbol '2' is expressed in terms of other symbols, in this case as '$1 + 1$' and this is expressed in the equation '$2 = 1 + 1$'.

Theorems of the Elementary Number Systems

The LEAD Process provides all the steps required to prove theorems. The Theory Section of each chapter is used to extend our knowledge of a number system by proving which equations are 'true' equations in that number system. To **prove a theorem is true** in a number system involves demonstrating that the equation representing the theorem is derivable from the axioms and definitions of that number system.

Applications of the LEAD Process

In this section, we will work through examples that show how to apply the language, axioms, definitions and theorems learnt in each chapter. These applications reinforce what has been taught in that chapter and ensure the student develops a clear understanding of the content that has been presented. This section should also provide students with 'aha' moments when they realise for the first time how a familiar mathematical result can follow on simply from using the LEAD process of explaining an elementary number system.

Objectives of this Book

The objectives to be achieved by developing mastery in the foundations of secondary school Mathematics are:

- To accelerate your learning in all branches of Mathematics
- To provide you with a simple hands-on explanation of a best-practice learning methodology in the basic languages of Mathematics (i.e. learning how to learn)
- To ensure you comprehend what is required to develop mastery in a subject area (i.e. learning the meaning of meaning)
- To develop a clear understanding of mathematical logic (i.e. learning the meaning of 'true' and 'false')
- To teach you how to model (and master) the world around you by applying these principles more broadly so as to become a subject-matter expert in other fields of interest.

These objectives are expected to 'spill over' to other areas of learning by increasing self-confidence within students in their inherent ability to learn more; by adding to the skill set necessary to overcome learning obstacles; and by fostering a passion for enquiry and understanding.

The most valuable outcome from reading the text and working through the mathematical examples presented in this book would be for secondary school students to become empowered learners and create opportunities to fully develop their potential and make a positive contribution to the world.

CHAPTER 1

THE FOUNDATIONS OF NUMBER SYSTEMS

Developing Mastery in Arithmetic

The approach we shall be adopting in this book to develop mastery in secondary school Mathematics follows the SME Learning Method. This methodology consists of three steps that make learning Mathematics (1) **S**imple, (2) **M**eaningful and (3) **E**xtendable (**SME**).

These learning steps are described in more detail below:

Step One – Simple (Genesis)

The MSSM Program begins with the easiest, smallest and most familiar set of concepts in Mathematics that represent a simple physical system – namely, the Physical Counting System. Modelling this simple physical system is the ideal place for students to commence on the optimal learning path in Mathematics. This starting point is consistent with our philosophy of avoiding 'challenges' by exposing the simplicity of Mathematics immediately. This is so the student won't be able to identify with the saying: 'The reason Mathematics is so *difficult*, is because it is so *simple*'! Our goal is to start with this simple model of the Physical Counting System and create 'meaning' in a way that ensures our mathematical development remains simple.

Step Two – Meaningful (The Meaning of Meaning)

The MSSM Program introduces basic symbols that will be used to represent concepts that describe properties of this simple physical system. The relationship between these symbols will then be formulated by modelling basic behaviours of this system and, thereby allow us to assign meaning to these symbols. Once our fundamental symbols have been assigned meaning and our rules of reasoning have been established, we can define other concepts that relate to other properties of this system. At this point, the student will understand how 'meaning' is created in Mathematics (and, indeed, in any language).

Step Three – Extendable (Learning to Learn)

The third element of the MSSM Program is to show how this approach allows the previous knowledge to be extended by small incremental steps. These incremental steps build knowledge in the same way that a principal deposit in a bank collects interest and becomes an exponentially improving investment. Learning becomes a repeatable process where all the elements are known and understood, and the success achieved in the initial steps motivates and drives the process and fosters creativity, intuition and confidence to continue building the body of knowledge called Mathematics.

We will use this SME Learning Method to build the elementary number systems starting from the simplest number system. This simplest number system is the formalisation of elementary Arithmetic called the **Natural Number System**. This number system is a simple **model** of the Physical Counting System which we learn about at the beginning of primary school.

Overview of the Physical Counting System

The **Physical Counting System** consists of processes that provide the answer to the question: 'How many objects are there?' in the following three situations:

1. There are a group of objects arranged in some random pattern

2. There are two groups of objects which we treat as a single large group of objects

3. There are multiple groups of objects that have the same number of objects in each group.

The Physical Counting System diagram illustrates the first situation where we have some groups of objects that we wish to count by a systematic process. The process we normally use is the process of '**counting**' by associating a group of objects with the Natural Numbers, i.e. the sequence of numbers $(0, 1, 2, ...)$.

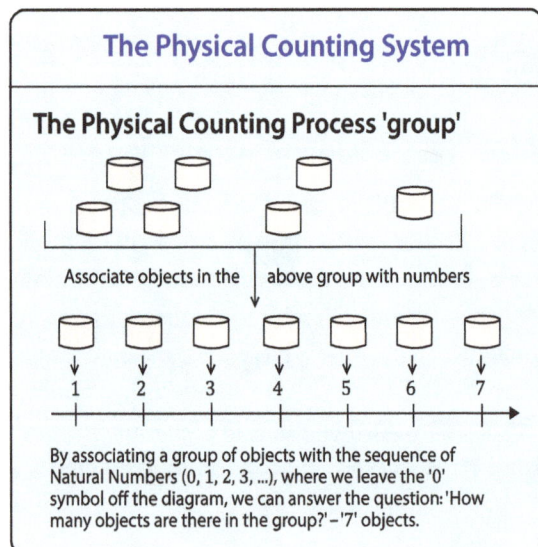

The Physical Counting System

The Physical Counting Process 'group'

Associate objects in the | above group with numbers

1   2   3   4   5   6   7

By associating a group of objects with the sequence of Natural Numbers (0, 1, 2, 3, ...), where we leave the '0' symbol off the diagram, we can answer the question: 'How many objects are there in the group?' – '7' objects.

This counting process is illustrated in the first diagram where we see that the objects at the top of the diagram don't appear to be arranged in any convenient pattern to allow counting. The simplest pattern in which to arrange these objects is along a straight line marked with equally-spaced Natural Numbers, in their normal sequential order. The last number to the right along this line which is associated with an object is then referred to as the number of objects in the group. Hence, this is the process of counting objects in a group when applied to any arrangement of objects.

Next, we wish to look at the situation where we have two groups of objects and we wish to count the total number of objects if we combined these two groups into one group. As you already know, you select either group and associate this group of objects with the line of Natural Numbers as we did in the first situation above. Then, we select the other group of objects and start associating them with the natural numbers to the right of the last number which already has an object associated with it. After doing this, the last natural number to the right which has an object associated with it, is called the number of objects in the two groups of objects combined.

In this second situation above, this process of counting two groups of objects is referred to as the '**addition**' process. That is, we can consider the addition process to be adding the numbers associated with each group to find out the number of objects in the combined group of objects. This process is illustrated in the Modelling the Physical Counting System diagram below.

Finally, we wish to examine the situation where we are combining multiple groups of objects which have the same number of objects in each group. A simple example is to combine three groups each of which have two objects each. This is equivalent to adding the first two groups of objects together, as we did in the second situation above, and then adding this result to the third group. This process of adding multiple identical groups of objects to get the total number of objects is referred to as the '**multiplication**' process.

The above processes of counting, addition and multiplication with the natural numbers is referred to as the **Physical Counting System**. The answers given by the Physical Counting System are irrefutable and are not reliant on the type of objects, the number symbols used or the person who carries out these processes. The Physical Counting System is the starting point for all future developments in this book.

The limitation of this Physical Counting System is clearly its ability to deal with large numbers and complex calculations. We try to compensate for this limitation by introducing the Multiplication Tables, algorithms for additions and multiplications, and lots of examples and exercises. However, a better approach is required for secondary school students. This approach involves creating a complete **model** of the Physical Counting System and then modelling the other elementary number systems.

This process of modelling using Mathematical symbols also starts in primary school where we assign the number '1' to each object and introduce the addition symbol '+' to indicate we are finding the number of objects in a group of objects. Similarly, we introduce the multiplication symbol '×' to indicate we have multiple groups of objects with the same number of objects in each group and we want to find the total number of objects when we combine these groups. This initial modelling is illustrated in the following diagram:

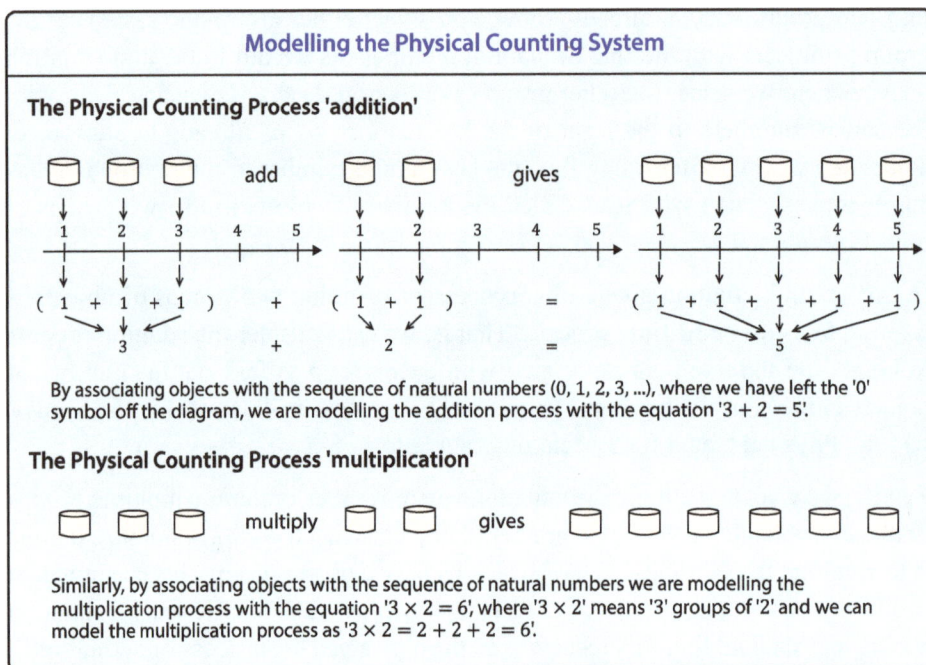

The complete model of the Physical Counting System is our first elementary number system – the Natural Number System. As outlined in the Introduction Section of this book, our approach to modelling the Natural Number System, as well as the other elementary number systems, is by developing a Language, Elementary Properties, Axioms, Definitions and Theorems for each of the four elementary number systems.

The modelling of the Physical Counting System as the Natural Number System is a way of representing the physical counting processes using the elementary number system components summarised in the right-hand column of the following diagram:

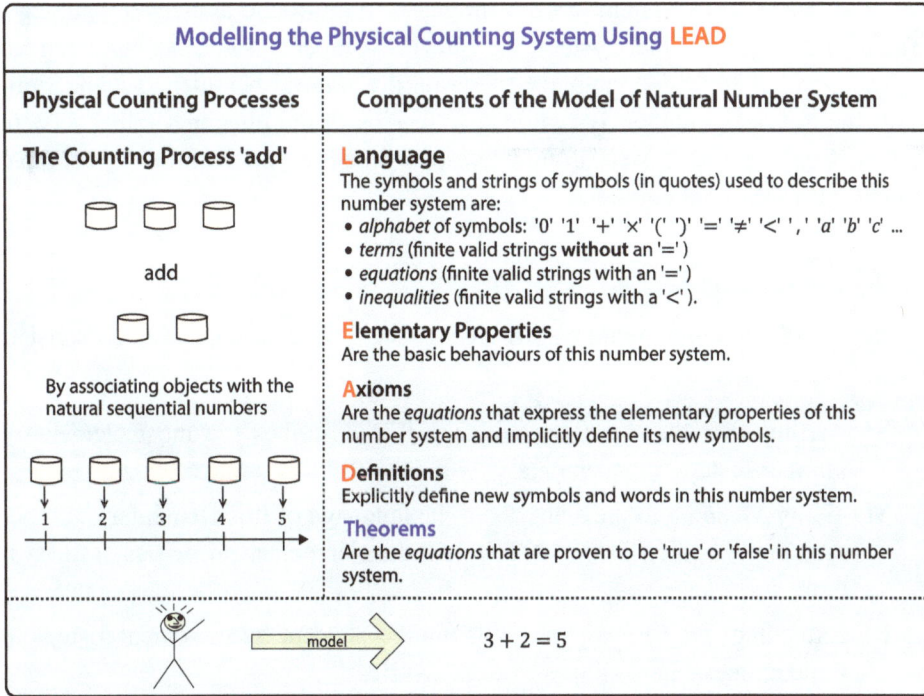

Modelling the Physical Counting System Using LEAD

Physical Counting Processes	Components of the Model of Natural Number System
The Counting Process 'add' add By associating objects with the natural sequential numbers 1 2 3 4 5	**L**anguage The symbols and strings of symbols (in quotes) used to describe this number system are: • *alphabet* of symbols: '0' '1' '+' '×' '(' ')' '=' '≠' '<' ', ' 'a' 'b' 'c' ... • *terms* (finite valid strings **without** an '=') • *equations* (finite valid strings with an '=') • *inequalities* (finite valid strings with a '<'). **E**lementary Properties Are the basic behaviours of this number system. **A**xioms Are the *equations* that express the elementary properties of this number system and implicitly define its new symbols. **D**efinitions Explicitly define new symbols and words in this number system. **T**heorems Are the *equations* that are proven to be 'true' or 'false' in this number system.

model ⟹ $3 + 2 = 5$

This section has described the main aim of this book which is to develop the models of the elementary number systems as the optimum way (using the SME approach) of achieving mastery in secondary school Mathematics.

The Four Elementary Number Systems

The four elementary number systems discussed in this book are the Natural, the Integer, the Rational and the Real Number system. These four systems are shown in order of simplicity from **bottom-to-top** in the diagram below:

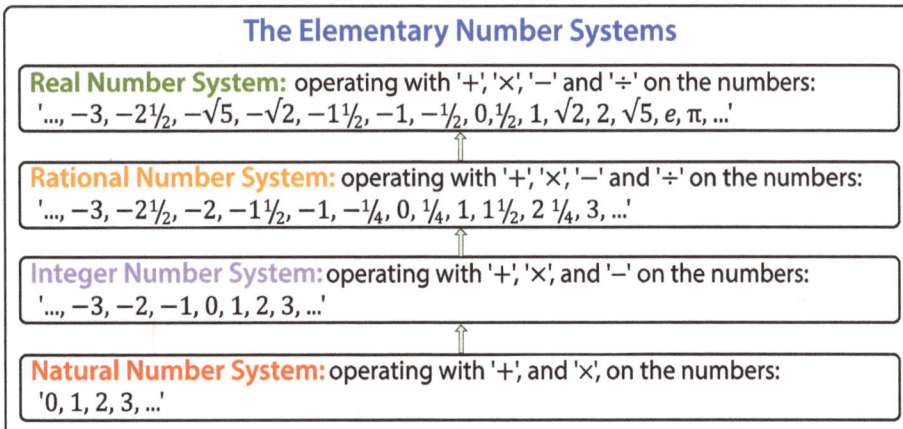

The Elementary Number Systems

Real Number System: operating with '+', '×', '−' and '÷' on the numbers:
'..., $-3, -2\frac{1}{2}, -\sqrt{5}, -\sqrt{2}, -1\frac{1}{2}, -1, -\frac{1}{2}, 0, \frac{1}{2}, 1, \sqrt{2}, 2, \sqrt{5}, e, \pi, ...$'

Rational Number System: operating with '+', '×', '−' and '÷' on the numbers:
'..., $-3, -2\frac{1}{2}, -2, -1\frac{1}{2}, -1, -\frac{1}{4}, 0, \frac{1}{4}, 1, 1\frac{1}{2}, 2\frac{1}{4}, 3, ...$'

Integer Number System: operating with '+', '×', and '−' on the numbers:
'..., $-3, -2, -1, 0, 1, 2, 3, ...$'

Natural Number System: operating with '+', and '×', on the numbers:
'$0, 1, 2, 3, ...$'

The Natural Number System has the simplest alphabet and associated language. The next simplest number system, the Integer Number System, has an extended alphabet and an extended language based on the Natural Number System. Similarly, the Rational Number System is an extension of the Integer Number System and the Real Number System is an extension of the Rational Number System that includes the irrational numbers.

SME Learning Method Applied to Arithmetic

As described at the beginning of this chapter the SME Learning Method consists of:

S: starting with the **S**implest language that describes the most elementary number system, i.e. elementary Arithmetic

M: giving **M**eaning to the symbols of the language of this elementary number system through equations that capture the elementary properties of the system

E: **E**xtending this language and this number system by incremental steps to build up to the next simplest number system.

In this way, the three steps of the SME Learning Method when applied to elementary Arithmetic are:

Step One: **Start with the simplest language that describes a model of the most elementary number system.**

This is achieved by starting with the Natural Number System since the majority of children can count even before they go to school.

When we first begin learning Arithmetic in primary school, the numbers we use are the 'Natural Numbers'; also known as the collection of counting numbers ('0, 1, 2, 3, 4, ...'). So in primary school we begin learning a new language with a new mathematical 'alphabet'.

In this way, we can claim to have successfully met the requirement of Step 1, above.

Next, we progress to the most important step of all and the fundamental question that underpins effective learning, namely: 'How do we give **meaning** to the symbols that make up the alphabet of our language?' That is, how will we meet the requirement of Step Two?

Step Two: **Give meaning to the symbols (the alphabet) of the language through equations that capture the elementary properties of the number system.**

The answer to this fundamental question is: **Identify the smallest collection of basic properties of an elementary number system (when formalised in our language, these properties are called axioms) that enable us to derive all the other properties of this number system.** This smallest collection of basic properties – where only the symbols of the alphabet are used – demonstrates that it is the relationship of these symbols to each other that creates meaning within this elementary number system.

Once we have created meaning for these basic symbols, we can assign meaning to other symbols in relation to these basic symbols. After we have successfully given our basic symbols their 'meaning' by using our language to model the properties of the Physical Counting System, we need to be able to show how to extend this language incrementally. This is the directive expressed in the next step.

Step Three: **Extend the language (and associated alphabet) by an incremental step to model the next simplest system.**

By using Step Three and including new symbols into the alphabet, we will be able to progress from the Natural Number System through to the Real Number System.

These steps immediately raise a question in students' minds. After adopting this approach, how will we know that it is producing best-practice learning in Arithmetic? The answer is: the student will be able to demonstrate mastery of these languages by clearly articulating what they are thinking, while hearing, seeing and doing arithmetic activities! Like riding a bicycle, these new languages of Arithmetic will become embedded in the student's reasoning processes.

Once the basic logic of Arithmetic is understood and its simplicity apparent, students will be able to accelerate their learning – not only in Arithmetic, but in other branches of Mathematics as well.

CHAPTER 2

THE NATURAL NUMBER SYSTEM

Overview of the Natural Number System

In this chapter, we begin our study of the elementary number systems by starting with the simplest and most familiar number system – the Natural Number System. Our study of the Natural Number System will involve the manipulation of numbers and variables in Arithmetic and Algebra using the operations of **addition** (denoted by '+') and **multiplication** (denoted by '×') applied to the set of natural numbers {0, 1, 2, ...} to model the Physical Counting System. In order to keep the notation simple – both within this chapter and throughout the MSSM Program – we will drop the use of quotes around mathematical symbols when they are already surrounded by braces or parentheses.

Developing mastery in this most basic mathematical system – the Natural Number System – lays a solid foundation not only for the study of Arithmetic and basic Algebra, but also for the foundations of secondary school Mathematics.

There are many practical reasons why we learn 'by rote' in the Physical Counting System, for example, the multiplication tables we learn in primary school. By the end of this chapter, we will have observed that these multiplication tables are derived from axioms, definitions and theory based on the Natural Number System.

The obvious value of Mathematics (and Arithmetic in particular) lies in its ability to model both simple and complex behaviours in our environment which we have to deal with in our everyday lives. In this chapter, we will use the Natural Number System to model the Physical Counting System and to demonstrate the basic elements of problem solving. This problem-solving skill will continue to be developed as we progress through each chapter.

Pictorial Representation of the Natural Number System

It can sometimes be helpful to provide a simple pictorial representation of the natural numbers. We start by representing some of the natural numbers – which include zero – as labels of 'equally-spaced points' along a horizontal line known as the number line.

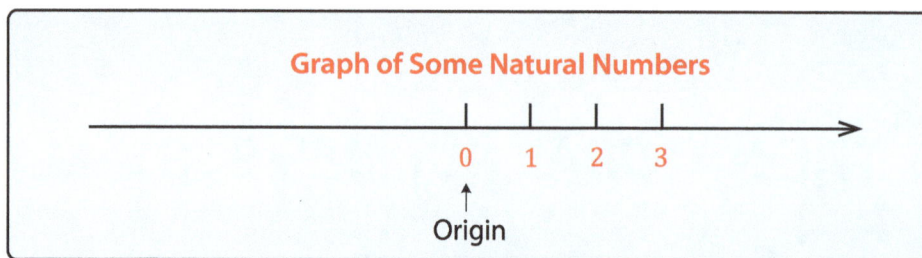

Figure 2.1

It is important to recognise that only those points equally spaced to the right of the arbitrary point labelled '0' can be assigned natural number coordinates. We will have to wait for a later extension to the Natural Number System before we can assign number coordinates to the remaining points.

Background and Context of the Natural Number System

To understand the Natural Number System (which is the simplest area of Mathematics) we are confronted with the fundamental question: 'What is meaning?' In other words, how do we ascribe 'meaning' to a collection of symbols and words? This problem is evident in our first encounter with a dictionary when we look up the 'meaning' of a word, only to find that it is explained in terms of other words which may also have no meaning for us. This is particularly apparent when we are trying to learn a new language.

In the Natural Number System, we are primarily focused on the operations of addition and multiplication on the set of symbols {0, 1, 2, ...} which we call the set of Natural Numbers. Before we can even start to manipulate these numbers, we need a language with which to talk about the properties of the Natural Number System. This language must be based on the alphabet which we identified in the Introduction to this book.

Approach to the Natural Number System

In this section our aim will be to assign meaning to the alphabetic symbols of the language used in the Natural Number System. To achieve this outcome, this language will be used to express the elementary properties of this number system in the form of equations. These elementary properties refer to the behaviour of the operators '+', '×' and also '=' when they are used to combine numbers in equations. We will then use these equations to create the axioms of the Natural Number System. These axioms and subsequent definitions give meaning to the symbols of the alphabet.

Language of the Natural Number System

At this point, we begin to develop the language that we will use throughout this chapter. As with all languages, we require a set of symbols which make up the alphabet in order to construct **terms** (i.e. equivalent to 'words' in the English language) and **equations** (i.e. equivalent to 'sentences' in the English language). You will recall that our language consists of: an alphabet, terms and equations.

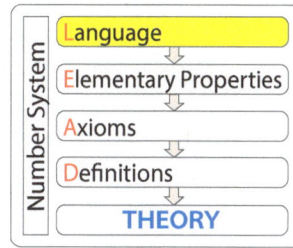

The alphabet that comprises the Natural Number System starts with the seven basic symbols: '0', '1', '+', '×', '(', ')' and '=' as set out in Table 2.1 below.

2.1 The **Alphabet** of the Natural Number System	
Symbols	**Description**
'0', '1'	'0' and '1' are called the numbers **zero** and **one** respectively.
'+'	'+' is called the **plus** sign or the **addition** operator.
'×'	'×' is called the **times** sign or the **multiplication** operator.
'='	'=' is called the **equal** sign or the **equal** operator.
'(', ')'	'(' and ')' are called the **left** and **right parenthesis** respectively.

Table 2.1

Although from primary school we understand intuitively what the seven basic symbols of the alphabet mean by association, we need a method that assigns formal meaning to these symbols. We give meaning to these symbols by writing down properties of these symbols so that they model the Physical Counting System.

However, before we can write down these properties we need to extend the alphabet by introducing the necessary **variable** symbols. There are an unlimited number of symbols we can add to our alphabet as variables. Some of the standard symbols used as variables are:

$a, b, c, ..., x, y, z$ ... which are symbols from the Latin alphabet

$\alpha, \beta, \gamma, ...$ ... which are symbols from the Greek alphabet

$v', v'', v''', ...$ ... which are symbols containing superscripts

$a_1, a_2, a_3, ...$ ... which are symbols containing subscripts.

Generally speaking, the natural number variables are represented by 'a' through to 'f'. If more alternatives than these variables are required, subscripts are used (i.e. 'a_1' through to 'f_1'). Either way, there are an infinite number of variables available in the number systems, so mathematicians generally use a common smaller set to make axioms, definitions and theorems more readable.

Once these variables are included in our alphabet, we have sufficient language to begin to develop our model of the Physical Counting System.

Before we start to give meaning to the symbols of the alphabet we need to summarise the language we will use to classify finite valid strings of symbols from the alphabet (see Table 2.2 below).

2.2 The Language of the Natural Number System	
In this table, let 'a', 'b', 'c' and 'd' be variables representing natural numbers.	
1. Arithmetic Terms	An individual natural number, or sum and/or product of natural numbers, is called an arithmetic term (or just term). In this way, '5', '5×2', '$5 + 7$', '$5 \times (2 + 3)$' and '$(2 \times 3) + 4 \times (5 + 9) + (8 \times 7 \times 16)$' are all finite valid strings of symbols called arithmetic terms.
2. Arithmetic Equations	When we have arithmetic terms on either side of an '$=$' sign, we call this string an arithmetic equation (or just equation). For example, '$1 = 1$', '$2 \times 4 = 3 + (5 + 0)$' and '$(2 \times 3) + 4 \times (5 + 9) + (8 \times 7 \times 16) = 958$' are all finite valid strings of symbols called arithmetic equations.
3. Algebraic Terms and Equations	If any of the above terms or equations contains one or more variables, then we refer to them as algebraic terms or algebraic equations. For example, 'a', '$a + 2 \times 3$', '$5 + a \times b \times c$', '$3 \times a \times 3 + 7 \times a \times c \times d$' are all finite valid strings of symbols called algebraic terms. The examples, '$a = 1$', '$a \neq 0$', '$a + 2 = 5$', '$a + a \times b + 3 \times c \times d \times e = a \times b \times c \times d$' are all finite valid strings of symbols called algebraic equations.

Table 2.2

A key part of our language for the Natural Number System is the role played by the binary operators for addition and multiplication. The order in which we apply these operators has a critical impact on the results these operators give.

Order of Operations

Where there is a combination of addition(s) and/or multiplication(s) in a term, we will get different results, depending on the order in which we apply the addition and/or multiplication operators. This concept is easily demonstrated by evaluating the arithmetic term: '2 + 3 × 4'. As there are no parentheses in this term to unambiguously determine the order of application of these operators, we don't know whether to do the addition or multiplication first and, hence, we have two options for evaluating this term. First, we can begin by placing parentheses around the numbers to be added and then proceed as follows:

$$(2 + 3) \times 4 = 5 \times 4$$

Alternatively, if we place parentheses around the numbers to be multiplied, we get:

$$2 + (3 \times 4) = 2 + 12$$

There are two ways that we could remove this ambiguity. First, we could always insist on adding parentheses to terms and equations to avoid any ambiguity, or second, we could introduce a convention that ensures one type of operator is always evaluated before another to ensure that no ambiguities occur.

Mathematicians decided on the second course of action – initiating the Order of Operations which permits us to significantly reduce the use of parentheses thereby making terms and equations much more readable. The general rules for evaluating terms and equations using the Order of Operations are set out in Table 2.3 below.

2.3 Order of Operations for the Natural Number System	
The process used to evaluate arithmetic and algebraic terms is determined by the order of operations according to the following steps:	
1.	**Parentheses** have highest priority in a term and control the order of evaluation. Parentheses may be removed by evaluating inside the parentheses **or** by using the Distributive Axiom (described below).
2.	Next, we need to evaluate **exponents** (i.e. an **index** to which a base number is raised. For example, in the term '3^2', '2' is the index).
3.	Where there are no parentheses or exponents, then we evaluate **multiplication** '×' from left to right.
4.	Finally, we evaluate **addition** '+' from left to right.

Table 2.3

Note:

Parentheses may be removed when evaluating terms as long as no ambiguity occurs. For example: '$(a) + 2 = a + 2$'.

Order of Operations – PEMA

Evaluate the following term using Order of Operations:

$$Term = (3 \times 5) + 2^3 \times 7$$

Parentheses	$= (3 \times 5) + 2^3 \times 7$
Exponents	$= 15 + 2^3 \times 7$
Multiplication	$= 15 + 8 \times 7$
Addition	$= 15 + 56$
Answer	$= 71$

To make it easier to remember the Order of Operations for the Natural Number System, we can use the acronym '**PEMA**' which stands for:

- Parentheses

- Exponents

- Multiplication

- Addition.

This acronym is dependent on the number system being used; in later chapters in this book it will be extended to describe the order of operations in the various number systems. *Note*: In some schools the acronym BODMAS is used.

The Eight Elementary Properties of the Natural Number System

Now that we have developed the language for the Natural Number System, we can start to identify the elementary properties of the Natural Number System by assuming that:

Number System
- Language
- Elementary Properties
- Axioms
- Definitions
- THEORY

1. Any number from the set of Natural Numbers {0, 1, 2, 3, ...} can be assigned to the variables 'a', 'b' and 'c'

2. These variables can be combined with the operators '$+$', '$\times$' and/or '$=$'

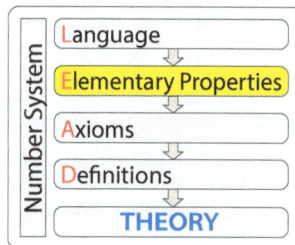

3. These variables can be combined with the set of Natural Numbers to form terms and equations.

When counting, there are eight elementary properties of the Physical Counting System that we want to model in the Natural Number System. These properties are described below.

1. Closure Property

In order to be able to use all the subsequent properties, we will first state the underlying property that guarantees Closure in the Natural Number System. When counting with two natural numbers, we always generate a unique (single) natural number.

This property is modelled in our Natural Number System by the statement that: 'When combining any two natural numbers using '+' and '×', we always end up with a unique natural number'. Since variables can be assigned any natural number, this property also applies to variables. This outcome is so fundamental that it is designated as the first property and is called the Closure Property.

2. Commutative Property for Addition

The order in which we add the total number of objects for two groups of objects does not change the answer to the question: 'How many objects are there?'

This property is modelled in our Natural Number System by the following equation:

$$a + b = b + a$$

So, when adding two natural numbers, the order in which we add these numbers does not alter the end result of the addition.

3. Associative Property for Addition

If we have to add the numbers of objects for any three groups of objects, then we can add the numbers for two groups of objects first and then add the third number. The order in which we add these numbers together for these three groups of objects does not change the final outcome.

This property is modelled in our Natural Number System by the following equation:

$$(a + b) + c = a + (b + c)$$

So, when adding three natural numbers, we can just write '$a + b + c$', which is now unambiguous.

4. Identity Property for Addition

If we have one object and we add no extra objects to it, then we will get the answer that there is still only one object.

This property is modelled in our Natural Number System by the following equation:

$$1 + 0 = 1$$

In this way, adding the natural number '0' to the natural number '1' does not change the value of the number '1'.

5. Commutative Property for Multiplication

The order in which we multiply the numbers for two groups of objects does not change the final result when counting. That is, we do not change the answer to the question: 'How many objects are there?' if we decide to multiply the numbers of objects in these two groups in a different order.

This property is modelled in our Natural Number System by the following equation:

$$a \times b = b \times a$$

In this way, we can note that when multiplying two natural numbers, the order in which we multiply these numbers does not alter the end result.

6. Associative Property for Multiplication

If we have to multiply the numbers of objects for any three groups of objects, then we can multiply the numbers for two groups of objects first and then multiply the third number. The order in which we multiply the numbers together does not alter the final result. This property is modelled in our Natural Number System by the following equation:

$$(a \times b) \times c = a \times (b \times c)$$

So, when multiplying three natural numbers we can just write '$a \times b \times c$' which is now unambiguous.

7. Identity Property for Multiplication

If we want one copy of one object, then we get one object. This property is modelled in our Natural Number System by the following equation:

$$1 \times 1 = 1$$

So, multiplying the natural number '1' by itself does not change its value.

8. Distributive Property

When we multiply the number of objects in one group of objects by the **sum of** the number of objects in two groups of objects, then we get the same answer as if we just multiply the number of objects in the one group of objects by both of the numbers of objects in the two groups of objects and add the results. This property is often called the 'distribution of multiplication over addition'. This property is modelled in our Natural Number System by the equation:

$$a \times (b + c) = a \times b + a \times c$$

This property ensures that 'multiplication' is just 'multiple additions'.

For example:

$$2 \times (3 + 4) = 2 \times 3 + 2 \times 4 \qquad \text{... by Distributive Property}$$
$$= 3 + 3 + 4 + 4 \qquad \text{... by multiple copies of '3', '4'}$$
$$\therefore \quad 2 \times (3 + 4) = (3 + 4) + (3 + 4) \qquad \text{... by multiple copies of '3 + 4'}$$

(The undefined symbol '$\therefore$' in the last equation is the standard abbreviation for 'therefore'.) In this way, this property captures the concept of 'multiplication' as 'multiple additions'.

If we assume that the numbers assigned to the variables 'a', 'b' and 'c' can be constructed from the symbols '1' and '+', for example '$2 = 1 + 1$' and '$3 = 1 + 1 + 1$' etc., then we observe that only seven basic symbols '0', '1', '+', '×', '(', ')', '=' and the variables are required to state the above eight elementary properties.

The Axioms of the Natural Number System

The above eight elementary properties – when expressed as equations – are the **axioms** of the Natural Number System. These axioms give meaning to the basic symbols of the natural number alphabet and are summarised in Table 2.4 on the following page.

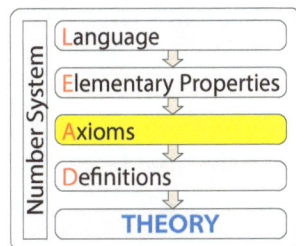

Number System
- Language
- Elementary Properties
- **Axioms**
- Definitions
- **THEORY**

2.4 The Axioms of the Natural Number System		
Axiom Type	**Addition '+'**	**Multiplication '×'**
Closure	1. The binary operators '+' and '×' (i.e. '$a + b$' and '$a \times b$') each give a natural number result for every natural number assigned to the variable 'a' and to the variable 'b'.	
Commutative	2. $\quad a + b = b + a$	3. $\quad a \times b = b \times a$
Associative	4. $(a + b) + c = a + (b + c)$	5. $(a \times b) \times c = a \times (b \times c)$
Identity	6. $\quad 1 + 0 = 1$	7. $\quad 1 \times 1 = 1$
Distributive	8. $\qquad a \times (b + c) = a \times b + a \times c$	

Table 2.4

In Table 2.4 above, the equations express how the symbols are related to each other and, as a result, these symbols are said to be 'implicitly-defined' by this process. Like the '+' and '×' operators, the equal symbol is an operator. However, in this case it does not provide an outcome of a number but, instead, gives a final result of 'true' or 'false'. Although in this table we use the '=' symbol, this table does not give the properties of the '=' symbol itself.

The '=' symbol compares two arithmetic terms and determines if those two terms result in the same natural number. If they do, then the '=' operator returns a 'true' value; if not, then the '=' operator returns a 'false' value.

Before we can use the Axioms of the Natural Number System (shown in Table 2.4) we must create some 'properties of reasoning' which are the elementary properties of the '=' symbol. These properties of reasoning are formally known as the **Properties of Logic**. First, we declare that the Axioms of the Natural Number System are 'true' equations because they model the Physical Counting System. Second, the Properties of Logic permit us to manipulate the 'true' equations so that the resulting equation(s) still model the Physical Counting System.

In order to make it easier to remember the Properties of Logic, we observe that there is a similarity between the properties of the number system and the Properties of Logic. We can highlight the existence of this similarity by placing the equivalent name from Table 2.4 after the appropriate "The Properties of Logic" name below.

The Six Elementary Properties of Logic

The six basic properties of 'true' and 'false', and the '=' sign are listed on the following page:

1. Declaration Property (Closure)

We declare that every arithmetic equation must be either a **true** or a **false** equation. We also declare that the Axioms of the Natural Number System (listed in Table 2.4) are **true** equations for any natural numbers assigned to the variables in those equations. That is, these equations accurately model the Physical Counting System.

2. Symmetry Property (Commutative)

Similarly, we would expect the equal sign to model the Physical Counting System with the property: given 'a' and 'b' are two variables (and given the two equations '$a = b$' and '$b = a$') then either both equations are **true** or both equations are **false**.

We can write this as:

'$a = b$' then '$b = a$'

3. Transitive Property (Associative)

Once again, we would expect the equal sign to model the Physical Counting System insofar as the property: given 'a', 'b' and 'c' are three variables and '$a = b$' and '$b = c$' are **true** equations, then '$a = c$' will also be a **true** equation. Likewise, we would expect that if one of these equations is **false** and the other **true**, then the third equation must be **false**.

We can write this as:

'$a = b$' and '$b = c$', then '$a = c$'

4. Not-Equal-To Property (Identity)

This property states that in the Physical Counting System the basic numbers '0' and '1' are different numbers, so that '$0 = 1$' is a **false** equation.

We can summarise this property as follows:

'$0 = 1$' is false

5. Deduction Property (Distributive)

In the Physical Counting System, we would expect the '$=$' symbol to interact with the addition and multiplication operators so that if we add or multiply two **true** equations, then the result will also be a **true** equation. Furthermore, if the result is a **true** equation and one of the input equations is **true**, then the other input equation must also be true (note the exception below).

We can write this as:

'a = b' and 'c = d', then 'a + c = b + d'
'a = b' and 'c = d', then 'a × c = b × d'

However, if one of the input equations for the multiplication operation (i.e. the second equation above) is '0 = 0', then we would always expect to get a **true** equation. Hence, we must **exclude** this case so that the general statements described above will apply.

6. Mathematical Induction Property

Finally, we would expect as a result of modelling the Physical Counting System that the process of reasoning we use to prove a sequence of equations is **true** will make use of the proof by the Mathematical Induction Process. The Mathematical Induction Process will be explained later in this chapter.

The Axioms of Logic

In summary, the properties of the equal symbol that model the Physical Counting Process can be defined as a set of axioms which we call the Axioms of Logic. These axioms are the same axioms for all the elementary number systems and are set out in Table 2.5 below.

2.5 The Axioms of Logic (for the Elementary Number Systems)	
Given 'a', 'b', 'c' and 'd' are variables, then the Axioms of Logic are:	
Declaration	1. Every arithmetic equation is either true or false. The Axioms of the Number Systems are declared as true so as to model the Physical Counting System and its extensions.
Symmetry	2. In the statement: Given 'a = b' then 'b = a': If one equation in quotes is true, then the other is also true; if one equation is false, then the other equation is also false.
Transitive	3. In the statement: Given 'a = b' and 'b = c', then 'a = c': If any two of the three equations in quotes are true, then the third equation is also true; if one equation is true and another false, then the third equation is also false.
Not-Equal-To	4. The arithmetic equation: '0 = 1' is false.

2.5 The Axioms of Logic (for the Elementary Number Systems)	
Deduction	5. There are two 'Given' statements in this axiom: Given '$a = b$' and '$c = d$', then '$a + c = b + d$'. Given '$a = b$' and '$c = d$', then '$a \times c = b \times d$'. For each 'Given' statement: if any two of the three equations in quotes are **true**, then the third equation is also **true**; if any one equation is **true** and another **false**, then the third equation in quotes is also **false**. **Exception**: the input equation '$0 = 0$' is excluded from the second 'Given' statement (multiplication).
Mathematical Induction	6. Given a sequence of mathematical equations, ('equation 1', 'equation 2', ...), if we have the following two conditions: - 'equation 1' in the sequence is **true**; and - given an arbitrary equation in the sequence, 'equation k', and the subsequent equation 'equation $k + 1$' is **derivable** from 'equation k' (i.e. using axioms and theorems); then we can declare every equation in this sequence is **true**.

Table 2.5

Notes:

1. Table 2.5 contains the Axioms of Logic and gives the 'Truth' Table for how we can combine equations so as to preserve their '**true**' or '**false**' value and hence 'truly' model the Physical Counting System.

2. In the Deduction Axiom stated in Table 2.5 above, if we consider '$0 = 0$' and '$0 = 1$' as the two input equations, then when we multiply these two equations together, we will get '$0 \times 0 = 0 \times 1$' which is a true equation. In this case, we have two true equations and one false equation. Therefore, we always treat the case of the input equation '$0 = 0$' as an exception in our Deduction Axiom for multiplication.

3. Having now established the meaning of the equal symbol '=' in Table 2.4 and Table 2.5, we can now define the not-equal symbol '≠' by the following statement:

Given 'a' and 'b' are two variables, then:

Given '$a = b$' is a **true** equation then '$a \neq b$' is a **false** equation.

Given '$a = b$' is a **false** equation then '$a \neq b$' is a **true** equation.

Another important mathematical property of the equal sign is known as the 'Reflexive Property'. This property is associated with the '=' sign and for a variable 'a' is simply stated as: '$a = a$'. This is a derived property that follows from the Transitive and Symmetry Axioms. This property will be proven as a theorem later in the Theory Section of this chapter and, hence, is not an Axiom of Logic.

Definitions of the Natural Number System

Now that we have discussed the seven basic symbols of the alphabet, namely '0', '1', '+', '×', '(', ')' and '=', in terms of their relationships to one another, it is time to define some new symbols relating to these basic symbols.

Number System

- Language
- Elementary Properties
- Axioms
- Definitions
- THEORY

In this section, we will create new symbols and words that describe the Natural Number System. In the Theory Section, we will explore the relationship between these new symbols, the axioms and definitions and the words and symbols in the basic alphabet.

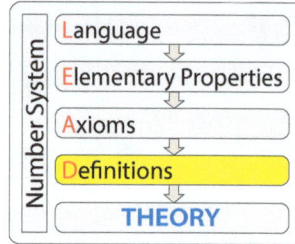

Definition 2.1   Extension of the Basic Alphabet

The basic set of implicitly-defined symbols – {0, 1, +, ×, (,), =} – is now extended to include a further eight natural number symbols to supplement the existing number symbols '0' and '1'. These new natural numbers are created from combinations of the symbols '1', '+' and '=' using the Closure Axiom.

These new natural number symbols are explicitly defined by the following equations being declared as true equations:

$$2 = 1 + 1$$
$$3 = 2 + 1$$
$$\vdots$$
$$9 = 8 + 1$$

Having defined natural numbers up to the number '9', we now require a system that allows us to use natural numbers larger than the number '9'. Without even being aware of it, most of us use 'decimal notation' on a daily basis to express these larger numbers. This notation will be defined in the definition below.

Definition 2.2   Decimal Notation

Decimal notation for natural numbers allows us to write numbers as multiples of '10' so that the first digit on the right-hand end of a string of digits represents the number of **units**, the second digit from the right-hand end represents the number of **tens**, the third digit represents the number of **hundreds**, and so on.

———————————————

If the digits in the decimal number are designated to be 'd_0', 'd_1', 'd_2', ... , 'd_m', then the decimal number 'd_m ... $d_2 d_1 d_0$' is defined by:

$$d_m \ldots d_2\, d_1\, d_0 = d_m \times 10 \ldots 0 + \ldots + d_2 \times 100 + d_1 \times 10 + d_0 \times 1$$

Note:

In the term '$d_m \times 10 \ldots 0$', the symbols '10 ... 0' represent the number '1' followed by 'm' zeros. Decimal notation will be explained in more detail in Chapter 6.

In the case of decimal notation, it is common practice to add a new symbol to the alphabet to improve readability of large decimal numbers (as described in Definition 2.3 below).

Definition 2.3   Comma Separator Symbol

The **comma separator** symbol ',' is explicitly defined to improve the readability of long strings of digits in numbers in decimal notation. The comma separator can be inserted at three digit intervals from the right-hand side of a string of decimal digits.

———————————————

For example, the number '10000000000' is more easily read by inserting the comma separator symbol so that it is written as: '10,000,000,000'. It is much easier to read this number as ten billion when the commas are used.

The next point is to define the concept of the 'value' of an arithmetic term. This is the single number generated when we simplify an arithmetic term using the axioms.

Definition 2.4   Value of an Arithmetic Term

The **value** of an arithmetic term is the natural number that this term evaluates to when using the Axioms of the Natural Number System.

For example, the two terms '3 + 5' and '2 × 4', which have no symbols in common, have the same value of '8'.

Similarly, when variables (or numbers) are being multiplied by themselves, it is helpful to have an abbreviation for multiple products of a variable (or number). To do this, **indices** are introduced to summarise otherwise tedious repetitive multiplications of the same variable (or number). Although we touch on this concept here, indices will be dealt with in detail in Chapter 5.

The notation for indices is set out in Definition 2.5 below:

Definition 2.5 **Index Notation 'a^n'**

Given a natural number variable 'a' (known as the **base**) and a natural number variable 'n' (known as the **index**) the combination of symbols 'a^n' is defined as:

$$a^n = a \times a \times \ldots \times a \quad \ldots \text{ where '}n\text{' is assigned '1, 2, ...'}$$

$$a^0 = 1 \qquad\qquad\qquad \ldots \text{ where '}n\text{' is assigned '0' and '}a \neq 0\text{'}$$

Notes:

1. The index is alternatively called the **power** or **exponent** of a number.

2. When we use words to describe the term 'a^n' we say: "The variable 'a' is raised to the power of 'n' which is the index".

An example of index notation occurs where we are required to simplify '35' multiplied by itself three times. In index notation this is written as: '35 × 35 × 35 = 35^3'.

Index notation also allows us to write out decimal numbers which use a base of '10' in the following way:

$$1 = 10^0$$

$$10 = 10^1$$

$$100 = 10^2$$

$$1000 = 10^3, \text{ etc.}$$

A good example of index notation for a number written with a base of ten is the number ten billion. In Definition 2.4 above, we observed that ten billion was written as '10,000,000,000'. However, index notation allows us to write the number '10,000,000,000' as '10^{10}'.

By using Definitions 2.2 and 2.4, we can now demonstrate the use of index notation to abbreviate decimal notation. This involves writing numbers as a string of digits expressed as multiples of '10'.

For example, we can express the number '4317' in two different forms, as:

$$4317 = 4 \times \mathbf{1000} + 3 \times \mathbf{100} + 1 \times \mathbf{10} + 7 \times 1 \text{ ... in decimal notation}$$

$$\therefore \quad 4317 = 4 \times \mathbf{10^3} + 3 \times \mathbf{10^2} + 1 \times \mathbf{10^1} + 7 \times \mathbf{10^0} \text{ ... in index notation}$$

With the ability to write numbers using index notation, we can now give a name to the factors (numbers), i.e. '4', '3', '1' and '7' in the term above that multiply the indexed terms '10^3', '10^2', '10^1' and '10^0', respectively. These factors are called 'coefficients' and their formal definition is provided below.

Definition 2.6   Coefficients

A term is called a coefficient if (and only if) it satisfies the following two conditions:

1. It is used as a factor for multiplying a number or a variable raised to a natural number index; **and**

2. It is a term that does **not** contain that indexed number or variable.

Multiple coefficients are usually expressed in subscript notation as: 'c_0', 'c_1', 'c_2' etc., where:

c_0 ... coefficient of the number or variable to the power of zero

c_1 ... coefficient of the number or variable to the power of one

c_2 ... coefficient of the number or variable to the power of two.

Note:

In the general term '$(2 \times a + 1) \times a^3$', the term '$(2 \times a + 1)$' is **not** a coefficient because it contains the indexed variable 'a'.

Using the definition of a term from the Introduction to this book, we can now use index notation and the definition of a coefficient (Definition 2.6) to write out a general algebraic term with coefficients. An algebraic term with coefficients is made up of terms which contain coefficients, along with products of the variable 'a' expressed in index notation.

Definition 2.7   Algebraic Term with Coefficients

Let the terms 'c_0', 'c_1', 'c_2' up to 'c_n' be coefficients. Also, let the terms 'a^0', 'a^1', 'a^2', up to 'a^n' be the products of the variable 'a' in index notation. We can now define the **algebraic term with coefficients** (in the variable 'a') up to the power of 'n' to be:

$$c_0 \times a^0 + c_1 \times a^1 + \ldots + c_{n-1} \times a^{n-1} + c_n \times a^n$$

Note:

When using index notation to represent decimal numbers, we start with the highest index first when we want to write the number as a decimal string.

Example:

Using the above example number '4317', the algebraic term with coefficients has specific coefficients that correspond to this number. They are:

$$c_0 \times a^0 + c_1 \times a^1 + \ldots + c_{n-1} \times a^{n-1} + c_n \times a^n$$
$$= c_n \times a^n + c_{n-1} \times a^{n-1} + \ldots + c_1 \times a^1 + c_0 \times a^0$$

... by starting with the highest index

$$4317 = 4 \times 10^3 + 3 \times 10^2 + 1 \times 10^1 + 7 \times 10^0$$

In the above equation, the coefficients are '$c_3 = 4$', '$c_2 = 3$', '$c_1 = 1$' and '$c_0 = 7$'.

In order to complete the language of the Natural Number System, we need to discuss two more definitions. In the pictorial representation of the set of Natural Numbers shown in Figure 2.1, the numbers '0', '1', '2', etc. are used to label 'equally-spaced points' which are visualised as the intersection of a long thick horizontal line with the smaller thin vertical line segments. The numbers associated with the 'equally-spaced points' along a number line are given an alternative name in our next definition.

Definition 2.8   Coordinate of a Point

A number that is used to label a point along the number line is called the **coordinate** of the point it labels.

With this definition of the word 'coordinate' in mind, we would say the origin has the coordinate '0', that is, the number zero. The first equally-spaced point to the right of the origin has the coordinate of '1', that is, the number one. We assume that this labelling process of points from left to right along the number line can occur indefinitely.

Generally speaking, in Mathematics (and in this book in particular) we often require a mathematical symbol that allows us to represent a repeating pattern of numbers, terms or equations. This mathematical symbol is called the 'ellipsis' and is our next definition.

Definition 2.9   Ellipsis Symbol '...'

A repeating pattern of numbers, terms or equations is represented by three full stops in a row. This is called an **ellipsis symbol** '...' and is sometimes referred to as 'dot-dot-dot'. An alternative ellipsis symbol is ':' which is called the vertical ellipsis symbol and is used to represent a sequence of terms of equations displayed vertically.

Now that we have covered the basic symbols of the alphabet, axioms and definitions for the Natural Number System, we can complete our description of its alphabet. The alphabet for the Natural Number System is summarised in Table 2.6 below.

2.6 The **Alphabet** of the Natural Number System	
Symbols	**Meaning from:**
'0', '1', '+', '×', '(', ')'	The Axioms of the Natural Number System
'=', '≠'	The Axioms of Logic
'2', '3', '4', '5', '6', '7', '8', '9'	Definition 2.1
','	Definition 2.3
'...'	Definition 2.9
'a', 'b', 'c', ..., 'a_1', 'a_2', 'a_3', ...	Common variables to which we assign natural numbers
'a^n'	Definition 2.5
'c_n'	Definition 2.6

Table 2.6

This alphabet for the Natural Number System is the starting point for the alphabet for the Integer Number System described in the next chapter.

Theory of the Natural Number System

To develop the Theory of the Natural Number System, we need to investigate the simple consequences of applying the Axioms of Logic to manipulate the Axioms of the Natural Number System. We commence our investigation of the Natural Number System by showing that the simplest equations we can write down, which aren't axioms or definitions, follow on from the axioms and definitions above. The simplest types of equations that can be proved to be true are:

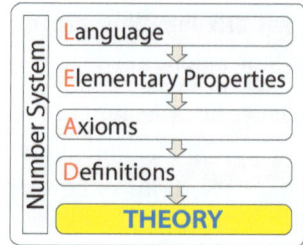

$$0 = 0$$

$$1 = 1$$

$$0 + 0 = 0$$

$$1 \times 0 = 0$$

$$0 \times 0 = 0$$

$$1 + 1 \neq 0$$

$$1 + 1 \neq 1$$

As the first two equations are closely linked, we will prove them as our first theorem.

Theorem 2.1   '0 = 0' and '1 = 1' Theorem

The simplest equations to prove are: '$0 = 0$' and '$1 = 1$'.

Proof

We will need to use the Identity Axiom for Addition, '$1 + 0 = 1$', the Symmetry Axiom for '=', the Transitive Axiom and the Deduction Axiom to prove this theorem. Since '0' is related to '1' by the equation '$1 + 0 = 1$', we can apply the Symmetry Axiom as follows:

$$1 = 1 + 0 \qquad \text{... by Symmetry Axiom for '='}$$

$$1 + 0 = 1 \qquad \text{... by Identity Axiom for '+'}$$

$$\therefore \quad 1 = 1 \qquad \text{... by Transitive Axiom}$$

By using the Transitive Axiom again we get:

$$1 + 0 = 1 + 0 \qquad \text{... by Transitive Axiom}$$

But: $$1 = 1 \qquad \text{... by first half of the proof}$$

$$\therefore \qquad\qquad 0 \ = \ 0 \qquad\qquad ... \text{ by Deduction Axiom}$$

That is, since the two equations '1 + 0 = 1 + 0' and '1 = 1' are true equations and we form '1 + 0 = 1 + 0' by adding '1 = 1' and '0 = 0', then '0 = 0' is also a true equation by the Deduction Axiom. In this way, the theorem is proved.

In the next theorem, we prove the next simplest result in a similar way to the above theorem.

Theorem 2.2 '0 + 0 = 0' Theorem

A simple arithmetic equation to prove is: '0 + 0 = 0'.

Proof

We will be using the Identity Axiom for Addition, '1 + 0 = 1', Theorem 2.1 and the Deduction Axiom to prove this theorem.

$$1 + 0 \ = \ 1 \qquad\qquad ... \text{ by Identity Axiom for '+'}$$

And: $\qquad\qquad 0 \ = \ 0 \qquad\qquad$ ... by Theorem 2.1

$\therefore \qquad 1 + 0 + 0 \ = \ 1 + 0 \qquad$ ... by the Deduction Axiom

But: $\qquad\qquad 1 \ = \ 1 \qquad\qquad$ ... by Theorem 2.1

$\therefore \qquad\qquad 0 + 0 \ = \ 0 \qquad\qquad$ ... by the Deduction Axiom.

That is, by adding '1 = 1' and '0 + 0 = 0' we get the true equation:

'1 + 0 + 0 = 1 + 0'. Hence, by the Deduction Axiom '0 + 0 = 0' must be a true equation. Therefore, the theorem is proved.

The next theorem proves that the more challenging equation, '1 × 0 = 0', is also a true equation.

Theorem 2.3 '1 × 0 = 0' Theorem

The simple arithmetic equation we want to prove is: '1 × 0 = 0'.

Proof

To prove this theorem, we need to utilise the Identity Axiom for Addition '$1 + 0 = 1$', the Identity Axiom for Multiplication, '$1 \times 1 = 1$', the Distributive Axiom, the Deduction Axiom and the Transitive Axiom.

	$1 + 0 = 1$	... by Identity Axiom for '+'
And:	$1 = 1$	... by Theorem 2.1
$\therefore$	$(1 + 0) \times 1 = 1 \times 1$	... by Deduction Axiom
	$1 \times 1 + 0 \times 1 = 1 \times 1$	... by Distributive Axiom
	$1 \times 1 + 0 \times 1 = 1 \times 1$	Result 1.

But:	$1 \times 1 = 1$	... by Identity Axiom for '×'
And:	$0 = 0$	... by Theorem 2.1
$\therefore$	$1 \times 1 + 0 = 1 + 0$	... by Deduction Axiom
	$= 1$	... by Identity Axiom for '+'
	$= 1 \times 1$	... by Identity Axiom for '×'
$\therefore$	$1 \times 1 + 0 = 1 \times 1$	Result 2.

Now using the Transitive Axiom on Results 1 and 2 above, gives:

	$1 \times 1 + 0 \times 1 = 1 \times 1 + 0$	... by Results 1 and 2
$\therefore$	$0 \times 1 = 0$	... by Deduction Axiom

In this way, the theorem has been proved.

————————————

The next theorem proves the fifth of our simple equations, '$0 \times 0 = 0$' is also a true equation.

Theorem 2.4   '$0 \times 0 = 0$' Theorem

The simple arithmetic equation to prove is: '$0 \times 0 = 0$'.

Proof

To prove this theorem, we need to utilise the Identity Axiom for Addition '$1 + 0 = 1$', Theorem 2.1, Theorem 2.3, the Deduction Axiom and the Transitive Axiom.

$$1 + 0 = 1 \qquad \text{... by Identity Axiom '+'}$$

$$\therefore \qquad (1 + 0) \times (1 + 0) = 1 \times 1 \qquad \text{... by Deduction Axiom}$$

$$1 \times 1 + 1 \times 0 + 0 \times 1 + 0 \times 0 = 1 \times 1 \qquad \text{... by Distributive Axiom}$$

$$1 \times 1 + 1 \times 0 + 0 \times 1 + 0 \times 0 + 0$$
$$= 1 \times 1 + 0 \qquad \text{... by adding '0 = 0'}$$

$$\therefore (1 \times 0) + (0 \times 1) + (0 \times 0) + 0 = 0 \qquad \text{... by Deduction Axiom}$$

But: $$0 = 0 + 0 \qquad \text{... by Theorem 2.2}$$

$$0 + 0 = 0 + 0 + 0 + 0 \quad \text{... by Deduction Axiom}$$

$$0 = 0 + 0 + 0 + 0 \quad \text{... by Transitive Axiom}$$

$$(1 \times 0) + (0 \times 1) + (0 \times 0) + 0 = 0 + 0 + 0 + 0 \quad \text{... by Transitive Axiom}$$

But: $$1 \times 0 = 0 \qquad \text{... by Theorem 2.3}$$

$$0 \times 1 = 0 \qquad \text{... by Symmetry Axiom}$$

$$0 = 0 \qquad \text{... by Theorem 2.1}$$

Having used these three equations and the Deduction Axiom, we are able to remove all symbols from the equation '$(1 \times 0) + (0 \times 1) + (0 \times 0) + 0 = 0 + 0 + 0 + 0$' to end up with:

$$0 \times 0 = 0$$

Hence, '$0 \times 0 = 0$' is a true equation and the theorem is proved.

The next theorem proves that the last two of our seven simple equations are true. From this theorem, we will demonstrate that the term '$1 + 1$' is a new natural number since it is not equal to '0' or '1'.

Theorem 2.5   '$1 + 1$' is a New Natural Number Theorem

The two arithmetic equations '$1 + 1 \neq 1$' and '$1 + 1 \neq 0$' are true equations.

Proof

To prove this theorem, we need to utilise the Transitive Axiom as it applies to a true and a false equation.

$$1 = 1 \qquad \text{(true)} \quad \text{... by Theorem 2.1}$$

$$1 = 0 \qquad \text{(false)} \quad \text{... by Not-Equal-To Axiom}$$

$$\therefore \qquad 1+1 \ = \ 1+0 \qquad \text{(false)} \quad \text{... by Deduction Axiom}$$

However, '$1 + 0 = 1$' as shown from the Identity Axiom for Addition. So we can rewrite the above equation using the Transitive Axiom as follows:

$$1+1 \ = \ 1 \qquad \text{(false)} \quad \text{... by Transitive Axiom}$$
$$\therefore \qquad 1+1 \ \neq \ 1 \qquad \text{(true)} \quad \text{... by definition of '$\neq$'}$$

To prove the second half of the theorem, we use Theorem 2.1, '$1 = 1$', added to itself using the Deduction Axiom to get '$1 + 1 = 1 + 1$'. We then multiply '$1 + 1 = 1 + 1$' by the false equation '$1 = 0$' as follows:

$$1+1 \ = \ 1+1 \qquad \text{(true)} \quad \text{... by Deduction Axiom}$$
$$1 \ = \ 0 \qquad \text{(false)} \quad \text{... by Not-Equal-To Axiom}$$
$$1 \times (1+1) \ = \ 0 \times (1+1) \quad \text{(false)} \quad \text{... by Deduction Axiom}$$
$$\therefore \qquad 1 \times 1 + 1 \times 1 \ = \ 0 \times 1 + 0 \times 1 \ \text{(false)} \quad \text{... by Distributive Axiom}$$
$$\therefore \qquad 1+1 \ = \ 0+0 \qquad \text{(false)} \quad \text{... by Transitive Axiom}$$

Using the Transitive Axiom on the two equations '$1 + 1 = 0 + 0$' and '$0 + 0 = 0$' (Theorem 2.2), then we get:

$$1+1 \ = \ 0 \qquad \text{(false)} \quad \text{... by reasoning above}$$
$$\therefore \qquad 1+1 \ \neq \ 0 \qquad \text{(true)} \quad \text{... by definition of '$\neq$'}$$

Therefore, the combination of natural numbers '$1 + 1$' must be a new natural number not equal to either of the natural numbers '0' or '1'. In Definition 2.1, we defined the new symbol of the alphabet '2' by the equation '$2 = 1 + 1$'. This theorem has now been proved and justifies the introduction of Definition 2.1.

Up until this point in the chapter, we have shown that the simplest equations of the Natural Number System can be derived from the Axioms of the Natural Number System and the Axioms of Logic. It is now appropriate to prove that the Reflexive Property, i.e. '$a = a$' also holds for the Natural Number System. This will require us to use the Mathematical Induction Axiom which is the last axiom of the Axioms of Logic. We now derive the Reflexive Property as our next theorem.

Theorem 2.6   Reflexive Theorem for Natural Numbers

If 'a' is a natural number variable, then it has the reflexive property so that:

$$a = a$$

Proof

In order to show that '$a = a$' for a variable 'a', it is appropriate to start with the results of Theorem 2.1. In that theorem, we proved the following two key results:

$$0 = 0$$

$$1 = 1$$

In this way, when we assign 'a' the values of '0' and '1', the equation '$a = a$' is true.

Our next step is to prove that these equations are not only true for just '0' and '1' but are also true for any natural number. This proof will require the use of the Mathematical Induction Axiom.

The Mathematical Induction Process (the Mathematical Induction Axiom)

The Mathematical Induction Process takes particular cases that are known to be true (e.g. '$1 = 1$') and then reasons, that under certain conditions, all the remaining cases must also be true.

Assume there is an unlimited sequence of equations, i.e. ('equation 1', 'equation 2', 'equation 3', ...) where the ellipsis symbol '...' implies this sequence of equations has no last equation. We can explicitly write out these equations as:

equation 1: $\qquad$ $1 = 1$

equation 2: $\qquad$ $2 = 2$

$$\vdots$$

equation k: $\qquad$ $k = k$

equation $k + 1$: $(k + 1) = (k + 1)$

$$\vdots$$

1. First, we set out to prove that 'equation 1' in the sequence is a true statement. This equation is true from Theorem 2.1 (i.e. '$1 = 1$').

2. Second, we will select an arbitrary equation in the sequence of equations and assume this arbitrary equation is true for the natural number 'k', i.e. for 'equation k'. Hence, '$k = k$' is assumed to be a true equation.

3. Third, we prove (if possible) that the subsequent equation, 'equation $k + 1$', is true if the arbitrary equation, 'equation k', is true. We can do this in the following way:

By combining 'equation k', i.e. '$k = k$', and 'equation 1', i.e. '$1 = 1$', and using the Deduction Axiom, we reach:

$$k + 1 = k + 1 \qquad \ldots \text{ by Deduction Axiom}$$

This proves that if 'equation k' is a true equation, then its subsequent equation, 'equation $k + 1$' is also true.

4. Fourth, if we let the arbitrary equation '$k = k$' be 'equation 1' (which we have proven to be true by Theorem 2.1), then its subsequent equation, 'equation 2' (i.e. '$1 + 1 = 1 + 1$', or alternatively, '$2 = 2$') must also be true.

5. Fifth, if we now let the arbitrary equation be 'equation 2', then 'equation 3' (i.e. '$3 = 3$') must also be a true equation.

6. Sixth, using the Mathematical Induction Axiom guarantees that we will be able to proceed through the sequence of equations while being assured that all equations in this sequence are true.

In this way, we have shown that for any given natural number 'k', the equation '$k = k$' is a true arithmetic equation. For the natural number variable 'a', we can declare that the algebraic equation '$a = a$' is a true equation since for every natural number assigned to 'a', we get a true equation. This completes the proof of the theorem and illustrates the use of the Mathematical Induction Axiom.

———————————

We now want to extend the Reflexive Theorem described above to apply to terms as well.

Theorem 2.7   Reflexive Theorem for Terms

If 't' is a general term, then this term also has the reflexive property so that:

$$t = t$$

Proof

A term 't' is a finite valid string of symbols constructed from natural numbers and variables using the operators '$+$' and '$\times$'.

By assigning natural numbers to each (and every) variable in a term 't', we get an arithmetic term. By then applying the Closure Axiom to this term, we are assured that the term 't' evaluates to a natural number. This statement will be true for each (and every) assignment of natural numbers to the variables belonging to 't'.

However, Theorem 2.6 assures us that '$a = a$' for every natural number 'a'. Therefore, we can declare that for each (and every) assignment of natural numbers to the variables of 't', the equation '$t = t$' is a true equation. In this way, the theorem is proved.

Given that coefficients are terms, we can state the basic properties we expect these terms to have before we start to manipulate these coefficients more generally.

Theorem 2.8 Properties of Terms Theorem

If three general terms (constructed from natural numbers and/or variables) are represented by 't_1', 't_2' and 't_3', then these terms should also obey the Axioms of the Natural Number System.

Proof

From the Axioms of the Natural Number System, we already know that every natural number and variable satisfy these axioms. A general term 't' is constructed from a finite valid string of additions and multiplications of natural numbers and variables.

By assigning a natural number to each (and every) variable in a term 't', and by using the Closure Axiom, we are assured that the general term 't' evaluates to a natural number. This statement is true for each (and every) assignment of a natural number to every variable belonging to the term 't'.

If the general term 't' always evaluates to a natural number after each (and every) variable is assigned a number, then it follows that 't' has the properties of the natural numbers derived from its component natural numbers and variables. Therefore, all terms will satisfy the Axioms of the Natural Number System.

In a similar manner, we can also apply the Axioms of Logic to these terms.

Since terms obey the Axioms of the Natural Number System and the Axioms of Logic, then every theorem that expresses the basic properties of the Natural Number System using variables will also be true for terms.

This concludes the proof of the theorem.

The Transitive Axiom is a basic method of replacing (or substituting) a variable in one simple equation, e.g. '$a = b$', with a variable from another simple equation, e.g. '$b = c$', where they have a variable in common, i.e. 'b'. That is, the variable 'b' occurs in both equations and can be eliminated by the Transitive Axiom to give a single equation, e.g. '$a = c$'. It is tedious using this axiom in such a restricted form, so in our next theorem we extend this 'substitution' process to permit substitution in general.

We will use Theorem 2.7 and the algebraic term with coefficients '$c_0 \times a^0 + c_1 \times a^1 + ... + c_n \times a^n$' to show that if '$a$' and '$b$' are variables such that '$a = b$', then in this algebraic term in 'a' we can substitute 'b' without altering the value of this term. This is called the substitution theorem and we prove this in the following theorem.

Theorem 2.9   Substitution Theorem

Assume we are given an algebraic term in the variable 'a' up to the power of 'n' with the coefficients 'c_0', 'c_1', 'c_2' up to 'c_n'.

In this algebraic term with coefficients – namely '$c_0 \times a^0 + c_1 \times a^1 + ... + c_n \times a^n$' – if '$a = b$', then '$b$' can be substituted for 'a' everywhere in this term and not change its value. In this way:

$$c_0 \times a^0 + c_1 \times a^1 + ... + c_n \times a^n = c_0 \times b^0 + c_1 \times b^1 + ... + c_n \times b^n$$

Proof

From Theorem 2.7 we know that all the equations equating coefficients to themselves, namely: '$c_0 = c_0$', '$c_1 = c_1$', ... , '$c_n = c_n$' are true equations. Likewise, from Theorem 2.8 we know that terms also obey the Axioms of the Natural Number System.

Given that '$a = b$' is true, and '$a^{n+1} = b^{n+1}$' is derivable from '$a^n = b^n$' by multiplying by '$a = b$' using the Deduction Axiom, then by Mathematical Induction the equation '$a^n = b^n$' is true for all assignments to the index variable 'n'. Therefore, the following equations must be true:

$$c_0 \times a^0 \;=\; c_0 \times b^0 \qquad \text{... by Deduction Axiom}$$

$$c_1 \times a^1 \;=\; c_1 \times b^1 \qquad \text{... by Deduction Axiom}$$

$$c_2 \times a^2 \;=\; c_2 \times b^2 \qquad \text{... by Deduction Axiom}$$

$$\vdots$$

$$c_n \times a^n \;=\; c_n \times b^n \qquad \text{... by Deduction Axiom}$$

We can now add the equations shown above in pairs to get true equations, so the following general equation must also be true by the Deduction Axiom:

$$c_0 \times a^0 + c_1 \times a^1 + ... + c_n \times a^n = c_0 \times b^0 + c_1 \times b^1 + ... + c_n \times b^n$$

Therefore, if '$a = b$', we can effectively substitute 'b' for 'a' in the above algebraic term '$c_0 \times a^0 + c_1 \times a^1 + \ldots + c_n \times a^n$' and not change the value of that term.

From Theorem 2.8, we can also declare that for the general terms 't' and 's', if '$t = s$' then we can substitute 's' for 't' in the algebraic term '$c_0 \times t^0 + c_1 \times t^1 + \ldots + c_n \times t^n$' and not change the value of that term i.e. '$c_0 \times s^0 + c_1 \times s^1 + \ldots + c_n \times s^n$'. Therefore, we can substitute for terms and not just variables in our Substitution Theorem.

In this way, the theorem has been proved.

_____

Note:

When substituting 's' for 't' in a term such as 't^2', where '$s = 1 + 1$', then 's' must be written with parentheses as '$(1 + 1)$' so that '$s^2 = (1 + 1)^2$'.

Just as we have extended the Transitive Axiom of Logic to the Substitution Theorem, we now extend our Identity Axiom for Addition – namely '$1 + 0 = 1$' – to apply to natural numbers, variables and terms and not just the number '1'.

Theorem 2.10   Identity Theorem for Natural Number Addition

If 'a' is a natural number variable, then '0' is the additive identity for this variable, so that:

$$a + 0 = a$$

Proof

We proceed to prove this theorem by mathematical induction by assigning the natural number 'k' to the variable 'a'. Here we are trying to show that the infinite sequence of equations listed below are, in fact, true equations:

equation 1:	$1 + 0 = 1$
equation 2:	$2 + 0 = 2$
	$\vdots$
equation k:	$k + 0 = k$
equation $k + 1$:	$(k + 1) + 0 = (k + 1)$

To carry out the standard proof by mathematical induction, we begin with 'equation k', that is: '$k = k + 0$'.

1. First we have to prove that 'equation 1' in the sequence is a true equation. This equation is true by the Identity Axiom for Addition.

2. Second, we will select an arbitrary equation in the sequence of equations and assume this arbitrary equation is true for the natural number 'k', i.e. 'equation k': '$k = k + 0$' is a true equation.

3. Third, we prove that the subsequent equation – 'equation $k + 1$' – is true if the arbitrary equation, 'equation k' is true. We do this in the following way:

   By combining 'equation k', i.e. '$k = k + 0$', and the true equation '$1 = 1$' and using the Deduction Axiom, we reach the following result:

$$k + 1 \;=\; k + 0 + 1 \qquad \text{... by Deduction Axiom}$$
$$\;=\; k + 1 + 0 \qquad \text{... by Commutative Axiom '+'}$$
$$\therefore \quad k + 1 \;=\; (k + 1) + 0$$

   This proves that if 'equation k' is a true equation, then so is its subsequent equation, 'equation $k + 1$'.

4. Fourth, if we let the arbitrary equation ('$k = k + 0$') be 'equation 1' (which we know is true from the Identity Axiom for Addition) then its subsequent equation, 'equation 2' (i.e. '$2 = 2 + 0$') must also be true.

5. Fifth, we now let the arbitrary equation be 'equation 2', so 'equation 3' ('$3 = 3 + 0$') must also be a true equation.

6. Sixth, using the Mathematical Induction Axiom guarantees that we can proceed through the sequence of equations and be confident that all equations in this sequence are true.

In this way, we are able to demonstrate that for any given natural number 'k', the equation '$k = k + 0$' is a true equation. For the natural number variable 'a', we can declare that the equation '$a = a + 0$' is a true equation since for every natural number assigned to 'a', we get a true equation. This completes the proof of the theorem.

From Theorem 2.8 we can also declare that for a general term 't' the equation '$t = t + 0$' must also be a true equation.

We now extend our Identity Axiom for Multiplication – namely '$1 \times 1 = 1$' – to apply to natural numbers, variables and terms, and not just the number '1'.

Theorem 2.11   Identity Theorem for Natural Number Multiplication

Let 'a' represent a natural number variable, then '1' is the multiplicative identity for 'a', that is:

$$a \times 1 = a$$

Proof

Using the proof by Mathematical Induction Process that we performed in Theorem 2.10, we can carry out the same process on the following equations:

equation 1:	$1 \times 1 = 1$
equation 2:	$2 \times 1 = 2$
	$\vdots$
equation k:	$k \times 1 = k$
equation $k + 1$:	$(k + 1) \times 1 = (k + 1)$

If the arbitrary equation, 'equation k', is true, then the subsequent equation, 'equation $k + 1$', must also be a true equation by applying the Deduction Axiom to the two equations '$1 \times 1 = 1$' and '$k \times 1 = k$' to get '$k \times 1 + 1 \times 1 = k + 1$' and then using the Distributive Axiom to get '$(k + 1) \times 1 = k + 1$'.

Since the first equation is true by the Identity Axiom for Multiplication, (i.e. '$1 \times 1 = 1$'), then its subsequent equation, 'equation 2', must also be true, (i.e. '$2 \times 1 = 2$'). Therefore, the Mathematical Induction Axiom allows us to declare that all subsequent equations must also be true.

In Theorem 2.3 we showed that '$1 \times 0 = 0$' and from the result above, we have '$k \times 1 = k$' is true for '$k = 1, 2, 3, ...$'. Therefore, '$k \times 1 = k$' is true for **all** natural numbers including zero. Therefore, the theorem has been proved for all natural numbers. From Theorem 2.8, we can also declare that for a general term 't' the equation '$t = t \times 1$' must also be a true equation.

––––––––––––––––––––

We now extend the two special results '$1 \times 0 = 0$' (Theorem 2.3) and '$0 \times 0 = 0$' (Theorem 2.4) to the general case of '$a \times 0 = 0$' for a variable 'a' using Theorems 2.9, 2.10 and 2.11.

Theorem 2.12   Null Multiplication of Natural Numbers Theorem

Given 'a' is a natural number variable, then:

$$a \times 0 = 0$$

Proof

We prove this theorem by starting with the assumption that 'a' is not assigned '0' and then using the two results '$1 + 0 = 1$' and '$a = a$' as well as proof by deduction. Using the Deduction Axiom, we multiply these two equations together to get:

$$a \times 1 = a \times (1 + 0) \qquad \text{... by Deduction Axiom}$$

$$\therefore \quad a \times 1 = a \times 1 + a \times 0 \qquad \text{... by Distributive Axiom}$$

$$\therefore \quad a = a + a \times 0 \qquad \text{... by subst. '}a \times 1 = a\text{'}$$

$$\therefore \quad a + 0 = a + a \times 0 \qquad \text{... by subst. '}a = a + 0\text{'}$$

$$\therefore \quad 0 = a \times 0 \qquad \text{... by Deduction Axiom}$$

In this way, the theorem has been proved. From Theorem 2.8, we can also declare that for a general term 't' the equation '$t \times 0 = 0$' must also be a true equation.

We can now summarise all of the theorems outlined above in simple table format to illustrate the progress we have made so far (see Table 2.7 below).

2.7 **Summary** of Natural Number System Theory		
1. *Reflexive Result:*	$0 = 0$	... Theorem 2.1
	$1 = 1$	... Theorem 2.1
	$a = a$	... Theorem 2.6
	$t = t$	... Theorem 2.7
2. *Multiplicative Results:*	$1 \times 1 = 1$	... Identity Axiom for '×'
	$1 \times 0 = 0$	... Theorem 2.3
	$0 \times 0 = 0$	... Theorem 2.4
	$a \times 0 = 0$	... Theorem 2.12
	$t \times 0 = 0$	... Theorem 2.12
	$a \times 1 = a$	... Theorem 2.11
	$t \times 1 = a$	... Theorem 2.11

2.7 **Summary** of Natural Number System Theory	
3. *Additive Results:*	$1 + 1 = 2$ ... Definition 2.1
	$1 + 0 = 1$ ... Identity Axiom for '+'
	$0 + 0 = 0$ ... Theorem 2.2
	$a + 0 = a$ ... Theorem 2.10
	$t + 0 = t$ ... Theorem 2.10
4. *General Results:*	If '$a = b$' then 'b' can be substituted for 'a' everywhere it occurs in a term and not result in the value of that term being changed. This is also the case for terms in '$t = s$'. ... Theorem 2.9

Table 2.7

Having established the results shown in the table above, we are now in a position to prove some simple (but very useful) theorems. The next step is for us to extend our Natural Number System by proving the simplest of counting results.

Theorem 2.13   Simple as '$2 + 2 = 4$'

The natural numbers '2' and '4' are related by '$2 + 2 = 4$'.

Proof

We begin with the left-hand side of this equation:

$$2 + 2 = 2 + (1 + 1) \qquad \text{... by Definition 2.1}$$
$$= (2 + 1) + 1 \qquad \text{... by Associative Axiom '+'}$$
$$= 3 + 1 \qquad \text{... by Definition 2.1}$$
$$= 4 \qquad \text{... by Definition 2.1}$$
$$\therefore \qquad 2 + 2 = 4$$

This concludes the proof of one of the simplest theorems in the Natural Number System. No doubt, you've heard the saying: 'That is as simple as: '$2 + 2 = 4$'. As we can see from the above reasoning, the result may be simple but it still requires a proof!

The next theorem is a very useful identity and follows from two applications of the Distributive Axiom. It is commonly known as the Binomial Expansion. We can now prove this identity holds true.

Theorem 2.14   Binomial Expansion

Let 'a', 'b', 'c' and 'd' be natural number variables, then:

$$(a + b) \times (c + d) = a \times c + a \times d + b \times c + b \times d$$

Proof

This proof requires two applications of the Distributive Axiom as follows:

$$(a + b) \times (c + d) = (a + b) \times c + (a + b) \times d \quad \text{... by Distributive Axiom}$$

$$= a \times c + b \times c + a \times d + b \times d$$

... by Distributive Axiom

$$\therefore \quad (a + b) \times (c + d) = a \times c + a \times d + b \times c + b \times d$$

In this way, we have proved a very useful identity as a theorem.

––––––––––––––––––––

This completes the introduction of the standard language and proofs of the basic theorems for the Natural Number System. Having established these axioms and definitions and proven the theorems, we now have a model for our Physical Counting System that lays the foundations for all later chapters in this book.

Applications of the Natural Number System

In this section, we will cover some simple applications of the theory of the Natural Number System that are considered to be standard syllabus questions. These applications will highlight how to manipulate simple arithmetic and algebraic terms and equations in such a manner that is normally not explained but is essential to understanding the logical processes used in mathematical reasoning.

The first application is intended to illustrate that we can prove a simple result if we can use an existing theorem. We can consider this application as an extension of the proof of Theorem 2.13.

Application 2.1

Evaluate the arithmetic term: '$2 + 3$'.

This evaluation doesn't have to be done from first principles (i.e. using axioms, definitions and theorems) because we have already proved Theorem 2.13 above. Hence, we start with the term '2 + 3' and simplify it as follows:

$$2 + 3 = 2 + (2 + 1) \quad \text{... by Definition 2.1 '3 = 2 + 1'}$$
$$= (2 + 2) + 1 \quad \text{... by Associative Axiom for '+'}$$
$$= (4) + 1 \quad \text{... by Theorem 2.13}$$
$$= 4 + 1 \quad \text{... by removing parentheses}$$
$$\therefore \quad 2 + 3 = 5 \quad \text{... by Definition 2.1 '5 = 4 + 1'}$$

Therefore, the arithmetic term '2 + 3' has the value of '5'.

Now we work out the equivalent of Theorem 2.13 for the multiplication operator.

Application 2.2

Evaluate the arithmetic term: '2 × 2'.

This evaluation will be done from first principles to demonstrate the theory behind the standard multiplication tables. We begin with the term '2 × 2' and simplify it as follows:

$$2 \times 2 = 2 \times (1 + 1) \quad \text{... by Definition 2.1}$$
$$= 2 \times 1 + 2 \times 1 \quad \text{... by Distributive Axiom}$$
$$= 2 + 2 \quad \text{... by Theorem 2.11}$$
$$\therefore \quad 2 \times 2 = 4 \quad \text{... by Theorem 2.13}$$

Therefore, the arithmetic term '2 × 2' has the value of '4'.

From this application we can see that it is straightforward (although somewhat tedious) to prove our standard multiplication tables that we learnt in primary school from first principles. In this way, while we may not have been aware of it, our multiplication tables are really a set of basic theorems of the Natural Number System. This demonstrates that our Natural Number System model is generating the practical results found in the Physical Counting System. The next application shows how we can make multiple additions simpler by using our basic axioms.

Application 2.3

Evaluate the arithmetic term: '27 + 39 + 73'.

This evaluation becomes much simpler if we use the Commutative Axiom for '27 + 39 + 73' as follows:

$$
\begin{aligned}
27 + 39 + 73 &= 27 + (39 + 73) &&\text{... by inserting parentheses} \\
&= 27 + (73 + 39) &&\text{... by Commutative Axiom for '+'} \\
&= (27 + 73) + 39 &&\text{... by Associative Axiom for '+'} \\
&= 100 + 39 &&\text{... by simplifying terms}
\end{aligned}
$$

$\therefore \quad 27 + 39 + 73 = 139$

Therefore, the arithmetic term '27 + 39 + 73' has the value of '139'.

———————————————

The next application will show us how we can simplify what seems to be a complex multiplication by using our axioms.

Application 2.4

Evaluate the arithmetic term: '27 × 73'.

This evaluation becomes much simpler if we use the definition of '27' and '73' as decimal numbers and then apply the Distributive Axiom as follows:

$$
\begin{aligned}
27 \times 73 &= (20 + 7) \times (70 + 3) &&\text{... by Definition 2.2} \\
&= 20 \times 70 + 7 \times 70 + 20 \times 3 + 7 \times 3 \\
& &&\text{... by Theorem 2.14} \\
&= 1400 + 490 + 60 + 21 &&\text{... by simplifying terms} \\
&= 1890 + 81 &&\text{... by simplifying terms}
\end{aligned}
$$

$\therefore \quad 27 \times 73 = 1971$

Therefore, the arithmetic term '27 × 73' has the value of '1971'. Of course, it is no accident that the numbers in the second last row of the proof above – namely '81' and '1890' – are the two numbers you would get if you used the long multiplication algorithm.

———————————————

The next application provides an example of collecting like terms in an algebraic term to produce an algebraic term written with coefficients.

Application 2.5

Given the algebraic term '$a \times b + 5 + (1 + a)\, a^2 + a \times 3 + 2 + a^2$', write this term as an algebraic term with coefficients.

We begin by collecting terms which have the variable 'a' raised to the same index.

$a \times b + 5 + (1 + a) \times a^2 + a \times 3 + 2 + a^2$

$= a \times b + 5 + 1 \times a^2 + a \times a^2 + a \times 3 + 2 + a^2$ ... by Distributive Axiom

$= a \times b + 5 + a^2 + a^3 + 3 \times a + 2 + a^2$ ... by simplifying terms

$= 1 \times a^3 + 2 \times a^2 + (3 + b) \times a + 7 \times a^0$ ... by simplifying terms

Therefore, the coefficients of the algebraic term in the variable 'a' are:

$$c_0 = 7$$
$$c_1 = 3 + b$$
$$c_2 = 2$$
$$c_3 = 1$$

This application is often called 'collecting like terms' in an algebraic term.

———————————

We have now completed the Application Section and observed the connection between the axioms and our modelling of the Physical Counting System. As a result, students will have observed how to reason from axioms to find the answers to typical questions that might feature in the Natural Number System syllabus.

Summary of the Natural Number System

At this point, we have covered the objectives of this chapter and so can now summarise our findings up to now. In terms of the Natural Number System, the most important definitions that have been introduced so far are:

1. Define the **eight natural number symbols**, '2, 3, 4, 5, 6, 7, 8, 9', that extend the set of implicitly-defined number symbols of the alphabet – namely '0' and '1' – with these eight explicitly-defined symbols.

2. Define '**decimal notation**' to allow us to write numbers as multiples of '10' so that the first digit on the right-hand end of a string of digits represents the number of **units**, the second digit from the right-hand end represents the number of **tens**, the third digit represents the number of **hundreds**, and so on.

3. Define the '**value**' of an arithmetic term as the natural number that this term evaluates to when using the Axioms of the Natural Number System.

4. Define '**index notation**' for a natural number variable 'a' (known as the base) and a natural number variable 'n' (known as the index) as the combination of symbols 'a^n' as: '$a^n = a \times a \times ... \times a$' where there are '$n$' factors of '$a$'.

5. Define a '**coefficient**' as a term if (and only if) it satisfies the following two conditions:

    1. It is used as a factor for multiplying a number or a variable raised to a natural number index; and

    2. It is a term that does **not** contain that indexed number or a variable.

6. Define the '**algebraic term with coefficients**' in the variable 'a' up to the power of 'n' to be:

$$c_0 \times a^0 + c_1 \times a^1 + ... + c_{n-1} \times a^{n-1} + c_n \times a^n$$

This chapter has provided an explanation of the operations of addition and multiplication of natural numbers. It has shown how the operations of long addition and long multiplication (known as algorithms) we first learnt in Arithmetic in primary school follow simply and logically from what has been discussed in this chapter.

These algorithms for addition and multiplication are an abbreviation of the many steps necessary to perform the same calculation using all the basic logical steps demonstrated in this chapter. For this reason, these algorithms are like applied arithmetic and do not participate in the normal proofs and formal procedures of the Natural Number System. In fact, when formatting mathematical solutions, it is normal to place the algorithms on the right-hand side of the page to indicate they are not participating in the normal logical flow of the solution to a mathematical problem.

The **Axioms of the Natural Number System** are summarised in Table 2.8 on the following page.

2.8 The Axioms of the Natural Number System		
Type	**Addition '+'**	**Multiplication '×'**
Closure	1. Each binary operator '+' and '×', i.e. '$a + b$' and '$a \times b$', produces a natural number for each natural number assigned to the variable 'a' and to the variable 'b'.	
Commutative	2. $\quad a + b = b + a$	3. $\quad a \times b = b \times a$
Associative	4. $(a + b) + c = a + (b + c)$	5. $(a \times b) \times c = a \times (b \times c)$
Identity	6. $\quad 1 + 0 = 1$	7. $\quad 1 \times 1 = 1$
Distributive	8. $\qquad\qquad a \times (b + c) = a \times b + a \times c$	

Table 2.8

The Axioms of Logic (for the Elementary Number Systems) as summarised in Table 2.9 below, will also be used in all the other elementary number systems we will be studying in the MSSM Program.

The Axioms of Logic (for the Elementary Number Systems) are:

2.9 The Axioms of Logic (for the Elementary Number Systems)	
Given 'a', 'b', 'c' and 'd' are variables, then the Axioms of Logic are:	
Declaration	1. Every arithmetic equation is either **true** or **false**. The Axioms of the Number Systems are declared as **true** so as to model the Physical Counting System and its extensions.
Symmetry	2. In the statement: Given '$a = b$' then '$b = a$': If one equation in quotes is **true**, then the other is also **true**; if one equation is **false**, then the other equation is also **false**.
Transitive	3. In the statement: Given '$a = b$' and '$b = c$', then '$a = c$': If any two of the three equations in quotes are **true**, then the third equation is also **true**; if one equation is **true** and another **false**, then the third equation is also **false**.
Not-Equal-To	4. The arithmetic equation: '$0 = 1$' is **false**

2.9 The Axioms of Logic (for the Elementary Number Systems)	
Deduction	5. There are two 'Given' statements in this axiom: Given '$a = b$' and '$c = d$', then '$a + c = b + d$'. Given '$a = b$' and '$c = d$', then '$a \times c = b \times d$'. For each 'Given' statement: if any two of the three equations in quotes are **true**, then the third equation is also **true**; if any one equation is **true** and another **false**, then the third equation in quotes is also **false**. **Exception**: The input equation '$0 = 0$' is excluded from the second 'Given' statement (multiplication).
Mathematical Induction	6. Given a sequence of mathematical equations, ('equation 1', 'equation 2', ...), if we have the following two conditions: - 'equation 1' in the sequence is **true**; and - given an arbitrary equation in the sequence, 'equation k', and the subsequent equation 'equation $k + 1$' is **derivable** from 'equation k' (i.e. using axioms and theorems); then we can declare every equation in this sequence is **true**

Table 2.9

The real importance of this study of the **Natural Number System** is that it demonstrates how language, elementary properties, axioms, definitions and theorems underpin our theory of the Natural Number System, and how this theory relies on a model of the Physical Counting System. When a student is able to correctly apply the axioms, definitions and theorems that have been discussed in this chapter to a broad range of physical systems, they will have achieved mastery in the Natural Number System.

In the next chapter, we will use all of the methods we have so-far investigated in relation to the Natural Number System to extend our knowledge to the theory of the **Integer Number System**.

CHAPTER 3

THE INTEGER NUMBER SYSTEM

Overview of the Integer Number System

In this chapter we extend our language of the Natural Number System to include the 'negative numbers'. The resulting set of numbers is called 'Integer Numbers' or, more commonly, the 'Integers'. Once again, our study begins by applying the operations of addition and multiplication to this new expanded set of integer numbers and building on this to derive the extended body of knowledge known as the Integer Number System.

Our purpose is to further develop mastery in the language of the number systems by extending our model of counting to encompass the concept of negative numbers to the left of zero on the number line. In this chapter we will model these negative numbers by creating the **additive inverses** of the set of natural numbers. For example, for every natural number assigned to 'a', we will create an additive inverse denoted by '$-a$' such that we can prove by adding 'a' to '$-a$' we get '0'. This concept is illustrated in the statement: '$a + -a = 0$'.

We will create these additive inverses by extending the natural number alphabet to include the additive inverse for the number '1' (namely '-1') and ensure that this new number complies with all the same Axioms of the Natural Number System while, at the same time, complying with one extra axiom: '$1 + -1 = 0$'. The same set of variables we used in the Natural Number System will now be assigned integer numbers to complete the alphabet for the Integer Number System.

Pictorial Representation of the Integers

It can be helpful to provide a pictorial representation of our model of this extended Physical Counting System. In Chapter Two we started with the Natural Number System as coordinates for 'equally-spaced points' along the number line to the right-hand side of the number zero. We will now include the 'negative numbers' (which are called 'negative integers') on this number line. This will allow us to assign coordinates to 'equally-spaced points' along the number line to the **left of zero**.

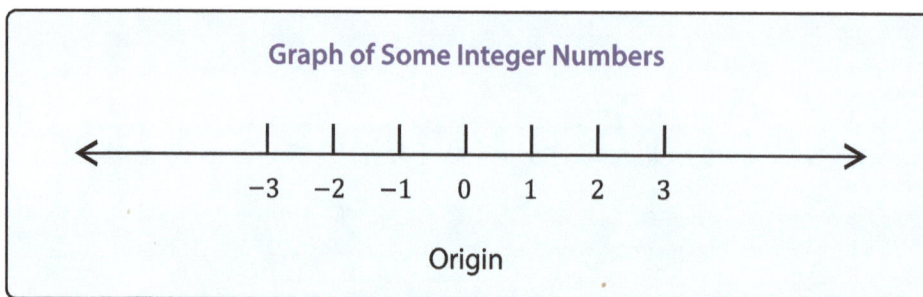

Figure 3.1

Once again, we recognise that only those points that are equally spaced away from the arbitrary point labelled '0' can be assigned integer numbers as coordinates. The remaining points will have to wait until a further extension to the Integer Number System before we can assign these points a set of numbers as coordinates.

Background and Context of the Integer Number System

In Chapter Two we saw that the symbols '0, 1, 2, ... , 9' were the number symbols of the Natural Number System. Extending the Natural Number System to the Integer Number System only requires the introduction of one new number symbol, '−1', to represent the new number we call **negative one**.

As was the case in Chapter Two, we will assume that we know intuitively what we want this symbol '−1' to represent and after we have defined the 'negative integers' using the symbol '−1', we will extend the Axioms of the Natural Number System to assign negative integers their meaning. Although two distinct symbols are used, the symbol '−1' is a new symbol in its own right.

In the Definition Section below, we will add another new symbol to our alphabet: the symbol '−', which we call the '**negative**' symbol. This new symbol will be used in this chapter in the following three distinct ways:

1. The 'negative' symbol attached to the front of an integer to make it a negative integer symbol (see Definition 3.1)

2. The minus 'unary' operator to create the additive opposite of an integer (see Definition 3.4)

3. The 'subtraction' operator to take one integer away from another integer (see Definition 3.5).

Now that we have established the set of symbols for the Integer Number System, it only remains to assign these new symbols meaning using the same approach used in Chapter Two. This approach is described in detail in the next section.

Approach to the Integer Number System

In this section our aim will be to assign meaning to the alphabetic symbols of the language of the Integer Number System. To achieve this outcome, the language will be used to express the elementary properties of this number system in the form of equations. These elementary properties refer to the behaviour of the operators '+', '×' and also '=' when they are used to combine numbers in equations. We will then use these equations to create the Axioms of the Integer Number System. These axioms and definitions give meaning to the symbols of the alphabet.

Language of the Integer Number System

At this point we begin to extend the alphabet developed in the Natural Number System to include the '**negative one**' symbol. This will give us the basic symbols of the alphabet required for the Integer Number System as shown in Table 3.1 below.

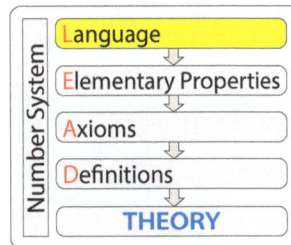

3.1 The **Alphabet** of the Integer Number System	
Symbols	**Meaning from:**
'0', '1', '+', '×', '(', ')'	The Axioms of the Natural Number System
'=', '≠'	The Axioms of Logic
'2', '3', '4', '5', '6', '7', '8', '9'	Definition 2.1
','	Definition 2.3
'...'	Definition 2.8
'a', 'b', 'c', ... , 'a_1', 'a_2', 'a_3', ...	Variables to which we assign integer numbers
'-1'	'-1' is the number called **negative one**. It derives its meaning from the equation: $$1 + -1 = 0$$

Table 3.1

Before we assign meaning to the new symbol '-1' of the alphabet, we must first incorporate the new symbol into the language we established in Chapter Two. This extension to the language is included in Table 3.2 on the following page:

3.2 The Language of the Integer Number System	
1. **Arithmetic Terms**	An individual integer or sum and/or product of integers is called an **arithmetic term** or just a **term**. Hence, '5' or '5 × −1', '5 + 7 × −1', or '5 × 3 + −1' are all finite valid strings of symbols called **arithmetic terms**.
2. **Arithmetic Equations**	When we have arithmetic terms on either side of an '=' or '≠' symbol, we call this equation an **arithmetic equation** or just an **equation**. For example, '3 + −1 = 2', '2 × 4 = 3 + (6 + −1)', '2 × −1 = −2' and '0 ≠ −1', are all finite valid strings of symbols called **arithmetic equations**.
3. **Algebraic Terms and Equations**	If any of the above terms or equations contains one or more variables, we then refer to them as **algebraic** terms or equations, respectively. For example, 'a', '$a + 2 \times -1$', '$-1 + a \times b \times c$', '$3 \times a \times b + -1 \times a \times c \times d$' are all finite valid strings of symbols called **algebraic terms**. '$a = -1$', '$a \neq 0$', '$a + 2 = -1$', '$a + a \times b + -1 \times c \times d \times e = a \times b \times c \times d$' are all finite valid strings of symbols called **algebraic equations**.

Table 3.2

As is the case for the Natural Number System, the addition and multiplication operators play a key role in the language of the Integer Number System. However, we will also be able to define a new operator, called the '**subtraction**' operator, for this new language using the negative numbers. Once again, the order in which we apply these operators has a critical impact on the outcomes of these operators.

Order of Operations

As we discovered in Chapter Two, there is an inherent ambiguity in the order in which we could apply operators if no parentheses are used to determine their order. The same Order of Operations convention that we applied to the Natural Number System is also extended to the Integer Number System and accommodates the new symbol '−1'.

To evaluate general integer terms, we use the Order of Operations as shown in Table 3.3 on the following page.

3.3 Order of Operations for the Integer Number System

The process used to evaluate arithmetic and algebraic terms in the Integer Number System is determined by the order of operations and proceeds according to the following steps:

1.	The **parentheses** have highest priority and control the order of evaluation. Parentheses may be removed by evaluating inside the parentheses **or** by using the Distributive Axiom.
2.	Next evaluate **exponents** (i.e. an integer index to which a base number is raised).
3.	Where there are no parentheses, then we evaluate **multiplication** '×' from left to right.
4.	Finally, evaluate **addition** '+' and **subtraction** '−' from left to right.

Table 3.3

The Order of Operations for the Integer Number System can be summarised using the acronym '**PEMAS**' which stands for:

- Parentheses
- Exponents
- Multiplication
- Addition
- Subtraction.

Order of Operations – PEMAS

Evaluate the following term using Order of Operations:

$$Term = (3 \times 5) + 2^3 \times 7 - 12$$

Parentheses $= (3 \times 5) + 2^3 \times 7 - 12$

Exponents $= 15 + 2^3 \times 7 - 12$

Multiplication $= 15 + 8 \times 7 - 12$

Addition $= 15 + 56 - 12$

Subtraction $= 71 - 12$

Answer $= 59$

The acronym is dependent on the number system being used and has been extended in this chapter to include the subtraction operation.

When counting with integers, we retain the eight elementary properties of the Natural Number System and then add one new property, the Inverse Property, that relates specifically to our new symbol, '−1'.

The Ninth Elementary Property of the Integer Number System

Now that we have developed the language for the Integer Number System, we can express the additional elementary property of this system.

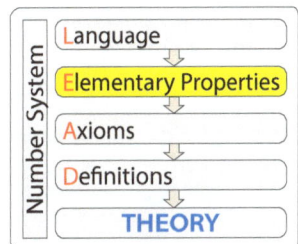

Number System: Language → Elementary Properties → Axioms → Definitions → THEORY

Inverse Property of Addition

This property will guarantee us that for every natural number there is another number (called the **inverse** number) such that when we **add** a number to its inverse, we get the number '0'. This inverse number is also called the **additive inverse** as it is the inverse of a number with respect to the addition operation.

This property is modelled in our Integer Number System by adding the new integer number '−1' to the integer number '1' to get the zero integer '0'. In symbols, we model this property using the equation:

$$1 + -1 = 0$$

This new property is called the '**Inverse Axiom for Addition**' which is Axiom '9' in Table 3.4 below.

The Axioms of the Integer Number System

By starting with the Axioms of the Natural Number System and then adding this new property as an axiom, we have extended the set of axioms available to us to use. This new set of axioms now applies to the broader set of numbers called the Integers. We refer to them as the Axioms of the Integer Number System.

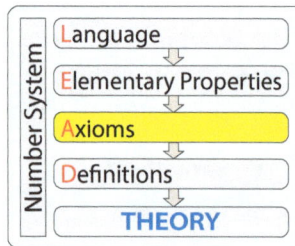

Number System

Language ⇩
Elementary Properties ⇩
Axioms ⇩
Definitions ⇩
THEORY

If we assume 'a', 'b' and 'c' are three integer variables that can be assigned any integer number, then the Axioms of the Natural Number System are now extended in scope and apply to all integer numbers. The Axioms of the Integer Number System are set out in Table 3.4 below.

3.4 The Axioms of the Integer Number System		
Axiom Type	**Addition '+'**	**Multiplication '×'**
Closure	1. The binary operators '+' and '×' (i.e. '$a + b$' and '$a \times b$') each give a unique integer number for every integer number assigned to the variable 'a' and to the variable 'b'.	
Commutative	2. $a + b = b + a$	3. $a \times b = b \times a$
Associative	4. $(a + b) + c = a + (b + c)$	5. $(a \times b) \times c = a \times (b \times c)$
Identity	6. $1 + 0 = 1$	7. $1 \times 1 = 1$
Distributive	8. $a \times (b + c) = a \times b + a \times c$	

3.4 The **Axioms** of the Integer Number System		
Inverse	9.　　$1 + -1 = 0$	Not applicable: there isn't a multiplicative inverse for the integers.

Table 3.4

According to our new axiom – the Inverse Axiom for Addition – the integer '1' has an additive inverse (denoted by '-1') that is related to it.

The Axioms of Logic

Before we can reason correctly from the Axioms of the Integer Number System, we again need to use the Axioms of Logic which provide us with the properties of reasoning that we learnt about in Chapter Two (refer to Table 2.5). We can now apply these same Axioms of Logic to the Integer Number System with the following notes.

Notes:

1. The variables 'a', 'b', 'c' and 'd' represented natural number variables in Chapter Two, however, in Chapter Three these variables now represent integer variables.

2. In the Deduction Axiom we expressed the additive property using the equation '$a + c = b + d$'. There is no need to use a minus symbol '$-$' to express the situation of subtracting two integers as this case is covered in Definition 3.5 below.

Definitions of the Integer Number System

Now that we have assigned meaning to the symbol '-1' in the Axioms of the Integer Number System, we will use this symbol to create new definitions in this number system.

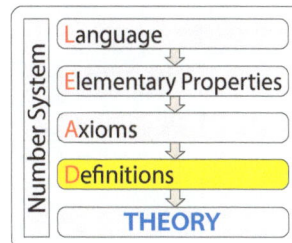

In the Theory Section we will explore the relationship between this new symbol, the other symbols in our alphabet, the axioms and definitions.

As stated in the Background and Context Section above, we shall define three different ways we can use the negative symbol '$-$'. The first of these is to create the full set of negative integers.

Definition 3.1   Negative Integers

An unlimited collection of **negative integers** can be generated using our new symbol '-1'. To create the negative of a natural number (excluding the number '1'), let 'k' be any natural number {2, 3, 4, ... }, and multiply it by '-1' as follows:

$$-k = -1 \times k$$

This new set of numbers of the form '$-k$' formed with the integer '-1' can be written as: {..., $-4, -3, -2, -1$}, and is called the set of **negative integers**. Multiplication by '-1' will generate an integer as a result of the Closure Axiom in the **Axioms of the Integer Number System** above. Later on in the Theory Section we will prove that '$-0 = 0$' so '-0' is not a different integer from '0'.

———————————

In the above definition, the equation '$-k = -1 \times k$' indicates that the negative symbol '$-$' is a part of the whole symbol for a negative integer. For example, in the case of the integer '-2', the negative symbol '$-$' is firmly attached to the symbol '2' so as to create the new symbol '-2'.

In some mathematical texts, the negative integers are written in a different format as '^{-}k'. Nevertheless, it continues to have the same meaning as defined above in that '$^{-}k = -k$'. In this book we will continue to use the '$-k$' format to express negative integers since it has become the international standard.

Now that we have defined negative integers, we can define the full set of 'integers'.

Definition 3.2   Integers

There is an unlimited collection of integers consisting of two sets of numbers. These sets are:

1.   The set of **natural numbers** which consists of: {0, 1, 2, 3, ...}
2.   The set of **negative integers** which consists of: {..., $-3, -2, -1$}.

When we combine these two sets we call them the **integers** and represent them as: {..., $-3, -2, -1, 0, 1, 2, 3, ...$}.

———————————

We take a subset of these integers and call them the **positive integers**.

Definition 3.3   Positive Integers

There is an unlimited set of natural numbers (**excluding** zero) i.e. {1, 2, 3, ...}, called the set of **positive integers**.

Note:

When dividing the integers into the subsets of negative and positive integers, the number '0' is neither a negative integer nor is it a positive integer. It does in fact become a subset on its own and is simply known as the set of zero, i.e. {0}.

In Chapter Two, we learnt that the addition '+' and multiplication '×' operators that we have been using so far are called **binary operators** because they combine **two** natural numbers to give a unique natural number result.

In the same way there are binary operators, there are also **unary operators** that operate on a single number to produce another single number. In the case of integers, a unary operator can be applied to either positive or negative integers and produce either negative or positive integers, respectively. The minus unary operator acts on an integer and reflects it through the origin of the Integer Number System to get its 'opposite' integer.

In Definition 3.1, the negative symbol '−' was used as part of the definition of a negative number; the second use of the negative symbol occurs in Definition 3.4 below. In this definition, the negative symbol is used as a symbol in its own right to operate on any integer and is called the **negative unary** operator. We define the negative unary operator in the following definition.

Definition 3.4   Negative Unary Operator '–'

The action of the negative symbol '−' as the **unary operator** is defined for an integer variable 'a' by multiplying 'a' by '−1' as follows:

$$-(a) = -1 \times a$$

Note:

We write the negative unary operator as '$-(a)$' to emphasise its role as an operator; however, we could have just as easily written this as '$-a$'. In this way, you can see that the operator '−' is applied to whatever we assign to the variable 'a'.

For example, if we assign 'a' the number '3', we get '$-(3) = -1 \times 3 = -3$'. Therefore, the unary operator takes a positive integer and turns it into a negative integer.

The third use of the negative symbol '−' is introduced to simplify some cumbersome operations such as '$7 + -5$' or by using the Commutative Axiom it can also be written as '$-5 + 7$'. The way to streamline these cumbersome operations is to define a new operator – the '**subtraction**' operator – which is a binary operator and is similar to our 'addition' and 'multiplication' operators.

We now define the subtraction operator as follows.

Definition 3.5 Subtraction Operator '$-$' as a Binary Operator

For the integer variables 'a' and 'b' we define the combination of symbols '$a - b$', also called 'a' subtract 'b', 'a' minus 'b' or 'a' take away 'b' as follows:

$$a - b = a + -b$$

On the right-hand side of this equation, the combination of symbols '$-b$' represent the unary operator applied to a variable.

How do we know when the negative symbol '$-$' is being used as a subtraction operator, a negative unary operator or a negative sign? As is often the case with words in any language, their meaning can vary according to the context in the sentence in which they are being used. Mathematicians have decided to follow this convention and do the same with the negative symbol '$-$'.

An easy way to interpret multiple uses of adjacent negative symbols which have a number or variable to their immediate left is to treat the first negative symbol (starting from the left of the string of negative symbols) as a binary operator, the last as a negative sign, and every other negative symbol in between as a unary operator.

For example: how do we evaluate the arithmetic term '$3 - - - 2$'? We treat:

- the **first** negative symbol to the right of '3' as a binary operator: this becomes the **subtraction operator**, i.e '$3 - (- - 2)$'

- the **last** negative symbol as a **negative sign** and we attach it to the number '2' to get '-2'. We also place parentheses around these two symbols to avoid confusion and get '(-2)'

- the **second** minus symbol as a **minus unary operator**. Again we place a set of parentheses around this negative symbol to avoid ambiguity and get '$(-(-2))$'.

Now, we can finally evaluate this term as: '$3 - (-(-2)) = 3 - 2 = 1$', as we will show later that '$-(-2) = 2$'.

With these basic symbols of the alphabet, axioms and definitions, we can complete the alphabet for the Integer Number System. This alphabet is summarised in Table 3.5 below.

3.5 The Alphabet of the Integer Number System	
Symbols	**Meaning from:**
'0', '1', '+', '×', '(', ')'	The Axioms of the Natural Number System
'=', '≠'	The Axioms of Logic
'2', '3', '4', '5', '6', '7', '8', '9'	Definition 2.1
','	Definition 2.3
'...'	Definition 2.8
'a', 'b', 'c', ..., 'a_1', 'a_2', 'a_3', ...	Common variables to which we assign integer numbers
'-1'	The Axioms of the Integer Number System
'$-$'	Definitions 3.4 and 3.5

Table 3.5

This alphabet for the Integer Number System is the starting point for the alphabet for the Rational Number System described in Chapter Four.

We have now completed our approach to developing the Integer Number System as an extension of the Natural Number System. In the next section of this chapter we will explore the theory that follows on from this extension to our number system axioms.

Theory of the Integer Number System

We now have sufficient language, axioms and definitions to proceed with the proof of some basic theorems. To develop the Theory of the Integer Number System, we will investigate the simple consequences of applying the Axioms of Logic to manipulate the Axioms of the Integer Number System in order to prove all the basic theorems involving the integers '-1', '0' and '1'.

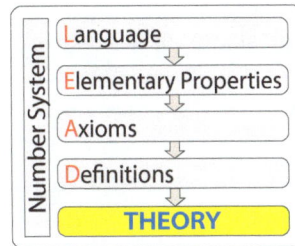

Number System

Language
⇩
Elementary Properties
⇩
Axioms
⇩
Definitions
⇩
THEORY

The first theorem we will prove is the simplest result;
namely, that '-1' satisfies the Reflexive Theorem just as '0' and '1' did in Chapter Two.

Theorem 3.1 '$-1 = -1$' Theorem

The arithmetic equation '$-1 = -1$' is a true equation.

Proof

We start by assuming that the integer '-1' satisfies the equation '$-1 = -1$'.

$$-1 = -1 \qquad \text{... by assumption}$$

$$1 = 1 \qquad \text{... by Theorem 2.1}$$

$$-1 + 1 = -1 + 1 \qquad \text{... by Deduction Axiom}$$

By using the Inverse Axiom for Addition and the Transitive Axiom on both sides of the equation, '$-1 + 1 = -1 + 1$', we get '$0 = 0$'. This last equation is true by Theorem 2.1 whose proof remains identical in the extended number system known as the Integer Number System.

Therefore, the original equation, i.e. '$-1 = -1$', is true by the Deduction Axiom. In this way, the theorem is proved.

The next theorem we will prove is the equivalent of the Identity Axiom for Addition using our new integer symbol, '-1'.

Theorem 3.2 Identity Theorem for Addition with '-1'

For the integer '-1', the following arithmetic equation is true:

$$-1 + 0 = -1$$

Proof

We start by assuming this equation is a true arithmetic equation:

$$-1 + 0 = -1 \qquad \text{... by assumption}$$

$$0 = 0 \qquad \text{... by Theorem 2.1}$$

$$-1 + (0 + 0) = -1 + 0 \qquad \text{... by Deduction Axiom}$$

But: $\qquad -1 = -1 \qquad \text{... by Theorem 3.1}$

$$0 + 0 = 0 \qquad \text{... by Theorem 2.2}$$

$$\therefore \quad -1 + (0 + 0) = -1 + 0 \qquad \text{... by Deduction Axiom}$$

This last equation is true by the Deduction Axiom. Therefore, the original assumption that '$-1 + 0 = -1$' must be true because it leads to a true equation using the Deduction Axiom. In this way, the theorem is proved.

———————————————

If we had had the Substitution Theorem at our disposal, the proof of the theorem outlined above could have been simplified by adding '$1 = 1$' to the first equation. However, we will not be investigating this theorem until Theorem 3.8 below and we will need more practice with using the Transitive Axiom until that point. Using the theorems shown above, we can now prove all the basic multiplicative identities using the integer '-1'.

Theorem 3.3   Theorems for Multiplication with '-1'

For multiplication by '-1', the following three arithmetic equations hold true:

(1) $-1 \times 1 = -1$

(2) $-1 \times 0 = 0$

(3) $-1 \times -1 = 1$

Proof

Part (1):

We will start with the first equation above: '$-1 \times 1 = -1$'. We will proceed by using mathematical deduction:

$$-1 + 1 = 0 \qquad \text{... by Inverse Axiom for '+'}$$

And: $\qquad\qquad 1 = 1 \qquad\qquad$ ... by Theorem 2.1

$\therefore \qquad 1 \times (-1 + 1) = 1 \times 0 \qquad$ ... by Deduction Axiom

However, according to Theorem 2.3, '$1 \times 0 = 0$'. Consequently, we can apply the Transitive Axiom to the above equation so that:

$$1 \times (-1 + 1) = 0 \qquad \text{... by Transitive Axiom}$$

$$1 \times -1 + 1 \times 1 = 0 \qquad \text{... by Distributive Axiom}$$

$$1 \times -1 + 1 \times 1 = -1 + 1 \qquad \text{... by Inverse Axiom for '+'}$$

From the Identity Axiom for Multiplication, Theorem 3.1 and the Deduction Axiom, we have '$-1 + 1 = -1 + 1 \times 1$'. Hence, we can write:

$$1 \times -1 + 1 \times 1 = -1 + 1 \times 1 \qquad \text{... by Transitive Axiom}$$

At this stage, we have the term '1×1' on both sides of this equation so we can 'remove' it using the Deduction Axiom. We are finally left with the result:

$$-1 \times 1 = -1$$

Hence, we have proved Part (1) of this theorem.

Part (2):

Next we will prove the second equation shown above: '$-1 \times 0 = 0$'. We will again use mathematical deduction.

	$1 + 0 = 1$	... by Identity Axiom for '$+$'
And:	$-1 = -1$	... by Theorem 3.1
$\therefore$	$-1 \times (1 + 0) = -1 \times 1$	... by Deduction Axiom
	$-1 \times 1 + -1 \times 0 = -1 \times 1$	... by Distributive Axiom
But:	$0 = 0$	... by Theorem 2.1
$\therefore$	$-1 \times 1 + -1 \times 0 + 0 = -1 \times 1 + 0$	... by Deduction Axiom
	$-1 \times 0 + 0 = 0$	... by Deduction Axiom
	$-1 \times 0 + 0 = 0 + 0$	... by Theorem 2.2
$\therefore$	$-1 \times 0 = 0$	... by Deduction Axiom

This completes the proof of Part (2) of this theorem.

Part (3):

Finally, we will prove the third equation shown above, i.e. '$-1 \times -1 = 1$'.

	$1 + -1 = 0$	... by Inverse Axiom for '$+$'
And:	$-1 = -1$	... by Theorem 3.1
$\therefore$	$-1 \times (1 + -1) = -1 \times 0$	... by Deduction Axiom
$\therefore$	$-1 \times 1 + -1 \times -1 = -1 \times 0$	... by Distributive Axiom
$\therefore$	$-1 \times 1 + -1 \times -1 = 0$	... by '$-1 \times 0 = 0$'
$\therefore$	$-1 \times 1 + -1 \times -1 = -1 + 1$	... by Inverse Axiom for '$+$'

We now want to replace '$-1 + 1$' on the right-hand side of the above equation with a term which includes '-1×1'.

We have:	$-1 = -1 \times 1$	... by Part (1)

$$1 = 1 \qquad \text{... by Theorem 2.1}$$

$$-1 + 1 = -1 \times 1 + 1 \qquad \text{... by Deduction Axiom}$$

Now we replace the '$-1 + 1$' on the right-hand side with '$-1 \times 1 + 1$' as follows:

$$-1 \times 1 + -1 \times -1 = -1 \times 1 + 1 \qquad \text{... by Transitive Axiom}$$

$$\therefore \qquad -1 \times -1 = 1 \qquad \text{... by Deduction Axiom}$$

Hence, we have proved that all three parts of this theorem are true.

Note:

Part 3 of the above theorem will lead us to the valuable outcome in Theorem 3.13 that "multiplying two negative numbers (or negative terms) together will result in a positive number (or positive term)".

Now that we have the basic multiplicative identities for the new symbol '-1', we can provide a simple expression that relates to subsequent negative numbers.

Theorem 3.4 Alternative Expression of Negative Integers Theorem

Negative numbers have already been defined in Definition 3.1 as '$-k = -1 \times k$'. They can also be expressed in the familiar form as:

$$-2 = -1 + -1$$

$$-3 = -2 + -1$$

$$\vdots$$

Proof

We start with the simplest case of '-2':

$$-2 = -1 \times 2 \qquad \text{... by Definition 3.1}$$

From Definition 2.1 ('$2 = 1 + 1$') and Theorem 3.1 ('$-1 = -1$'), we have:

$$-1 \times 2 = -1 \times (1 + 1) \qquad \text{... by Deduction Axiom}$$

$$= -1 \times 1 + -1 \times 1 \qquad \text{... by Distributive Axiom}$$

$$= -1 + -1 \qquad \text{... by Theorem 3.3(1)}$$

$$\therefore \qquad -2 = -1 + -1 \qquad \text{... by Transitive Axiom}$$

We can again use proof by the Mathematical Induction Process to show that every subsequent negative number can be derived from its predecessor by adding '−1'.

Now that we have established the simplest arithmetic equations using the integer number alphabet and the alternative expression of negative integers, we state and prove an extension of the Reflexive Theorem (Theorem 2.6), used for the set of natural numbers to now include the integer numbers.

Theorem 3.5   Reflexive Theorem for Integers

If 'a' is an integer number variable, then it has the reflexive property so that:

$$a = a$$

Proof

We know from Chapter Two that the algebraic equation '$a = a$' is true for natural numbers. In the Integer Number System we build on this result and show that the same equation holds for the set of negative integers.

To begin, we let 'k' represent a positive integer, so that we have:

$$k = k \qquad \text{... by Theorem 2.1}$$
$$-1 = -1 \qquad \text{... by Theorem 3.1}$$
$$\therefore \quad k \times -1 = k \times -1 \qquad \text{... by Deduction Axiom}$$
$$\therefore \quad -k = -k \qquad \text{... by Definition 3.1}$$

In this way, we have shown that for any given negative integer '$-k$', the equation '$-k = -k$' is a true equation. For the integer variable 'a', we can declare that the equation '$a = a$' is a true equation since for every integer number assigned to 'a', we get a true equation. This completes the proof of the theorem and extends the equivalent theorem on Natural Numbers.

We now want to extend the Reflexive Theorem described above to apply to terms also.

Theorem 3.6   Reflexive Theorem for Terms

If 't' is a general integer term, then this term also has the reflexive property so that:

$$t = t$$

Proof

A term 't' is a finite string of symbols constructed from integer numbers and variables using the operators '$+$' and '$\times$' (and also supported by parentheses).

Following the proof for natural number terms in Theorem 2.7, the equation '$t = t$' is also true for integer number terms as Theorem 3.5 assures us that '$a = a$' for every integer number 'a'. Therefore, this theorem is also true for the Integer Number System.

Now that we have the Reflexive Theorem for Terms, '$t = t$', we will develop several theorems that will help us operate with terms. First, we need to show how some of the basic axioms also apply to terms and not just integers and variables.

Theorem 3.7   Properties of Terms Theorem

If three general terms (i.e. constructed from integer numbers and/or variables) are given by 't_1', 't_2' and 't_3', then these terms also have the properties expressed in the Axioms of the Integer Number System.

Proof

Following the proof for natural number terms in Theorem 2.8, this proof follows exactly the same logic and, hence, this theorem is also true for the Integer Number System.

In Chapter Two we observed that it was tedious using the Transitive Axiom in its restrictive form to carry out number and variable substitutions; the same principle applies in this chapter. We will again state the Substitution Theorem as it is applied to integer numbers, variables and terms, and provide an abbreviated proof.

Theorem 3.8   Substitution Theorem

Let's assume we are given an algebraic term in an integer variable 'a' up to the power of 'n' with the integer coefficients 'c_0', 'c_1', 'c_2' up to 'c_n', and we are also given another integer variable 'b'.

In the general algebraic term with coefficients, namely '$c_0 \times a^0 + c_1 \times a^1 + ... + c_n \times a^n$', for each assignment of integers to 'a' and 'b' where '$a = b$', then 'b' can be substituted for every instance of 'a' in this term and not result in the value of this term changing.

In this way:

$$c_0 \times a^0 + c_1 \times a^1 + \ldots + c_n \times a^n = c_0 \times b^0 + c_1 \times b^1 + \ldots + c_n \times b^n$$

Proof

From Theorem 3.6 we know that all the equations equating coefficients (i.e. terms) to themselves (namely: '$c_0 = c_0$', '$c_1 = c_1$', ... , '$c_n = c_n$') are true equations. Also, from Theorem 3.7 we know that terms also obey the Axioms of the Integer Number System.

The remainder of this proof follows directly on from Theorem 2.9. If we assign values so that '$a = b$', we can substitute 'b' for 'a' in the above algebraic term with coefficients in the variable 'a' without changing the value of that term. In this way, the theorem is also true for the Integer Number System.

In the next theorem, we set out to prove an extension of the result '$a \times 0 = 0$' for the Natural Number System from Chapter Two where 'a' will now be an integer variable.

Theorem 3.9   Null Multiplication of Integers Theorem

Given 'a' is an integer number variable, then:

$$a \times 0 = 0$$

Proof

From Chapter Two we already know that when we assign any one of the numbers $\{0, 1, 2, 3, \ldots\}$ to the variable 'a' then '$a \times 0 = 0$' by Theorem 2.12. Consequently, using Definition 3.2(1) and Definition 3.3, we know this theorem is true for zero and the positive integers. At this stage, we need to prove this result is true for the negative integers as well.

Let the general negative integer be given by: '$-k = -1 \times k$' as outlined in Definition 3.1. We now wish to show that '$-k \times 0 = 0$' where 'k' is assigned a positive integer:

$$\begin{aligned}
-k \times 0 &= (-1 \times k) \times 0 && \text{... by Definition 3.1} \\
&= -1 \times (k \times 0) && \text{... by Associative Axiom '$\times$'} \\
&= -1 \times 0 && \text{... by Theorem 2.12} \\
\therefore \quad -k \times 0 &= 0 && \text{... by Theorem 3.3(2)}
\end{aligned}$$

Therefore, the equation '$a \times 0 = 0$' is true when 'a' is assigned positive integers, zero or negative integers and, thus, is true for all the integer numbers. Hence, the theorem has been proved.

Note:

From Theorem 3.7 we can also declare that for a general term 't' the equation '$t \times 0 = 0$' must also be a true equation.

———————————

The next theorem is an extension of the result '$a \times 1 = a$' for the Natural Number System and is called the Identity Theorem for Integer Multiplication.

Theorem 3.10   Identity Theorem for Integer Multiplication

Let 'a' represent an integer variable. The number '1' is the multiplicative identity for 'a', so that:

$$a \times 1 = a$$

Proof

From Chapter Two we already know that when we assign any one of the natural numbers $\{0, 1, 2, 3, ...\}$ to the variable 'a' then '$a \times 1 = a$' by Theorem 2.11. Consequently, we know this theorem holds true for zero and the positive integers by Definition 3.2(1) and Definition 3.3. So we now need to prove this result is true for the negative integers as well.

If we assume 'k' is a positive integer, we have: '$k \times 1 = k$' from Theorem 2.11. From Theorem 3.1, we also know that '$-1 = -1$' so by the Deduction Axiom we have:

$$-1 \times (k \times 1) = -1 \times k \qquad \text{... by Deduction Axiom}$$
$$(-1 \times k) \times 1 = -1 \times k \qquad \text{... by Associative Axiom '}\times\text{'}$$
$$-k \times 1 = -k \qquad \text{... by Definition 3.1}$$

Therefore, the integer variable 'a' in the equation '$a \times 1 = a$' is true when 'a' is assigned positive integers, zero or negative integers and, hence, is true for all the integer numbers. In this way, the theorem has been proved.

Note:

From Theorem 3.7 we can also declare that for a general term 't', the equation '$t = t \times 1$' must also be a true equation.

———————————

Now that we have established the above two theorems, it is straightforward to prove the generalisation of the Identity Axiom for Addition of Integers.

Theorem 3.11   Identity Theorem for Integer Addition

Let 'a' represent an integer variable. The number '0' is the additive identity for 'a', so that:

$$a + 0 = a$$

Proof

We will now use the Identity Axiom for Addition, '$1 + 0 = 1$', and the Reflexive Theorem for Terms (Theorem 3.6), '$a = a$' with the Deduction Axiom to get the following equation:

$$a \times (1 + 0) = a \times 1 \qquad \text{... by Deduction Axiom}$$

$$a \times 1 + a \times 0 = a \times 1 \qquad \text{... by Distributive Axiom}$$

However, '$a \times 1 = a$' by Theorem 3.10 above, and '$a \times 0 = 0$' by Theorem 3.9. Consequently, by substitution, we have:

$$a + 0 = a \qquad \text{... by Theorems 3.9 and 3.10}$$

Hence, this theorem – '$a + 0 = a$' – has been proved for all integers.

Note:

From Theorem 3.7 we can also declare that for a general term 't', the equation '$t + 0 = t$' must also be a true equation.

We can now use the above theorems to extend our Inverse Axiom for Addition (i.e. '$1 + -1 = 0$') to include integers, variables and terms.

Theorem 3.12   Inverse Theorem for Integer Addition

Let 'a' represent an integer variable. The variable '$-a$' is the additive inverse for 'a', so that:

$$a + -a = 0$$

Proof

If we start with the left-hand side of this equation, we have:

$$a + -a = a + -1 \times a \qquad \text{... by Definition 3.4}$$

$$= 1 \times a + -1 \times a \qquad \text{... by Theorem 3.10}$$

$$= a \times (1 + -1) \qquad \text{... by Distributive Axiom}$$

$$= a \times 0 \qquad \text{... by Inverse Axiom for '+'}$$

$$\therefore \qquad a + -a = 0 \qquad \text{... by Theorem 3.9}$$

So the theorem is true for the integer variable 'a' and thus for all integers. In this way, the theorem has been proved.

Note:

From Theorem 3.7, we can also declare that for a general term 't', the equation '$t + -t = 0$' must also be a true equation.

_____

The above theorem demonstrates that for every integer number 'a' there is an **opposite** integer number '$-a$' always related to it, so that every integer number and its **opposite** always add to zero. Therefore, 'a' and '$-a$' are additive opposites of each other.

We now set out to investigate the consequences of multiplying negative numbers.

Theorem 3.13   Multiplication of Negative Integers Theorem

If we assume 'a' and 'b' are integer variables, then the product of the negatives of these two variables is given by the following equation:

$$-a \times -b = a \times b$$

Proof

We prove this theorem deductively beginning with Definition 3.4 (i.e. '$-a = -1 \times a$') as follows:

$$-a \times -b = -1 \times a \times -1 \times b \qquad \text{... by Definition 3.4}$$

$$= -1 \times -1 \times a \times b \qquad \text{... by Commutative Axiom '$\times$'}$$

$$= 1 \times a \times b \qquad \text{... by Theorem 3.3(3)}$$

$$\therefore \qquad -a \times -b = a \times b \qquad \text{... by Theorem 3.10}$$

So the theorem is true for the integer variable 'a' and hence for all integers. Therefore, the theorem has been proved.

Note:

From Theorem 3.7 we can also declare that for a general term 't_1' and 't_2' the equation '$-t_1 \times -t_2 = t_1 \times t_2$' must also be a true equation.

———————————

This next theorem demonstrates that a double application of the minus unary operator to an integer does not change that variable.

Theorem 3.14   Double Negatives

If 'a' is an integer variable then:

$$-(-(a)) = a$$

Proof

We prove this equation is true from the previous equation in 'a' as follows:

$$
\begin{aligned}
-(-(a)) &= -1 \times -(a) && \text{... by Definition 3.4} \\
&= -1 \times -1 \times a && \text{... by Definition 3.4} \\
&= (-1 \times -1) \times a && \text{... by Associative Axiom `$\times$'} \\
&= 1 \times a && \text{... by Theorem 3.3(3)} \\
\therefore \quad -(-(a)) &= a && \text{... by Theorem 3.10}
\end{aligned}
$$

So the theorem is true for the integer variable 'a' and hence for all integers. Therefore, the theorem has been proved.

Note:

From Theorem 3.7 we can also declare that for a general term 't' the equation '$-(-(t)) = t$' must also be a true equation.

———————————

We have now completed the introduction of the standard language and proofs of the basic theorems for the Integer Number System. With these axioms, definitions and theorems, we now have a model for our Physical Counting System that includes negative numbers that lays the foundations for all later chapters in this book.

Applications of the Integer Number System

In this section we can refer to three comments from the Summary Section of Chapter Two that also apply to the Integer Number System. In summary these are:

1. The algorithms for long addition and long multiplication involving negative numbers are simple applications of the above theorems in this chapter.

2. These algorithms are applied arithmetic and do not participate in the normal proofs and formal procedures of the Integer Number System.

3. When formatting general mathematical solutions, it is normal to place the calculations using algorithms to the right-hand side of the page. This indicates that the algorithm is not participating directly in the normal logical flow to reach the solution.

In this section, we again cover some simple applications of the theory developed in this chapter that form syllabus questions for basic Arithmetic and Algebra. These applications highlight the simple manipulation of both arithmetic and algebraic terms and equations which contain negative numbers. The first application illustrates how we can extend the information presented in the previous chapter in a way that accommodates the negative symbol.

Application 3.1

Evaluate the arithmetic term: '$-27 + 13$'.

We evaluate the term '$-27 + 13$' using our axioms, definitions and theorems in a similar manner to which we carried out the same operations with natural numbers, as follows:

$$-27 + 13 = -1 \times 27 + 1 \times 13 \qquad \text{... by Definition 3.1}$$
$$= -1 \times 27 + -1 \times -1 \times 13 \qquad \text{... by Theorem 3.3(3)}$$
$$= -1 \times (27 + -1 \times 13) \qquad \text{... by Distributive Axiom}$$
$$= -1 \times (27 + -13) \qquad \text{... by Definition 3.1}$$
$$= -1 \times (27 - 13) \qquad \text{... by Definition 3.5}$$
$$= -1 \times 14 \qquad \text{... by simplifying}$$
$$\therefore \quad -27 + 13 = -14 \qquad \text{... by Definition 3.1}$$

In this way, the arithmetic term '$-27 + 13$' has the value of '-14'.

———————————

All steps are added to the above proof to illustrate one way of using the axioms, definitions and theorems to reach the correct answer. In practice, when a student is answering an examination question, they will need to use their judgment to achieve a balance between how many steps to include against the risk of making a mistake by leaving out too much working. The aim of those examination questions that require us to 'work from first principles' is to test our knowledge to determine whether we are able to add in all the missing steps.

Application 3.2

Evaluate the arithmetic term: '-27×13'.

We evaluate the term '-27×13' using our axioms, definitions and theorems in a similar manner to which we carried out the same operations with natural numbers, as follows:

$$-27 \times 13 = -1 \times 27 \times 13 \qquad \text{... by Definition 3.1}$$
$$= -1 \times (27 \times 13) \qquad \text{... by Associative Axiom '}\times\text{'}$$
$$= -1 \times (351) \qquad \text{... by simplifying}$$
$$\therefore \qquad -27 \times 13 = -351 \qquad \text{... by Definition 3.1}$$

In this way, the arithmetic term '-27×13' has the value of '-351'.

———————————

Once again, this application shows that we can isolate the negative symbol '$-$' in a way that makes the multiplication of integers easily derivable using our primary school algorithms.

The next application demonstrates how to use the order of operations to evaluate complex arithmetic terms involving negative and positive integers.

Application 3.3

Evaluate the arithmetic term: '$-(-28 \times 12 - 28 + 13)$'.

We evaluate the term '$-(-28 \times 12 - 28 + 13)$' using the order of operations and our usual axioms, definitions and theorems in a similar manner to which we carried out the same operations with the set of integer numbers, as follows:

$$-(-28 \times 12 - 28 + 13)$$
$$= -(-28 \times 12 - 28 \times 1 + 13) \qquad \text{... by Identity Axiom '}\times\text{'}$$
$$= -(-28 \times (12 + 1) + 13) \qquad \text{... by Distributive Axiom}$$

$$= -(-28 \times 13 + 13) \qquad \text{... by simplifying}$$

$$= -(-28 \times 13 + 1 \times 13) \qquad \text{... by Identity Axiom for '×'}$$

$$= -(13 \times (-28 + 1)) \qquad \text{... by Distributive Axiom}$$

$$= -(13 \times -27) \qquad \text{... by simplifying}$$

$$= -(-351) \qquad \text{... by Application 3.2}$$

$$= 351 \qquad \text{... by simplifying}$$

Therefore:

$$-(-28 \times 12 - 28 + 13) = 351 \qquad \text{... by Theorem 3.14}$$

Hence, the arithmetic term '$-(-28 \times 12 - 28 + 13)$' has the value of '351'.

In the next application we illustrate the use of the integers as **directed** numbers along the number line. In other words, we assume the integers are set out along a number line as described at the beginning of the chapter so that the positive integers are to the right of the origin and negative integers are to the left of the origin.

Application 3.4

Students sitting in a classroom are required to create an arithmetic term that describes a journey of another student who is at the front of the class taking one-metre steps along a number line across the front of that classroom. We assume steps taken by the student to the right from '0', as seen by the seated students, are positive and steps taken to the left from '0' are negative.

The description of the starting position and steps taken is as follows:

1.   The student starts at the negative five-metre marker on the floor.

2.   The student then travels to the right for seven steps and then stops.

3.   The student travels three meters to the right and stops again.

4.   The student finally travels six meters to the left.

Find an arithmetic term that describes the journey taken by this student and the student's finishing position.

The steps can be replaced by numbers where a positive sign is used for steps taken to the right of zero and a negative sign is used for steps taken to the left of zero. In this way, by replacing these steps we have:

Step 1. The starting position is indicated by '-5'

Step 2. Seven steps to the right are indicated by: '7'

Step 3. Three steps to the right are: '3'

Step 4. Six steps to the left are indicated by: '-6'.

The arithmetic term that describes the path taken by the student using integer numbers is: '$-5 + 7 + 3 - 6$'. The value of this term is the finishing position which is at '-1' metres along the number line.

The following application is another example which demonstrates the way in which negative numbers are used naturally in our everyday lives. It is important because it affects our standard of living in a 'negative' way if we do not understand how to manage our money to gain the best return on our investment.

Application 3.5

Assume you have one hundred dollars in a bank account at the beginning of the month and on the second day of that month you add ten dollars. If, on the last day of the month, you earn five dollars in interest and you withdraw ten dollars to cover your monthly entertainment expenses, how much money will you have on the first day of the next month?

We write out each transaction against that account using the following steps:

Step 1. $100 (opening balance of the account for the month)

Step 2. $10 (deposited money)

Step 3. $5 (earned interest)

Step 4. $$-10$ (withdrawn money).

The amount of money left in the account for the beginning of the subsequent month is illustrated by: '$100 + $10 + $5 + $$-10$' which is equal to: '$105'.

The next application demonstrates that a variable may contain a 'hidden' negative sign in a term when we assign a negative integer to that variable.

Application 3.6

Evaluate the algebraic term: '$-(a - b)$' where 'a' and 'b' are assigned the values '-8' and '-3' respectively.

We evaluate the term '$-(a - b)$' by substituting for 'a' and 'b' and using the order of operations and our usual axioms and definitions for integers as follows:

$$-(a - b) = -(-8 - (-3)) \qquad \text{... by subst. for '}a\text{' and '}b\text{'}$$

$$= -(-8 + 3) \qquad \text{... by Theorem 3.10}$$

$$= -(-5) \qquad \text{... by simplifying}$$

$$\therefore \qquad -(a - b) = 5 \qquad \text{... by Theorem 3.10}$$

Therefore, the algebraic term '$-(a - b)$' evaluates to '5' for these assignments to 'a' and 'b'.

_____

We have now completed our application of the Integer Number System to model actions that take place in our everyday lives. It is worth pointing out here that society does not necessarily consistently model the integers when describing objects.

For example, when you enter a lift on the ground floor of a building, according to the Integer Number System, it should be labelled the '0' floor and when you want to go down one floor you should just have to press a button labelled '-1'. Within our society, we have adopted such conventions as labelling a floor in a lift so it defaults to the function of that floor (for example, it might be called 'Car Park A' instead of '-1').

Summary of the Integer Number System

At this point, we have covered the objectives of this chapter and so can now summarise our findings up to now. In terms of the Integer Number System, the most important definitions that have been introduced so far are:

1. Define the **Negative Integers** using the negative one symbol '-1' and the natural numbers (excluding '0' and '1') to create an unlimited set of Negative Integers which can be written as: $\{\dots, -4, -3, -2, -1\}$.

2. Define the full set of **Integers** which consist of the set of natural numbers and the set of negative integers.

3. Define the **Positive Integers** as the set of natural numbers excluding zero.

4. Define the **negative unary operator** '$-$' as a unary operator to create the additive opposite of an integer number.

5. Define the **subtraction operator** '$-$' as a binary operator to allow us to take one integer away from another integer.

Within this chapter, we have explored the operations of addition and multiplication of integer numbers and seen how the operations of long addition and long multiplication are still applied when using negative numbers.

This chapter has also shown how we can extend the development of the most elementary number system – the Natural Number System – by extending the language, axioms, definitions and theorems of the Natural Number System to include negative numbers and then develop the theorems of the Integer Number System.

This chapter further highlights the process of assigning meaning to symbols by first using the Axioms of the Integer Number System and then using the Axioms of Logic to prove the basic theorems of the Integer Number System. The Axioms of the Integer Number System are summarised in Table 3.6 below:

3.6 The Axioms of the Integer Number System		
Axiom Type	**Addition '+'**	**Multiplication '×'**
Closure	1. The binary operators '+' and '×' (i.e. '$a + b$' and '$a \times b$') give a unique integer number for each integer number assigned to the variable 'a' and to the variable 'b'.	
Commutative	2. $a + b = b + a$	3. $a \times b = b \times a$
Associative	4. $(a + b) + c = a + (b + c)$	5. $(a \times b) \times c = a \times (b \times c)$
Identity	6. $1 + 0 = 1$	7. $1 \times 1 = 1$
Distributive	8. $a \times (b + c) = a \times b + a \times c$	
Inverse	9. $1 + -1 = 0$	Not applicable: there isn't a multiplicative inverse for the Integers.

Table 3.6

Note:

The additional axiom that has been included is the Inverse Axiom for Addition, namely '$-1 + 1 = 0$'. This axiom allows us to generate all the negative integers using '-1'.

The Axioms of Logic used to manipulate the Axioms of the Integer Number System in this chapter have not changed in form since Table 2.5 in Chapter Two. However, they can now be applied to the broader set of numbers – the Integers. The basic theorems of this chapter are the starting point for proving all other results in the Integer Number System.

This study of the Integer Number System reinforces the structure and processes that are performed in Mathematics. We have now extended the language to describe terms and equations used in the Integer Number System. When students are able to correctly apply the axioms, definitions and theorems discussed in this chapter to a broad range of physical systems they will be on the road to achieving mastery in the Integer Number System.

In Chapter Four we will continue to extend the language we have used up to this point in order to incorporate new symbols that define Rational Numbers as Fractions. We will also be using the same methods we covered in the theory of Natural Number Arithmetic and Integer Arithmetic to explore the nature of fractions and develop the theory of the **Rational Number System**.

CHAPTER 4

THE RATIONAL NUMBER SYSTEM USING FRACTIONS

Overview of the Rational Number System using Fractions

In this chapter we extend our language of the Integer Number System to include the set of 'multiplicative inverses' of all integers (excluding '0'). The arithmetic terms formed from the resulting set of numbers that includes these multiplicative inverses is called the set of 'rational numbers'. This set of rational numbers (with their associated alphabet and operations) is called the Rational Number System. In this chapter you will learn how to represent the rational numbers as 'fractions'.

Once again, our study begins by applying the operations of addition and multiplication to this new expanded set of numbers – the rational numbers – and building the extended body of knowledge we refer to as the Rational Number System. The operations of subtraction and division are then defined in terms of these addition and multiplication operations, respectively. By extending our model of counting to include the concept of numbers between the integer numbers along the number line, this process takes us further on our journey of developing mastery over the language of the number systems.

In Chapter Three we developed the set of integers by modelling the equally-spaced negative numbers to the left of '0' along the number line using the additive inverse of the set of natural numbers. In this chapter we will model new numbers between the integers using the **multiplicative inverse** of the set of integers. That is, we will model more of the number line by assigning new numbers to certain points marked along this line between the integer numbers. In this way, for every non-zero integer 'a', a corresponding multiplicative inverse is denoted by '$\frac{1}{a}$' such that we can multiply 'a' by '$\frac{1}{a}$' to get '1' (i.e. '$a \times \frac{1}{a} = 1$').

Once again, we will ensure these new numbers comply with the same axioms as the Integer Number System, but that they also comply with one extra axiom. The Rational Number System also includes an extended set of variables to those we used in the Integer Number System and this concept is outlined below.

Pictorial Representation of Rational Numbers using Fractions

At this point it is helpful to visualise what this process of enhancing the set of integers with the set of rational numbers is actually modelling. We begin this process by using the set of integers from Chapter Three as labels for a set of 'equally-spaced' points along the number line. In this chapter we will be adding numbers like '$\frac{1}{2}$' and '$1\frac{1}{3}$' which can be considered as labelling proportionally-spaced points between the set of points with integer values along this number line. This concept is illustrated in Figure 4.1 below.

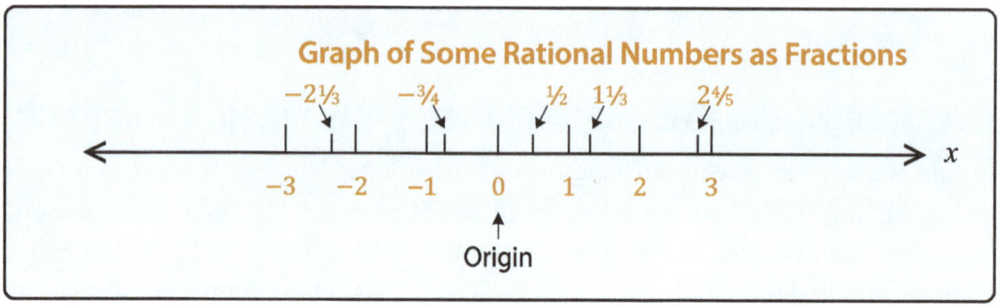

Figure 4.1

Once again, it is only those points that are proportionally spaced away from the arbitrary point labelled as the coordinate '0' that can be assigned fractions as coordinates. For example, '$2\frac{17}{24}$' corresponds to a proportionally-spaced point away from the origin, but '$\sqrt{2}$' (a non-rational number which will be defined in Chapter Eleven) does not correspond to a proportionally-spaced point. These remaining points will have to wait until our final extension to the Rational Number System which is called the Real Number System.

Background and Context of the Rational Number System using Fractions

To commence our study of the Rational Number System, we will need to introduce a new symbol called the **division symbol** that will allow us to represent Rational Numbers using Fractions. The division symbol can be represented by three possible variations: '÷', '−' or '/'. These three symbols for division are interchangeable and any of the symbols can be used to represent the same operation. In the examples throughout this chapter we will generally use the well-known division symbol '$\frac{1}{\square}$' as this has now become the standard form of representation.

In mathematical terms, the symbol '$\frac{1}{\square}$' is called the multiplicative inverse. After defining the 'Multiplicative Inverse of an Integer' using this symbol, we will extend our Axioms of the Integer Number System to assign meaning to the Rational Numbers using Fractions.

The use of the symbol '$\frac{1}{\square}$' can be considered as breaking up (or fracturing) the interval from '0' to '1' into equal pieces known as fractions. For example, when dividing the integer '1' by the integer '3', the number '1' is divided into '3' equal pieces with each piece being labelled '$\frac{1}{3}$', so that '$3 \times \frac{1}{3} = 1$'. By extending the set of Integers by including fractions, we arrive at the set of Rational Numbers using Fractions.

To continue on from the example above, if we treat '$\times 3$' as an operation that extends a number such as '7' to the number '21' (i.e. '$7 \times 3 = 21$'), then we can treat the operation '$\times \frac{1}{3}$' as the '**undo**' operation that reduces '21' back to the number '7' (i.e. '$21 \times \frac{1}{3} = 7$'). Therefore, when we consecutively apply the operations of '$\times 3$' and '$\times \frac{1}{3}$' to the number '7' (i.e. '$7 \times 3 \times \frac{1}{3} = 7$'), it is the same as multiplying '7' by '1'.

Now that we have established the set of symbols for the Rational Number System using Fractions, we need to assign meaning to these new symbols using the same approach we used in the previous chapters. This approach is outlined in the section below.

Approach to the Rational Number System using Fractions

In this section our aim will be to assign meaning to the alphabetic symbols of the language of the Rational Number System using Fractions. This language is an extension of the language we used in the Integer Number System as it applies to a further extended set of numbers – the Rational Numbers using Fractions. For the remainder of Chapter Four we will simply refer to 'Rational Numbers using Fractions' as **fractions**.

Language of The Rational Number System using Fractions

At this point we again extend the alphabet that was developed in the Integer Number System to include the '**multiplicative inverse**' symbol '$\frac{1}{\square}$'. This will give us the basic symbols of the alphabet required for the Rational Number System using Fractions as shown in Table 4.1 on the following page.

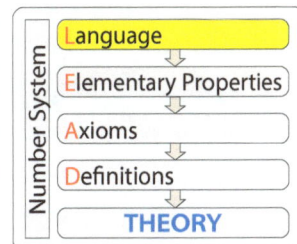

Number System

Language → Elementary Properties → Axioms → Definitions → **THEORY**

4.1 The **Alphabet** of the Rational Number System using Fractions	
Symbols	**Meaning from:**
'0', '1', '+', '×', '(', ')'	The Axioms of the Natural Number System
'=', '≠'	The Axioms of Logic
'2', '3', '4', '5', '6', '7', '8', '9'	Definition 2.1
','	Definition 2.3
'...'	Definition 2.8
'a', 'b', 'c', ... , 'a_1', 'a_2', 'a_3', ...	Variables to which we assign integer numbers
'−1'	The Axioms of the Integer Number System
'−'	Definitions 3.4 and 3.5
'$\frac{1}{\square}$'	'$\frac{1}{\square}$' is called the multiplicative inverse and derives its meaning from the equation: $a \times \frac{1}{a} = 1$
'x', 'y', 'z', ...	Variables to which we assign rational numbers using fractions

Table 4.1

In this extended language – the Rational Number System using Fractions – we now use variables to represent fractions. In this way, a variable such as 'x', 'y' or 'z' is a symbol of the alphabet to which we can assign any (and all) fractions from the set $\{..., -3, -2\frac{2}{3}, -2, -1\frac{1}{2}, -1, \frac{-1}{2}, 0, \frac{1}{4}, \frac{2}{3}, 1, 1\frac{7}{9}, 1\frac{5}{6}, 2, 3, ...\}$. However, we will continue to use 'a', 'b' and 'c' to represent integer number variables in this chapter.

This principle is summarised in Table 4.2 below.

4.2 The **Language** of the Rational Number System using Fractions	
In this table, 'a', 'b', 'c' and 'd' represent integer variables and 'x', 'y' and 'z' represent fraction variables.	
1. **Arithmetic Terms**	An individual fraction, or sum and/or product of fractions (each product is called a **factor** of the term) constructed from integer numbers, is referred to as an **arithmetic term** or simply a **term**. In this way, '$\frac{5}{1}$', '$\frac{5}{1} \times \frac{-2}{3}$', '$2 + \frac{3}{4}$' or '$\frac{5}{1} \times 2 \times \frac{-1}{3}$' are all finite valid strings of symbols in the Rational Number System using Fractions called **arithmetic terms**.

4.2 The **Language** of the **Rational Number System using Fractions**	
2. **Arithmetic Equations**	When we have arithmetic terms on either side of an '$=$' or '$\neq$' sign, we call this equation an **arithmetic equation** or simply an **equation**. For example, '$\frac{2}{3} \times -3 = -2$', '$\frac{2}{1} \times 4 = \frac{3}{1} + (\frac{12}{2} - 1)$' and '$\frac{3}{1} - 5 = \frac{-2}{1}$' are all finite valid strings of symbols in the Rational Number System using Fractions called **arithmetic equations**.
3. **Algebraic Terms and Equations**	If any terms or equations contain one or more variables, then we refer to them as **algebraic** terms or equations. For example, '$\frac{a}{1}$', '$\frac{a}{b}$', '$\frac{2}{c+3}$', '$\frac{a+2b+3c}{4d+13}$' and '$\frac{5c}{d}$' are all finite valid strings of symbols called **algebraic fractions** (or **algebraic terms**). However, 'x', '$x + \frac{2}{1} \times -3$', '$\frac{-5}{1} + x \times y \times z$', '$3 \times x \times y + \frac{-7}{1} \times x \times x \times y \times z$' are also **algebraic terms**. The equations, '$y = 1$', '$z \neq 0$', '$z + 2 = \frac{-5}{1}$', '$z - 3 \times z \times y \times x = 2 \times x + 5 \times z$' are all finite valid strings of symbols in the Rational Number System using Fractions called **algebraic equations**.

Table 4.2

To express the set of integers $\{\dots, -3, -2, -1, 0, 1, 2, 3, \dots\}$ in fraction notation, we write them as $\{\dots, \frac{-3}{1}, \frac{-2}{1}, \frac{-1}{1}, \frac{0}{1}, \frac{1}{1}, \frac{2}{1}, \frac{3}{1}, \dots\}$. However, when referring to integers as fractions, it is traditional practice in Mathematics to leave out the division by '1' to make rational number terms and equations more readable. This allows us to write fractions with a denominator of '1' in a more compact format. For example: $\{\dots, -3, -2\frac{2}{3}, -2, -1\frac{1}{2}, -1, \frac{-1}{4}, 0, \frac{1}{4}, \frac{2}{3}, 1, 1\frac{5}{6}, 2, 3, \dots\}$. We still treat all of these numbers in this set as fractions.

Once again, we start our investigation by assuming that each of the variables 'x', 'y' and 'z' can be assigned any rational number (in fraction form) and these can be combined with the operators '$+$' and/or '$\times$' to form terms and equations.

Order of Operations

As we discovered in previous chapters, there is an inherent ambiguity in the order in which we should apply operators if no parentheses are used to determine their order. The same **Order of Operations** convention that we applied to the Integer

Number System is also extended to include the set of fractions and to accommodate the new symbol for multiplicative inverses '$\frac{1}{\square}$'.

To evaluate these general terms composed of fractions, we use the Order of Operations as shown in Table 4.3 below.

4.3 Order of Operations for the Rational Number System using Fractions
The process used to evaluate arithmetic and algebraic terms is determined by the **order of operations** and proceeds according to the following steps:

1.	The **parentheses** have highest priority and control the order of evaluation. Parentheses may be removed by evaluating inside the parentheses **or** by using the Distributive Axiom.
2.	Next evaluate **exponents** (i.e. an integer index to which a base number is raised).
3.	Where there are no parentheses, then we evaluate **multiplication** '×' and **division** '÷' from left to right.
4.	Finally, evaluate **addition** '+' and **subtraction** '−' from left to right.

Table 4.3

The Order of Operations for the Rational Number System using Fractions can be remembered by using the acronym '**PEMDAS**' which stands for:

- Parentheses
- Exponents
- Multiplication
- Division
- Addition
- Subtraction.

Order of Operations – PEMDAS

Evaluate the following term using Order of Operations:

$$Term = (3 \times 5) + 2^3 \times 7 \times 2^{-1} \div 2 - 12$$

Parentheses	$= (3 \times 5) + 2^3 \times 7 \times 2^{-1} \div 2 - 12$
Exponents	$= 15 + 2^3 \times 7 \times 2^{-1} \div 2 - 12$
Multiplication	$= 15 + 8 \times 7 \times 2^{-1} \div 2 - 12$
Division	$= 15 + 28 \div 2 - 12$
Addition	$= 15 + 14 - 12$
Subtraction	$= 29 - 12$
Answer	$= 17$

The acronym is dependent on the number system being used and has been extended in this chapter to include the division operation.

When counting with rational numbers as fractions, we retain the nine elementary properties of the Integer Number System and then add one new property called the Inverse Property of Multiplication that relates specifically to our new symbol, '$\frac{1}{\square}$'.

The Tenth Elementary Property of the Rational Number System using Fractions

Now that we have developed the language for the Rational Number System using Fractions, we can identify the additional elementary property of this system.

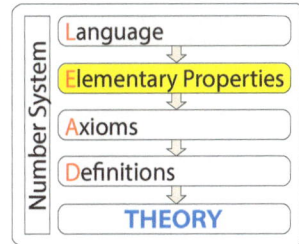

Inverse Property of Multiplication

This property will guarantee us that for every non-zero integer number there is another number (called the **inverse** number) such that when we multiply a number by its inverse, we get the number '1'. This inverse number is also called the **multiplicative inverse** as it is the inverse of a number with respect to the multiplication operation.

On our number line, this property models our ability to break a unit interval into a whole number of equal parts. For example, if we break up the interval between '0' and '1' into three equal lengths, then we would add the coordinate labels at the end of the first equal length (i.e. '$\frac{1}{3}$'), the end of the second equal length (i.e. '$\frac{2}{3}$') and the end of the third equal length (i.e. '$\frac{3}{3}$' or '1'). In symbols, we model this property using the following equation:

$$a \times \frac{1}{a} = 1$$

This new property shows how every integer number 'a' has a multiplicative inverse (denoted by '$\frac{1}{a}$') which is related to 'a' by the '**Inverse Axiom for Multiplication**'. This property forms axiom '10' which is modelled in Table 4.4 below.

The Axioms of the Rational Number System using Fractions

By starting with the Axioms of the Integer Number System and adding this new property, we have extended the set of axioms available to us to use. This new set of axioms now applies to the broader set of numbers called the Rational Numbers using Fractions.

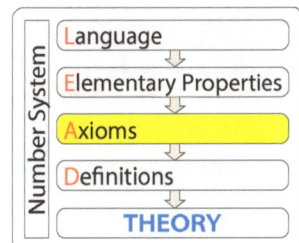

If we assume 'x', 'y', and 'z' are three rational number variables that can be assigned any rational number, then the Axioms of the Integer Number System are now extended in scope and apply to rational numbers. Also, we assume that 'a' is a variable that can be assigned any **integer** number. The Axioms of the Rational Number System using Fractions are set out in Table 4.4 on the following page.

4.4 The Axioms of the Rational Number System using Fractions		
Axiom Type	Addition '+'	Multiplication '×'
Closure	1. Let 'a' be a non-zero integer variable and 'x', 'y' and 'z' be rational number variables. This Closure Axiom states that the following operations: '$x + y$' and '$x \times y$' result in unique rational numbers.	
Commutative	2. $x + y = y + x$	3. $x \times y = y \times x$
Associative	4. $(x + y) + z = x + (y + z)$	5. $(x \times y) \times z = x \times (y \times z)$
Identity	6. $1 + 0 = 1$	7. $1 \times 1 = 1$
Distributive	8. $x \times (y + z) = x \times y + x \times z$	
Inverse	9. $1 + -1 = 0$	10. $a \times \frac{1}{a} = 1$

Table 4.4

According to our new axiom – the Inverse Axiom for Multiplication – every integer number variable 'a' (where 'a' cannot be assigned '0') has a multiplicative inverse (denoted by '$\frac{1}{a}$') that is always related to it.

The Axioms of Logic

Before we can deduce correctly from the Axioms of the Rational Number System using Fractions, we again need to use the Axioms of Logic which present the properties of reasoning that we learnt about in Chapter Two, Table 2.5. We will apply these same axioms to the Rational Number System using Fractions while, at the same time, observing the following note.

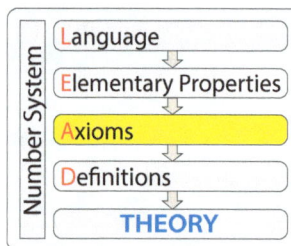

Note:

The variables 'a', 'b', 'c' and 'd' represented integer number variables in Chapter Three, however, in Chapter Four we supplement these variables with the variables 'x', 'y', 'z' and 'w' to indicate they are used as fraction variables.

Definitions of The Rational Number System using Fractions

Now that we have assigned meaning to the symbol '$\frac{1}{\square}$' in the Axioms of the Rational Number System using Fractions, we will use this symbol to create definitions for all the rational numbers.

In the Theory Section we will explore the relationship between this new symbol, the other symbols in our alphabet, the axioms and definitions.

After creating the multiplicative inverse of integers, we can now expand the set of integers to become the set of rational numbers.

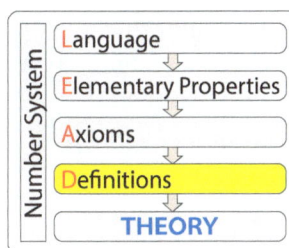

Number System	
Language	⇩
Elementary Properties	⇩
Axioms	⇩
Definitions	⇩
THEORY	

Definition 4.1   Rational Numbers

A **rational number** is defined to be any number formed by using any integer number from the first set multiplied by any multiplicative inverse from the second set below:

1.   The set of **integer numbers**: $\{... -3, -2 -1, 0, 1, 2, 3, ...\}$

2.   The set of **multiplicative inverses**: $\{..., \frac{1}{-3}, \frac{1}{-2}, \frac{1}{-1}, \frac{1}{1}, \frac{1}{2}, \frac{1}{3}, ...\}$.

———————————

We will demonstrate later in the Theory Section of this chapter that every arithmetic fraction can be written as a product by multiplying a number from the set of integers with a number from the set of multiplicative inverses (such as: '$-3 \times \frac{1}{2}$' or '$4 \times \frac{1}{-5}$') which are rational numbers.

The resulting rational numbers are cumbersome to read and write so we create a compressed notation (which we call 'fraction' notation) in our next definition.

Definition 4.2   Rational Numbers using Fraction Notation

Given 'a' and 'b' are integer variables and 'b' is not assigned '0', then we write the product of the rational numbers '$\frac{a}{1} \times \frac{1}{b}$', or equivalently '$a \times \frac{1}{b}$', in **fraction notation** as '$\frac{a}{b}$' so that:

$$\frac{a}{b} = a \times \frac{1}{b}$$

The symbol '$\frac{1}{b}$' can also be written as '$1/b$' or '$1 \div b$'.

———————————

Note:

We call the number on the top line of the fraction '$\frac{a}{b}$' the **numerator**. The number on the bottom line of this fraction is called the **denominator**.

In Definition 4.2 we wrote a fraction as one integer 'divided' by another integer. In the next definition we will create the 'reciprocal' of a fraction. In the Theory Section we shall discover that this 'reciprocal' of a fraction is the multiplicative inverse of this fraction.

Definition 4.3   Reciprocal of a Fraction

We define the **reciprocal** of a fraction '$\frac{a}{b}$' as this fraction with its numerator and denominator interchanged, i.e. '$\frac{b}{a}$' (where both 'a' and 'b' are integer variables and 'a' and 'b' cannot be assigned '0').

We now set out to define the division of one fraction by another fraction.

Definition 4.4   Fraction Divided by a Fraction – Complex Fraction

If 'a', 'b', 'c' and 'd' are any integer variables, where 'b', 'c' and 'd' cannot be assigned '0', then we define the fraction '$\frac{a}{b} \div \frac{c}{d}$' (called '$\frac{a}{b}$' **divided** by the fraction '$\frac{c}{d}$') as follows:

$$\frac{a}{b} \div \frac{c}{d} = \frac{a}{b} \times \frac{d}{c}$$

A fraction divided by another fraction is also called a **complex fraction**.

Note:

1. The division of two fractions '$\frac{a}{b} \div \frac{c}{d}$' is often written in two alternative notations as: '$\frac{\frac{a}{b}}{\frac{c}{d}}$' or '$\frac{a}{b} \times \frac{1}{\frac{c}{d}}$'.

2. This definition is often stated in the following way: 'To divide a fraction by another fraction, multiply the first fraction by the reciprocal of the second fraction'.

We will establish the connection between Definition 4.3 and Definition 4.4 in the Theory Section below. In the same way that integers can be positive or negative, we can define positive and negative fractions.

Definition 4.5   Positive Fractions

Let 'a' and 'b' be integer variables where 'a' and 'b' are assigned either **both** positive or **both** negative integers. We define a **positive fraction** as: '$\frac{a}{b}$'.

Note:

Later we will be able to show that '$\frac{2}{3} = \frac{-2}{-3}$' and hence both terms give a positive fraction.

We will now extend the use of the symbol '$-$' from Chapter Three to define negative fractions. Up until this stage, we wrote fractions as '$\frac{-2}{3}$' or '$\frac{2}{-3}$' but not as '$-\frac{2}{3}$'. However, when we label points to the left of the origin that are not negative integers, we refer to them as 'negative' fractions. As yet, we have not defined negative fractions; however, this is the purpose of our next definition.

Definition 4.6   Negative Fractions

If 'a' and 'b' are both positive integer variables, we can define the **negative fraction** '$-\frac{a}{b}$' as follows:

$$-\frac{a}{b} = -1 \times \frac{a}{b}$$

Note:

1. This definition is consistent with our definition of negative integers, namely Definition 3.1. For example, in the fraction '$-\frac{1}{2}$' we consider the negative symbol to be attached to the '$\frac{1}{2}$' symbol and we call it 'negative one-half'.

2. In Theorem 4.7 below, we prove that '$-\frac{1}{a} = \frac{1}{-a} = \frac{-1}{a}$' so that '$-\frac{1}{a}$' really is the multiplicative inverse of '$-a$', as well as being the additive inverse of '$\frac{1}{a}$'.

3. The fraction '$\frac{0}{1} = 0$' is the only fraction that is **neither** positive nor negative where 'a' and 'b' have been assigned '0' and '1', respectively.

We will now define another use of the '$-$' symbol as a unary operator '$-$' for fractions in general.

Definition 4.7   Unary Operator '$-$' Applied to a Fraction

The action of the negative symbol '$-$' as the **unary operator** is defined for a fraction '$\frac{a}{b}$' where the integer variables 'a' and 'b' can be assigned **any integer number** (except 'b' cannot be assigned '0') as follows:

$$-\left(\frac{a}{b}\right) = -1 \times \frac{a}{b}$$

Note:

In Definition 4.6 and Definition 4.7, we interpret the negative symbol '$-$' according to the context in which it appears. If the variables 'a' and 'b' are **not** both assigned a positive integer, then we must interpret the negative symbol as a unary operator and apply Definition 4.7. For example: '$-\frac{-2}{3} = -1 \times \frac{-2}{3}$'.

Now that we have established the definition of the unary operator '$-$', we can also define the familiar subtraction operator for fractions as follows.

Definition 4.8 Subtraction Operator '$-$' as a Binary Operator

For the fractions '$\frac{a}{b}$' and '$\frac{c}{d}$' where the integer variables 'a', 'b', 'c' and 'd' can be assigned **any integer numbers** (except 'b' and 'd' cannot be assigned '0'), the **subtraction operator** is defined by the combination of symbols '$\frac{a}{b} - \frac{c}{d}$' as follows:

$$\frac{a}{b} - \frac{c}{d} = \frac{a}{b} + -\frac{c}{d}$$

It is standard practice in Mathematics to classify fractions based on whether the denominator is greater than the numerator or vice versa. These classifications are known as '**proper fractions**' and '**improper fractions**', respectively.

Definition 4.9 Proper and Improper Fractions

A **proper fraction** is a fraction that lies between '-1' and '1' on the number line and is not equal to either '-1' or '1'. A fraction that is not equal to either '-1' or '1' and is not a proper fraction is called an **improper fraction**.

Examples of **proper fractions** include: '$\frac{2}{3}$', '$\frac{1}{7}$', '$-\frac{9}{10}$', '$\frac{1}{99}$', etc.

Examples of **improper fractions** are: '$\frac{3}{2}$', '$-\frac{99}{7}$', '$\frac{41}{40}$', etc.

When we reduce an improper fraction to an integer number and a proper fraction number, it is commonly called a '**mixed fraction**'. This practice is the motivation behind our next definition.

Definition 4.10 Mixed Fractions

A **mixed fraction** (or **mixed number**) is an improper fraction written in the form of an integer part and a fraction part with the '$+$' or '$-$' signs left out.

For example, we can write: '$2 + \frac{3}{5}$' (or the equivalent improper fraction '$\frac{13}{5}$') as the mixed fraction '$2\frac{3}{5}$' and '$-2 - \frac{3}{5}$' (or the equivalent improper fraction '$-\frac{13}{5}$') as the mixed fraction '$-2\frac{3}{5}$'.

Having established the above basic symbols of the alphabet, axioms and definitions, we can now complete the alphabet of the Rational Number System using Fractions. This alphabet is summarised in Table 4.5 below.

4.5 The **Alphabet** of the Rational Number System using Fractions	
Symbols	**Meaning from:**
'0', '1', '+', '×', '(', ')'	The Axioms of the Natural Number System
'=', '≠'	The Axioms of Logic
'2', '3', '4', '5', '6', '7', '8', '9'	Definition 2.1
','	Definition 2.3
'...'	Definition 2.8
'a', 'b', 'c', ... , 'a_1', 'a_2', 'a_3', ...	Variables to which we assign integer numbers
'−1'	The Axioms of the Integer Number System
'−'	Definitions 3.4 and 3.5
'$\frac{1}{\square}$'	The Axioms of the Rational Number System using Fractions
'x', 'y', 'z', ...	Variables to which we assign rational numbers as fractions
'$\frac{a}{b}$'	Definition 4.2
'÷'	Definition 4.2 and 4.4

Table 4.5

This alphabet for the Rational Number System using Fractions will be the starting alphabet for the Rational Number System using Index Form described in Chapter Five.

We have now completed our approach to developing the Rational Numbers using Fractions as an extension of the set of Integer numbers. In the next section of this chapter we will explore the theory that follows from this extension to our number system axioms.

Theory of the Rational Number System using Fractions

We now have sufficient language, axioms and definitions to proceed with the proof of some basic theorems. To develop the Theory of the Rational Number System using Fractions we will investigate the simple consequences of applying the Axioms of Logic to manipulate the Axioms of the Rational Number System using Fractions in order to prove all the basic theorems involving the multiplicative inverse of an integer.

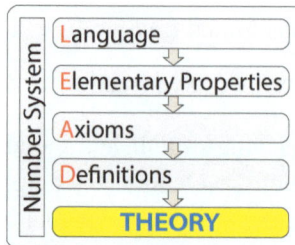

Number System
- Language
- Elementary Properties
- Axioms
- Definitions
- **THEORY**

The extended set of Rational Numbers using Fractions has an expanded set of theorems building on the theorems from previous chapters. We will now go on to prove several theorems that demonstrate the manipulation of fractions. The first and most important theorem to prove is that fractions also have the reflexive property.

Theorem 4.1 Reflexive Theorem for Fractions

If 'x' is a fraction variable, where we write '$x = \frac{a}{b}$' with 'a' and 'b' as integer variables and 'b' cannot be assigned '0', then the equation '$x = x$' is a true equation.

Proof

We have already proved that '$a = a$' for integer variables, so we only have to show that '$\frac{1}{b} = \frac{1}{b}$' for multiplicative inverses. We begin this process with the Reflexive Theorem for Integers as follows:

$$b = b \qquad \text{... by Theorem 3.5}$$

Assume: $$\frac{1}{b} = \frac{1}{b} \qquad \text{... by assumption}$$

∴ $$b \times \frac{1}{b} = b \times \frac{1}{b} \qquad \text{... by Deduction Axiom}$$

∴ $$1 = 1 \qquad \text{... by Inverse Axiom '×'}$$

Since this last equation is true by the Reflexive Theorem (Theorem 3.5), our original assumption that '$\frac{1}{b} = \frac{1}{b}$' must also be true. Therefore, when we multiply the equation '$a = a$' and '$\frac{1}{b} = \frac{1}{b}$' we arrive at:

$$a \times \frac{1}{b} = a \times \frac{1}{b} \qquad \text{... by Deduction Axiom}$$

∴ $$\frac{a}{b} = \frac{a}{b}$$

Hence, for every assignment of integers to the variables 'a' and 'b' (where '0' cannot be assigned to 'b') we have shown that '$x = x$' is a true equation for fraction variables. Therefore, the theorem has been proved.

_____

We can now extend the Reflexive Theorem outlined above so that it applies to terms as well as simple fractions.

Theorem 4.2   Reflexive Theorem for Terms

If 't' is a general term, then this term also has the reflexive property so that:

$$t = t$$

Proof

A term 't' is a finite string of symbols constructed from fractions and variables using the operators '$+$' and '$\times$' (supported by parentheses).

Following similar steps to the proof for integer number terms in Theorem 3.6, the equation '$t = t$' is also true for terms in the Rational Number System using Fractions because Theorem 4.1 assures us that '$x = x$' for every fraction assigned to 'x'. Therefore, this theorem is also true for the Rational Numbers using Fractions.

_____

Now that we have established the Reflexive Theorem for Terms (i.e. '$t = t$'), we will develop several theorems that help us operate with terms.

First, we need to show how some of the basic properties of the Rational Number System using Fractions also apply to terms and not just fractions and variables.

Theorem 4.3   Properties of Terms Theorem

If three general terms (constructed from fractions and/or variables) are given by 't_1', 't_2' and 't_3', then these terms also have the properties expressed in the Axioms of Rational Number System using Fractions.

Proof

Following similar steps to the proof for integer number terms in Theorem 3.7, this proof is deduced using exactly the same logic and, hence, this theorem is also true for the Rational Numbers using Fractions.

_____

In Chapter Two we observed that it was tedious using the Transitive Axiom in its restrictive form to carry out number and variable substitution. As a result, we will again state the Substitution Theorem as it applies to rational numbers using fraction variables and terms.

Theorem 4.4   Substitution Theorem

Assume we are given a general term in a rational number variable 'x' with coefficients 'c_0', 'c_1', 'c_2' up to 'c_n' which are terms. In this way, the general term can be written as:

$$c_0 \times x^0 + c_1 \times x^1 + ... + c_n \times x^n$$

In the case of all rational number assignments which make the equation '$x = y$' true, then the rational number variable 'y' can be substituted for 'x' everywhere in this general term and not change the value of that term so that:

$$c_0 \times x^0 + c_1 \times x^1 + ... + c_n \times x^n = c_0 \times y^0 + c_1 \times y^1 + ... + c_n \times y^n$$

Proof

From Theorem 4.2 we can conclude that all the coefficients which are terms have the reflexive property so that: '$c_0 = c_0$', '$c_1 = c_1$', ... , '$c_n = c_n$'.

The remainder of this proof follows directly on from Theorem 2.9 whereby if for the appropriate assignments, '$x = y$' is true, we can then substitute 'y' for 'x' in the above general algebraic term with coefficients in the variable 'x' and not change the value of that term. In this way, the theorem is also true for the Rational Numbers using Fractions.

───────────────

Now that we have the Reflexive Theorem using Fractions and the Substitution Theorem, we will use these theorems to simplify the following proofs.

The next result we are required to prove follows on directly from the Inverse Axiom for Multiplication. It leads us to the simplest outcome, namely that: '$\frac{1}{1} = 1$'.

Theorem 4.5   Trivial Identity Theorem

If 'a' is an integer variable and 'a' is not assigned the value '0', then the fraction '$\frac{a}{a}$' can be simplified to:

$$\frac{a}{a} = 1$$

Proof

If 'a' is assigned to a non-zero integer, then we have:

$$\frac{a}{a} = a \times \frac{1}{a} \qquad \text{... by Definition 4.2}$$

$$\therefore \qquad \frac{a}{a} = 1 \qquad \text{... by Inverse Axiom '×'}$$

Consequently, this theorem proves that when the numerator and the denominator are identical integer numbers (and not equal to zero), then the rational number '$\frac{a}{a}$' is '1'. In particular, this demonstrates that '$\frac{1}{1} = 1$'.

In the Approach Section of this chapter, we sought to write integers as fractions in order to demonstrate that all fractions could be written in the same format. The next theorem provides part of the justification for this approach.

Theorem 4.6   Integers as Fractions Theorem

If 'a' is an integer number variable, then 'a' can be written as the fraction '$\frac{a}{1}$', that is: $\frac{a}{1} = a$

Proof

First, from Theorem 4.5 we arrive at '$\frac{1}{1} = 1$' by assigning 'a' the value '1'. We now use this result to find '$\frac{a}{1}$' as follows:

$$\frac{a}{1} = a \times \frac{1}{1} \qquad \text{... by Definition 4.2}$$

$$= a \times 1 \qquad \text{... by Theorem 4.5}$$

$$= a \qquad \text{... by Identity Axiom '×'}$$

$$\therefore \qquad \frac{a}{1} = a$$

Therefore, any integer variable 'a' can be expressed as the fraction '$\frac{a}{1}$'.

In the Inverse Axiom for Multiplication (i.e. '$a \times \frac{1}{a} = 1$'), we observe that when we assign a negative integer to 'a', then the multiplicative inverse will have a negative number in the denominator. For example, the negative integer '-2' has a multiplicative inverse '$\frac{1}{-2}$'. However, we normally write the negative fraction '$\frac{1}{-2}$' as '$-\frac{1}{2}$'. The following theorem justifies this alternative way of writing the negative form of fractions.

Theorem 4.7   Applying the Unary Operator '−' to a Fraction Theorem

If 'a' and 'b' are integer variables and 'b' is not assigned '0', then the negative form of a fraction '$\frac{a}{b}$' (i.e. '$-\frac{a}{b}$') can be written as:

$$-\frac{a}{b} = \frac{-a}{b} = \frac{a}{-b}$$

Proof

This theorem is proved directly from Definition 4.7. We prove the first equation:

$$-\frac{a}{b} = -1 \times \frac{a}{b} \qquad \text{... by Definition 4.7}$$

$$= -1 \times a \times \frac{1}{b} \qquad \text{... by Definition 4.2}$$

$$= -a \times \frac{1}{b} \qquad \text{... by Definition 3.4}$$

$$\therefore \qquad -\frac{a}{b} = \frac{-a}{b} \qquad \text{... by Definition 4.2}$$

We now can prove the second equation using the Deduction Axiom and Reflexive Theorem.

Assume the equation '$\frac{-a}{b} = \frac{a}{-b}$' is true and apply deduction as follows:

$$\frac{-a}{b} = \frac{a}{-b} \qquad \text{... by assumption}$$

$$\frac{-a}{b} \times -b = \frac{a}{-b} \times -b \qquad \text{... by multiplying by '$-b$'}$$

$$-a \times \frac{1}{b} \times -b = a \times \frac{1}{-b} \times -b \qquad \text{... by Definition 4.2}$$

$$-a \times \frac{1}{b} \times -1 \times b = a \times \frac{1}{-b} \times -b \qquad \text{... by Definition 3.4}$$

$$-a \times -1 = a \qquad \text{... by Inverse Axiom '$\times$'}$$

Using Definition 3.4 and Theorem 3.3(3), this last equation can be shown to be a true equation. Therefore, by the Deduction Axiom, the equation '$\frac{-a}{b} = \frac{a}{-b}$' must be a true equation and we have now completed the proof of the theorem.

_____

The next theorem provides the extension of the result '$a \times 0 = 0$' for the Integer Number System in Chapter Three, so that it applies to fractions.

Theorem 4.8   Null Multiplication of Fractions Theorem

If 'x' is a fraction variable, then '0' is the null multiplier of 'x', so that:

$$x \times 0 = 0$$

Proof

The fraction variable 'x' can be written in the form '$x = \frac{a}{b}$', where 'a' and 'b' are integer variables and 'b' cannot be assigned '0'. We need to prove that '$\frac{a}{b} \times 0 = 0$'.

We commence this process by showing that the multiplicative inverse '$\frac{1}{b}$' satisfies the equation '$0 \times \frac{1}{b} = 0$'.

$$0 = 0 \times 1 \qquad \text{... by Theorem 2.3}$$

$$= 0 \times \frac{b}{b} \qquad \text{... by Theorem 4.5}$$

$$= (0 \times b) \times \frac{1}{b} \qquad \text{... by Definition 4.2}$$

$$= 0 \times \frac{1}{b} \qquad \text{... by Theorem 3.9}$$

$$\therefore \qquad 0 \times \frac{1}{b} = 0$$

It is now straightforward to show the more general result '$\frac{a}{b} \times 0 = 0$' is true as follows:

$$\frac{a}{b} \times 0 = a \times \frac{1}{b} \times 0 \qquad \text{... by Definition 4.2}$$

$$= a \times 0 \qquad \text{... by first half of proof}$$

$$= 0 \qquad \text{... by Theorem 3.9}$$

$$\therefore \qquad \frac{a}{b} \times 0 = 0$$

Therefore, the equation '$x \times 0 = 0$' gives a true equation when 'x' is assigned any fraction. Hence, the theorem has been proved.

Note:

From Theorem 4.3 we can also declare that for a general term 't' the equation '$t \times 0 = 0$' must also be a true equation.

The next theorem we set out to prove is an extension of the result '$a \times 1 = a$' for the Integer Number System from Chapter Three, so that it also applies to fractions.

Theorem 4.9   Identity Theorem for Fraction Multiplication

If 'x' is a fraction variable, then the multiplicative identity for 'x' is '1', so that:

$$x \times 1 = x$$

Proof

First, let 'a' and 'b' be integer variables (where 'b' cannot be assigned '0'). Now we have to prove: '$\frac{a}{b} \times 1 = \frac{a}{b}$'. We start with the left-hand side of this equation as follows:

$$\frac{a}{b} \times 1 = a \times \frac{1}{b} \times 1 \qquad \text{... by Definition 4.2}$$

$$= (a \times 1) \times \frac{1}{b} \qquad \text{... by Commutative Axiom '×'}$$

$$= a \times \frac{1}{b} \qquad \text{... by Theorem 3.10}$$

$$= \frac{a}{b} \qquad \text{... by Definition 4.2}$$

$$\therefore \qquad \frac{a}{b} \times 1 = \frac{a}{b}$$

Therefore, the equation '$x \times 1 = x$' results in a true equation when 'x' is assigned any fraction. In this way, the theorem has been proved.

Note:

From Theorem 4.3 we can also declare that for a general term 't', the equation '$t \times 1 = t$' must also be a true equation.

The next theorem we prove is an extension of the result '$a + 0 = a$' for the Integer Number System from Chapter Three, so that it also applies to all fractions.

Theorem 4.10   Identity Theorem for Fraction Addition

If 'x' is a fraction variable then the additive identity for 'x' is '0', so that:

$$x + 0 = x$$

Proof

We now take the Identity Axiom for Addition, '$1 + 0 = 1$', and the Reflexive Theorem, '$x = x$', and multiply these two equations together to get the following outcome:

$$x \times (1 + 0) = x \times 1 \qquad \text{... by Deduction Axiom}$$

$$x \times 1 + x \times 0 = x \times 1 \qquad \text{... by Distributive Axiom}$$

However, according to Theorem 4.8 and Theorem 4.9, using substitution we have:

$$x + 0 = x \qquad \text{... by Theorems 4.8 and 4.9}$$

Therefore, the equation '$x + 0 = x$' gives a true equation when 'x' is assigned any fraction. In this way, the theorem has been proved.

Note:

From Theorem 4.3 we can also declare that for a general term 't', the equation '$t + 0 = t$' must also be a true equation.

In the next theorem we prove an extension of the result '$a + -a = 0$' for the Integer Number System from Chapter Three, so that it applies to all fractions.

Theorem 4.11   Inverse Theorem for Addition of Fractions

If 'x' is a fraction variable, then '$-x$' is the additive inverse of 'x' so that:

$$x + -x = 0$$

Proof

From the definition of the unary operator '$-$' applied to a fraction we have:

$$
\begin{aligned}
x + -x &= x + -1 \times x &&\text{... by Definition 4.7} \\
&= 1 \times x + -1 \times x &&\text{... by Theorem 4.9} \\
&= (1 + -1) \times x &&\text{... by Distributive Axiom} \\
&= 0 \times x &&\text{... by Inverse Axiom '+'} \\
&= 0 &&\text{... by Theorem 4.8}
\end{aligned}
$$

$$\therefore \qquad x + -x = 0$$

Therefore, the equation '$x + -x = 0$' gives a true equation when 'x' is assigned any fraction. In this way, the theorem has been proved.

Note:

From Theorem 4.3 we can also declare that for a general term 't', the equation '$t + -t = 0$' must also be a true equation. Also note, from Definition 4.8, this theorem gives us the equation '$x - x = 0$'.

The next theorem we prove is an extension of the result '$-(-(a)) = a$' for the Integers from Chapter Three, so that it also applies to fractions.

Theorem 4.12 Double Negatives

If 'x' is a fraction variable, then the double negative of 'x' is just 'x' itself so that:

$$-(-(x)) = x$$

Proof

We start with the left-hand side of this equation as follows:

$$-(-(x)) = -1 \times -(x) \qquad \text{... by Definition 4.7}$$
$$= -1 \times -1 \times x \qquad \text{... by Definition 4.7 again}$$
$$= 1 \times x \qquad \text{... by Theorem 3.3(3)}$$
$$= x \qquad \text{... by Theorem 4.9}$$

$$\therefore \qquad -(-(x)) = x$$

Therefore, the equation '$-(-(x)) = x$' gives a true equation when 'x' is assigned any fraction. In this way, the theorem has been proved.

Note:

From Theorem 4.3 we can also declare that for a general term 't' the equation '$-(-(t)) = t$' must also be a true equation.

The next theorem establishes a simple rule for combining two fractions which do have the same denominator. For example, it will allow us to easily prove results such as: '$\frac{7}{3} + \frac{17}{3} = \frac{24}{3} = 8$'.

Theorem 4.13 Theorem for Addition of Fractions with Common Denominators

If 'a', 'b' and 'c' are integer variables (where 'b' cannot be assigned '0'), then we can add the fractions '$\frac{a}{b}$' and '$\frac{c}{b}$' as follows:

$$\frac{a}{b} + \frac{c}{b} = \frac{a+c}{b}$$

Proof

Given 'a', 'b' and 'c' are integer variables (where 'b' cannot be assigned '0'), then:

$$\frac{a}{b} + \frac{c}{b} = a \times \frac{1}{b} + c \times \frac{1}{b} \qquad \text{... by Definition 4.2}$$

$$= (a + c) \times \tfrac{1}{b} \qquad \text{... by Distributive Axiom}$$

$$= \tfrac{a + c}{b} \qquad \text{... by Definition 4.2}$$

$$\therefore \qquad \tfrac{a}{b} + \tfrac{c}{b} = \tfrac{a + c}{b}$$

This proves the theorem. The same proof holds for the subtraction of two fractions which do have the same common denominator.

_____

The next theorem sets out the equivalent result of multiplying two fractions, however, the denominators aren't required to be the same. For example, it will also allow us to easily get results like: '$\frac{7}{3} \times \frac{17}{2} = \frac{7 \times 17}{3 \times 2} = \frac{119}{6} = 19\frac{5}{6}$'.

Theorem 4.14   Theorem for Multiplication of Fractions

If 'a', 'b', 'c' and 'd' are integer variables (where 'b' and 'd' cannot be assigned '0'), then we multiply the fractions '$\frac{a}{b}$' and '$\frac{c}{d}$' as follows:

$$\frac{a}{b} \times \frac{c}{d} = \frac{a \times c}{b \times d}$$

Proof

The first step is to prove the simpler result '$\frac{1}{b} \times \frac{1}{d} = \frac{1}{b \times d}$'. We start by applying the Deduction Axiom to this equation as follows:

$$\tfrac{1}{b} \times \tfrac{1}{d} = \tfrac{1}{b \times d} \qquad \text{... by assumption}$$

$$(b \times d) \times \tfrac{1}{b} \times \tfrac{1}{d} = (b \times d) \times \tfrac{1}{b \times d} \qquad \text{... by multiplying by '}(b \times d)\text{'}$$

$$(b \times d) \times \tfrac{1}{b} \times \tfrac{1}{d} = 1 \qquad \text{... by Inverse Axiom '}\times\text{'}$$

$$b \times \tfrac{1}{b} \times d \times \tfrac{1}{d} = 1 \qquad \text{... by Commutative Axiom '}\times\text{'}$$

$$\therefore \qquad 1 \times 1 = 1 \qquad \text{... by Inverse Axiom '}\times\text{'}$$

Therefore, we have proved that '$\frac{1}{b} \times \frac{1}{d} = \frac{1}{b \times d}$' is a true equation.

We are now able to prove this result is also true for general fractions and not just for multiplicative inverses. We begin this process with the left-hand side of the equation stated in the theorem as follows:

$$\tfrac{a}{b} \times \tfrac{c}{d} = a \times \tfrac{1}{b} \times c \times \tfrac{1}{d} \qquad \text{... by Definition 4.2}$$

$$= a \times c \times \tfrac{1}{b} \times \tfrac{1}{d} \qquad \text{... by Commutative Axiom '}\times\text{'}$$

$$= a \times c \times \frac{1}{b \times d} \qquad \text{... by first half of this proof}$$

$$= \frac{a \times c}{b \times d} \qquad \text{... by Definition 4.2}$$

$$\therefore \quad \frac{a}{b} \times \frac{c}{d} = \frac{a \times c}{b \times d}$$

Consequently, this proves the theorem that in order to multiply two fractions we simply need to create a new fraction by multiplying the numerators and denominators of the original fractions.

A situation that occurs often is when a fraction has the same factor (called a 'common' factor) in both its numerator and denominator. For example, we would like to simplify results such as: '$\frac{2 \times 175}{3 \times 175} = \frac{2}{3}$'. The following theorem helps us to simplify such fractions.

Theorem 4.15   Cancelling a Factor Theorem

If 'a', 'b' and 'c' are integer variables (where 'b' and 'c' cannot be assigned '0'), then:

$$\frac{a \times c}{b \times c} = \frac{a}{b}$$

Proof

Given that 'a', 'b' and 'c' are integer variables, we start with the left-hand side of the equation above such that:

$$\frac{a \times c}{b \times c} = \frac{a}{b} \times \frac{c}{c} \qquad \text{... by Theorem 4.14}$$

$$= \frac{a}{b} \times 1 \qquad \text{... by Theorem 4.5}$$

$$\therefore \quad \frac{a \times c}{b \times c} = \frac{a}{b} \qquad \text{... by Theorem 4.9}$$

In this way, the theorem is proved.

The theorem outlined above gives rise to the necessity for additional language to describe the state of play. The terms on either side of the equation '$\frac{a \times c}{b \times c} = \frac{a}{b}$' in Theorem 4.15 are called **equivalent fractions**. For the term '$\frac{a \times c}{b \times c}$' we say that we have cancelled the common factor 'c' from the numerator and the denominator of this fraction to get '$\frac{a}{b}$'.

Next, we set out to prove the equivalent theorem to the Theorem for Multiplication of Fractions (Theorem 4.14) for the addition operator. In this case, this theorem requires that we create a common denominator for both of the fractions added together so that we can then use the Theorem for Addition of Fractions with Common Denominators (Theorem 4.13). For example, we will be able to prove a result such as: '$\frac{2}{3} + \frac{4}{5} = \frac{2 \times 5 + 4 \times 3}{3 \times 5} = \frac{22}{15} = 1\frac{7}{15}$'.

Theorem 4.16   Addition of Fractions with Different Denominators

If 'a', 'b', 'c' and 'd' are integer variables (where 'b' and 'd' cannot be assigned '0'), then we can add the fractions '$\frac{a}{b}$' and '$\frac{c}{d}$' as follows:

$$\frac{a}{b} + \frac{c}{d} = \frac{a \times d + c \times b}{b \times d}$$

Proof

Let 'a', 'b', 'c' and 'd' be integer variables (where 'b' and 'd' cannot be assigned '0'), then we begin with the left-hand side of the above equation. Our first task will be to create a common denominator so we can use Theorem 4.13. Therefore, we have:

$$\frac{a}{b} + \frac{c}{d} = \frac{a}{b} \times 1 + \frac{c}{d} \times 1 \qquad \text{... by Theorem 4.9}$$

However, to create a common denominator we must multiply the denominator of '$\frac{a}{b}$' by the denominator of the second fraction '$\frac{c}{d}$'. We do this by multiplying the first fraction '$\frac{a}{b}$' by '$\frac{d}{d} = 1$' and the second fraction '$\frac{c}{d}$' by '$\frac{b}{b} = 1$' as follows:

$$\frac{a}{b} + \frac{c}{d} = \frac{a}{b} \times \frac{d}{d} + \frac{c}{d} \times \frac{b}{b} \qquad \text{... by Theorem 4.5}$$

$$= \frac{a \times d}{b \times d} + \frac{c \times b}{b \times d} \qquad \text{... by Theorem 4.14}$$

$$= \frac{a \times d + c \times b}{b \times d} \qquad \text{... by Theorem 4.13}$$

$$\therefore \qquad \frac{a}{b} + \frac{c}{d} = \frac{a \times d + c \times b}{b \times d}$$

Consequently, this proves the theorem.

_____

Note:

When **subtracting** two fractions, the proof follows the same steps as in this theorem.

In Theorem 4.6 outlined above we demonstrated that all integers could be written in fraction form as '$\frac{a}{1}$' where 'a' is an integer variable. In Theorem 4.14 when we multiplied any two fractions we always got a fraction. Likewise, in Theorem 4.16, whenever we added any two fractions we always ended up with a fraction. These three theorems together prove that **any rational number can be written in fraction form**.

Now that we have proved theorems using addition, subtraction and multiplication of fractions, we need to prove a theorem using **division**, namely, the Inverse Theorem for Multiplication of Fractions. The inverse of a fraction using multiplication is also called the multiplicative inverse of a fraction. The multiplicative inverse of the fraction '$\frac{a}{b}$' is that fraction you assign to the symbol '□' in order to make '$\frac{a}{b} \times$ □ $= 1$' a true equation.

Theorem 4.17 Inverse Theorem for Multiplication of Fractions

The **multiplicative inverse** of any fraction '$\frac{a}{b}$' (where both 'a' and 'b' are integer variables and 'a' and 'b' cannot be assigned '0'), namely '$\frac{1}{\frac{a}{b}}$', satisfies the following equation:

$$\frac{a}{b} \times \frac{1}{\frac{a}{b}} = 1$$

Proof

From Definition 4.3 and Definition 4.4 we have:

$$\frac{a}{b} \times \frac{1}{\frac{a}{b}} = \frac{a}{b} \div \frac{a}{b} \qquad \text{... by Definition 4.4}$$

$$= \frac{a}{b} \times \frac{b}{a} \qquad \text{... by Definition 4.3}$$

$$= a \times \frac{1}{b} \times b \times \frac{1}{a} \qquad \text{... by Definition 4.2}$$

$$= a \times \frac{1}{a} \times \frac{1}{b} \times b \qquad \text{... by Associative Axiom '$\times$'}$$

$$= 1 \times 1 \qquad \text{... by Inverse Axiom '$\times$'}$$

$$\therefore \quad \frac{a}{b} \times \frac{1}{\frac{a}{b}} = 1 \qquad \text{... by Identity Axiom for '$\times$'}$$

In this way, the theorem has been proved.

_____

Note:

1. Our definition of division in Definition 4.4 guarantees us that a non-zero rational number divided by itself equals '1'.

2. It also follows from Theorem 4.17, that if we use a fraction variable 'x' with '$x = \frac{a}{b}$', we can write the result of this theorem as '$x \times \frac{1}{x} = 1$'. The above theorem proves that when we multiply a fraction by its multiplicative inverse we get '1'. This result is consistent with our creation of the multiplicative inverse of an integer 'a' – namely '$\frac{1}{a}$' – so that '$a \times \frac{1}{a} = 1$'. From the above theorem, we observe that the multiplicative inverse of a fraction is also the **reciprocal** of that fraction.

We have now completed the introduction of the standard language and proofs of the basic theorems for the **Rational Number System using Fractions**. With these axioms, definitions and theorems we now have an available model for our counting process with fractions.

Applications of the Rational Number System using Fractions

In this section we again explore some simple applications of the theory developed in this chapter (and that would be relevant as syllabus questions in Arithmetic and Algebra). These simple applications highlight the easy manipulation of arithmetic equations containing fractions.

The first application illustrates the origin of the common algorithm used to add general fractions. It also illustrates how we have to use Theorem 4.9, 4.5, 4.14 and 4.13 when we wish to derive a simple addition of fractions from 'first principles'.

Application 4.1

Evaluate the arithmetic term: '$-\frac{2}{3} + \frac{4}{5}$'.

We can evaluate the term '$-\frac{2}{3} + \frac{4}{5}$' by using our axioms, definitions and theorems as follows:

$$-\frac{2}{3} + \frac{4}{5} = -\frac{2}{3} \times 1 + \frac{4}{5} \times 1 \qquad \text{... by Theorem 4.9}$$
$$= -\frac{2}{3} \times \frac{5}{5} + \frac{4}{5} \times \frac{3}{3} \qquad \text{... by Theorem 4.5}$$
$$= -\frac{2 \times 5}{3 \times 5} + \frac{4 \times 3}{5 \times 3} \qquad \text{... by Theorem 4.14}$$
$$= -\frac{10}{15} + \frac{12}{15} \qquad \text{... by simplifying terms}$$
$$= \frac{-10 + 12}{15} \qquad \text{... by Theorem 4.13}$$

$$\therefore \quad -\frac{2}{3}+\frac{4}{5}=\frac{2}{15} \quad \text{... by simplifying terms}$$

Consequently, when we evaluate the term '$-\frac{2}{3}+\frac{4}{5}$', we get the fraction '$\frac{2}{15}$'.

As stated before Application 4.1, using the application of Theorem 4.14 in the manner described above can be referred to as: 'finding the answer from first principles'. However, if you were to just use Theorem 4.16, (i.e. the equation '$\frac{a}{b}+\frac{c}{d}=\frac{a\times d+c\times b}{b\times d}$'), then you have just used the algorithm. This is the short-cut we were taught in primary school. Therefore, proving a result from first principles implies you are using the steps in the proof of the original theorem.

Another algorithm we learnt in primary school is the algorithm for converting a mixed number to a fraction. As mixed numbers are just an abbreviated notation, all we have to do is to expand this notation before applying the axioms, definitions and theorems with fractions.

Application 4.2

Evaluate the arithmetic term with mixed fractions: '$-5\frac{2}{3}+3\frac{4}{5}$'.

We can evaluate the term '$-5\frac{2}{3}+3\frac{4}{5}$' by using our axioms, definitions and theorems.

$$
\begin{aligned}
-5\frac{2}{3}+3\frac{4}{5} &= -(5+\frac{2}{3})+3+\frac{4}{5} && \text{... by Definition 4.10}\\
&= -(\frac{5}{1}+\frac{2}{3})+\frac{3}{1}+\frac{4}{5} && \text{... by Theorem 4.6}\\
&= -\frac{5\times3+2\times1}{1\times3}+\frac{3\times5+4\times1}{1\times5} && \text{... by Theorem 4.16}\\
&= -(\frac{15+2}{3})+\frac{15+4}{5} && \text{... by simplifying terms}\\
&= -\frac{17}{3}+\frac{19}{5} && \text{... by simplifying terms}\\
&= \frac{-17}{3}+\frac{19}{5} && \text{... by Theorem 4.7}\\
&= \frac{-17\times5+19\times3}{3\times5} && \text{... by Theorem 4.16}\\
&= \frac{-28}{15} && \text{... by simplifying terms}\\
&= \frac{-15+-13}{15} && \text{... by rewriting '}-28\text{'}\\
&= \frac{-15}{15}+\frac{-13}{15} && \text{... by Theorem 4.13}\\
&= -\frac{15}{15}+-\frac{13}{15} && \text{... by Theorem 4.7}
\end{aligned}
$$

$$= -1 + -\frac{13}{15} \qquad \text{... by Theorem 4.5}$$

$$\therefore \qquad -5\tfrac{2}{3} + 3\tfrac{4}{5} = -1\tfrac{13}{15} \qquad \text{... by Definition 4.10}$$

Hence, when we evaluate the term '$-5\tfrac{2}{3} + 3\tfrac{4}{5}$', we get the mixed fraction '$-1\tfrac{13}{15}$'.

So once again, our axioms, definitions and theorems allow us to easily perform mathematical operations on any mixed numbers as well.

The next application illustrates the use of the Distributive Axiom in practical situations with numbers in mixed formats.

Application 4.3

Evaluate the arithmetic term '$\tfrac{3}{4} \times (-5\tfrac{2}{3} + 4)$' using the Distributive Axiom.

We can evaluate the term '$\tfrac{3}{4} \times (-5\tfrac{2}{3} + 4)$' by using our axioms, definitions and theorems as follows:

$$\tfrac{3}{4} \times (-5\tfrac{2}{3} + 4) = \tfrac{3}{4} \times (-5 - \tfrac{2}{3} + 4) \qquad \text{... by Definition 4.10}$$

$$= \tfrac{3}{4} \times (\tfrac{-5}{1} + \tfrac{-2}{3} + 4) \qquad \text{... by Theorem 4.6 \& 4.7}$$

$$= \tfrac{3}{4} \times ((\tfrac{-5 \times 3 + -2 \times 1}{1 \times 3}) + 4) \qquad \text{... by Theorem 4.16}$$

$$= \tfrac{3}{4} \times (\tfrac{-17}{3} + \tfrac{4}{1}) \qquad \text{... by simplifying terms}$$

$$= \tfrac{3}{4} \times \tfrac{-17}{3} + \tfrac{3}{4} \times \tfrac{4}{1} \qquad \text{... by Distributive Axiom}$$

$$= \tfrac{-17 \times 3}{3 \times 4} + \tfrac{4 \times 3}{1 \times 4} \qquad \text{... by Theorem 4.14}$$

$$= \tfrac{-17}{4} + \tfrac{12}{4} \qquad \text{... by Theorem 4.15}$$

$$= \tfrac{-17 + 12}{4} \qquad \text{... by Theorem 4.13}$$

$$= \tfrac{-5}{4} \qquad \text{... by simplifying terms}$$

$$= -\tfrac{4 + 1}{4} \qquad \text{... by Theorem 4.7}$$

$$= -(1 + \tfrac{1}{4}) \qquad \text{... by Theorem 4.13}$$

$$\therefore \qquad \tfrac{3}{4} \times (-5\tfrac{2}{3} + 4) = -1\tfrac{1}{4} \qquad \text{... by Definition 4.10}$$

In this way, when we evaluate the term '$\frac{3}{4} \times (-5\frac{2}{3} + 4)$', we get the mixed number '$-1\frac{1}{4}$'. Fewer steps would be required if we had evaluated the term '$-5\frac{2}{3} + 4$' first and then multiplied by '$\frac{3}{4}$'. However, the approach described above was shown in detail to illustrate the definitions and theorems discussed in this chapter in action.

———————————

Next, we explore an application demonstrating the use of the axioms, definitions and theorems to evaluate more complex looking terms involving the use of complex fractions (i.e. fractions in both the denominator and the numerator).

Application 4.4

Evaluate the arithmetic term: '$\dfrac{\frac{3}{4}}{\frac{5}{8}} \times -5\frac{2}{3} + 4$'.

We evaluate the term: '$\dfrac{\frac{3}{4}}{\frac{5}{8}} \times -5\frac{2}{3} + 4$', by using our axioms, definitions and theorems as follows:

$$\dfrac{\frac{3}{4}}{\frac{5}{8}} \times -5\frac{2}{3} + 4 = \frac{3}{4} \times \frac{8}{5} \times -5\frac{2}{3} + 4 \quad \text{... by Definition 4.4}$$

$$= \frac{3 \times 8}{4 \times 5} \times -5\frac{2}{3} + 4 \quad \text{... by Theorem 4.14}$$

$$= \frac{6 \times 4}{5 \times 4} \times -5\frac{2}{3} + 4 \quad \text{... by Theorem 4.14}$$

$$= \frac{6}{5} \times -5\frac{2}{3} + 4 \quad \text{... by Theorem 4.15}$$

$$= \frac{6}{5} \times -1 \times (\frac{5}{1} + \frac{2}{3}) + 4 \quad \text{... by Definitions 4.7 \& 4.10}$$

$$= -1 \times \frac{6}{5} \times (\frac{5 \times 3 + 2 \times 1}{1 \times 3}) + 4 \quad \text{... by Theorem 4.16}$$

$$= -1 \times \frac{6}{5} \times \frac{17}{3} + 4 \quad \text{... by simplifying terms}$$

$$= -1 \times \frac{2 \times 3 \times 17}{5 \times 3} + 4 \quad \text{... by Theorem 4.14}$$

$$= -1 \times \frac{2 \times 17}{5} + 4 \quad \text{... by Theorem 4.15}$$

$$= \frac{-34}{5} + \frac{4}{1} \quad \text{... by Theorem 4.7}$$

$$= \frac{-34 \times 1 + 4 \times 5}{5 \times 1} \quad \text{... by Theorem 4.16}$$

$$= \frac{-14}{5} \quad \text{... by simplifying terms}$$

$$= -\frac{10+4}{5} \qquad \text{... by Theorem 4.7}$$

$$\therefore \quad \frac{\frac{3}{4}}{\frac{5}{8}} \times -5\frac{2}{3} + 4 = -2\frac{4}{5}$$

Next, we present an application that demonstrates the use of the axioms, definitions and theorems to determine whether an arithmetic equation using complex fractions is true.

Application 4.5

Evaluate whether this arithmetic equation is true or false:

$$-\frac{\frac{9}{11}}{\frac{4}{13}} \times \frac{2}{3} \div \frac{\frac{5}{11}}{\frac{7}{13}} = -2\frac{1}{10}$$

We evaluate the terms on both sides of this equation using our axioms, definitions and theorems as follows:

$$-\frac{\frac{9}{11}}{\frac{4}{13}} \times \frac{2}{3} \div \frac{\frac{5}{11}}{\frac{7}{13}} = -2\frac{1}{10} \qquad \text{... given}$$

$$-\frac{9}{11} \times \frac{13}{4} \times \frac{2}{3} \times \frac{\frac{7}{13}}{\frac{5}{11}} = -2\frac{1}{10} \qquad \text{... by Definition 4.4}$$

$$-\frac{9}{11} \times \frac{13}{4} \times \frac{2}{3} \times \frac{7}{13} \times \frac{11}{5} = -2\frac{1}{10} \qquad \text{... by Definition 4.4}$$

$$-\frac{9^3}{11} \times \frac{13}{4_2} \times \frac{2}{3} \times \frac{7}{13} \times \frac{11}{5} = -2\frac{1}{10} \qquad \text{... by Theorem 4.15}$$

$$-\frac{3 \times 7}{2 \times 5} = -2\frac{1}{10} \qquad \text{... by simplifying terms}$$

$$-\frac{21}{10} = -2\frac{1}{10} \qquad \text{... by simplifying terms}$$

$$-\frac{20+1}{10} = -2\frac{1}{10} \qquad \text{... by writing '21 = 20 + 1'}$$

$$\therefore \quad -2\frac{1}{10} = -2\frac{1}{10} \qquad \text{... by Theorem 4.13}$$

Therefore, the original arithmetic equation was true.

Although the proofs shown above may appear to be longer than conventional proofs, only one operation per line is carried out in order to illustrate clearly which axiom, definition or theorem is being used.

We have now completed our application of the Rational Number System using Fractions to model actions that take place in our everyday lives.

Summary of the Rational Number System using Fractions

At this point, we have covered the key objectives of this chapter and so can summarise our findings up to now. In terms of the Rational Number System using Fractions, the most important definitions that have been introduced in this chapter are:

1. Define the **multiplicative inverse** of the integer variable 'a' as '$\frac{1}{a}$' (where 'a' cannot be assigned '0') so that:

$$a \times \frac{1}{a} = 1$$

2. Define a **fraction** where the integer variable 'a' is divided by the integer variable 'b' and 'b' cannot be assigned '0', as follows:

$$\frac{a}{b} = a \times \frac{1}{b}$$

3. Define the **division** of the fraction '$\frac{a}{b}$' by the fraction '$\frac{c}{d}$', where 'a', 'b', 'c' and 'd' are integer variables and 'b', 'c' and 'd' cannot be assigned '0' by:

$$\frac{a}{b} \div \frac{c}{d} = \frac{a}{b} \times \frac{d}{c}$$

Note:

A fraction of a fraction is called a **complex fraction**.

4. Define the operation of the negative symbol '$-$' as a **unary operator** on a general fraction '$\frac{a}{b}$', where 'a' and 'b' are integer variables and 'b' cannot be assigned '0', as:

$$-\frac{a}{b} = -1 \times \frac{a}{b}$$

5. Define the **subtraction** operator '$-$' between two fractions where 'a', 'b', 'c' and 'd' are integer variables and 'b' and 'd' cannot be assigned '0', as:

$$\frac{a}{b} - \frac{c}{d} = \frac{a}{b} + -\frac{c}{d}$$

6.   Define a **mixed fraction** (or mixed number) to be an abbreviation for the sum of an integer and a fraction. For example:

$$2\frac{3}{5} = 2 + \frac{3}{5}$$

The Axioms of the Rational Number System using Fractions that model the partitioning of a number line into equal portions are set out in Table 4.6 below. In this table, fraction variables and an integer variable are used to describe the multiplicative inverse.

4.6 The **Axioms** of the Rational Number System using Fractions		
Axiom Type	**Addition '+'**	**Multiplication '×'**
Closure	1. In this table the fraction variables are 'x', 'y' and 'z' and the non-zero integer variable is 'a'. The Closure Axiom states that the operations with the following fractions: 'x + y' and 'x × y' result in unique fractions.	
Commutative	2. $x + y = y + x$	3. $x \times y = y \times x$
Associative	4. $(x + y) + z = x + (y + z)$	5. $(x \times y) \times z = x \times (y \times z)$
Identity	6. $1 + 0 = 1$	7. $1 \times 1 = 1$
Distributive	8. $x \times (y + z) = x \times y + x \times z$	
Inverse	9. $1 + {-1} = 0$	10. $a \times \frac{1}{a} = 1$

Table 4.6

Once again, the Axioms of Logic used to manipulate the axioms in the table above have not changed in form since Table 2.5. However, they can now be applied to the broader set of numbers – the Rational Numbers using Fractions.

The basic theorems of this chapter are the starting point for proving all other results in the Rational Number System. When students are able to apply the axioms, definitions and theorems of this chapter to a broad range of physical systems, they will be on the road to achieving mastery in the Rational Number System using Fractions.

In Chapter Five we will continue to extend the language we have used up to this point by applying index notation instead of fraction notation to describe the rational numbers. We will also use the same methods we covered in the theory of the Rational Number System using Fractions to explore the theory of rational numbers using a different notation system.

CHAPTER 5

THE RATIONAL NUMBER SYSTEM USING INDEX FORM

Overview of the Rational Number System using Index Form

In Chapter Four we started our study of the Rational Number System using rational numbers in fraction form. In this chapter we extend our language of the Rational Number System by introducing an alternative way of expressing rational numbers using '**index form**', which is commonly abbreviated to the term '**indices**'.

Once again, our study begins by applying the operations of addition and multiplication to the Rational Numbers using Index Form. In particular, we will show that expressing rational numbers using index form is equivalent to the Rational Number System using Fractions. Using indices to represent rational numbers has the advantage of being able to express very large and very small numbers in a simple way.

Another advantage of using index form to express rational numbers is that it leads on to the natural extension of the definition of these indices to include all the integer numbers. After proving several theorems, we will discover the simple result that these indices behave as an Integer Number System in their own right. This new index system will be the key to our development and understanding of the Finite Decimal System and the Infinite Decimal System we will encounter in later chapters in this book.

In this chapter (as in Chapter Four), we are extending the Integer Number System by including multiplicative inverses of integers. Here, however, we will **rewrite fractions** using index form to express the **division** of one integer number by another integer number. For example, '3' divided by '4' as a fraction is written as '$\frac{3}{4}$', but when using index form, it is written as '3×4^{-1}'. This '-1' symbol in the term '4^{-1}' is a negative index. You will observe that this is a new use of the '-1' symbol from Chapter Three. It will be given new meaning in our axioms. Consequently, the negative index '$\square^{-1}$' is used to represent division in index form.

We will refer to a 'rational number in index form' as simply 'index form'; it is a number that can be written in the form '$a \times b^{-1}$' where 'a' and 'b' are integer variables and 'b' is not assigned zero. However, there are two abbreviations associated with the index form of a rational number:

1. If we assign '1' to the variable 'a', then the index form can be written as:

    '$1 \times b^{-1}$' or 'b^{-1}'   ... this will be justified later in the Theory Section

2. If we assign '1' to the variable 'b', then the index form can be written as:

    '$a \times 1^{-1}$' or 'a'   ... this will also be justified later in the Theory Section.

Although this chapter initially uses an index of '-1' to help represent rational numbers in index form, in the Definition Section we extend the index form so that not only '-1' but all integers can be used as indices.

In this chapter (as we did in Chapter Four), we will once again model the new numbers between the integers using the multiplicative inverse but, this time, using index form. In this way, for every non-zero integer assigned to the variable 'a', this integer has a multiplicative inverse denoted by 'a^{-1}' using the axiom '$a \times a^{-1} = 1$' (see Table 5.4).

Pictorial Representation of Rational Numbers using Index Form

In this chapter we will also visualise the Rational Numbers using Index Form. In Chapter Four, we followed this process for rational numbers as fractions and included numbers like '$\frac{1}{2}$', '$1\frac{1}{3}$' and '$2\frac{4}{5}$'. Now, we will express these fractions using indices to label the points between integer values along the number line. This concept is illustrated in Figure 5.1 below.

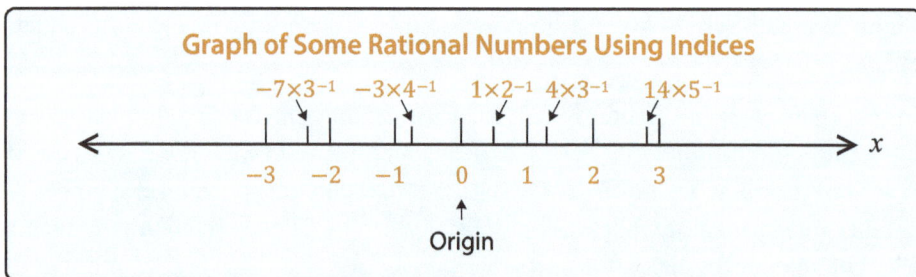

Figure 5.1

As index forms and fractions will be equivalent forms for representing rational numbers, our comment following this equivalent diagram in Chapter Four still applies. That is, those points that can't be labelled by coordinates from the Rational

Number System will have to wait until the later extension of our number system to the Real Number System.

Background and Context of the Rational Number System using Index Form

To continue our study of the Rational Number System, we now introduce an additional symbol for the division symbol that we introduced in Chapter Four, namely '$1 \div \square$'. This new index symbol '$\square^{-1}$' will allow us to represent the Rational Numbers using Index Form. This symbol is called the **index of negative one** symbol which is equivalent to '$1 \div \square$', '$\frac{1}{\square}$' and '$1/\square$' that we learnt about in Chapter Four, where the symbol '$\square$' can be replaced by any non-zero integer.

Once again, we will assume that we know intuitively what this symbol '$\square^{-1}$' means. Then, after defining the 'Multiplicative Inverse of an Integer' in index form using the symbol '$\square^{-1}$', we will again extend our Axioms of the Integer Number System to assign meaning to the Rational Numbers using Index Form. In the second half of the Theory Section below, we will extend the use of 'the index of negative one' to include general integer indices.

An example of a multiplicative inverse in index form for the number '$\frac{1}{3}$' is '3^{-1}'. We will show in the Theory Section that this means:

$$3^{-1} + 3^{-1} + 3^{-1} = 3 \times 3^{-1} = 1$$

Now that we have established the set of symbols for the Rational Number System using Index Form, it only remains to assign meaning to these new symbols using the same approach as given in the previous chapters. This approach is outlined in the section below.

Approach to the Rational Number System using Index Form

In this section our aim will be to assign meaning to the alphabetic symbols of the language of the Rational Number System using Index Form. This language is an extension of the language we used in the Integer Number System as it applies to this extended set of numbers – the Rational Numbers using Index Form. For the remainder of this chapter, we will simply refer to the 'Rational Numbers using Index Form' as the **index form**.

Language of the Rational Number System using Index Form

At this point, we again extend the alphabet that we developed in the Rational Number System using Fractions to include the 'multiplicative inverse' symbol '$\square^{-1}$'. This will give us the basic symbols of the alphabet required for the Rational Number System using Index Form as shown in Table 5.1 below.

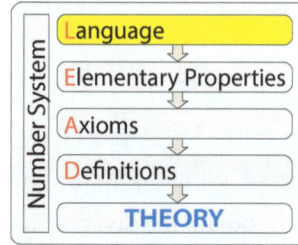

Number System
- Language
- Elementary Properties
- Axioms
- Definitions

THEORY

5.1 The Alphabet of the Rational Number System using Index Form	
Symbols	**Meaning from:**
'0', '1', '+', '×', '(', ')'	The Axioms of the Natural Number System
'=', '≠'	The Axioms of Logic
'2', '3', '4', '5', '6', '7', '8', '9'	Definition 2.1
','	Definition 2.3
'...'	Definition 2.8
'a', 'b', 'c', ..., 'a_1', 'a_2', 'a_3', ...	Common variables to which we assign Integer Numbers
'-1'	The Axioms of the Integer Number System
'$-$'	Definitions 3.4 and 3.5
'$\frac{1}{\square}$'	The Axioms of the Rational Number System using Fractions
'x', 'y', 'z', ...	Common variables to which we assign rational numbers using fractions
'$\frac{a}{b}$'	Definition 4.2
'$\div$'	Definition 4.4
'$\square^{-1}$'	'$\square^{-1}$' is called the index of negative one and derives its meaning from the property: $a \times a^{-1} = 1$ where 'a' cannot be assigned '0'.

Table 5.1

In this extended language, we again use variables to represent numbers in index form. That is, a variable such as 'x', 'y' or 'z' is a symbol of the alphabet to which we can assign any (and all) rational numbers using index form from the following set $\{..., -3, 8 \times (-3)^{-1}, -2, -1, 1 \times (-4)^{-1}, 0, 4^{-1}, 2 \times 3^{-1}, 1, 2, ...\}$. However, we will continue to use 'a', 'b' and 'c' to represent integer number variables in this chapter.

The use of rational numbers using index form is summarised in Table 5.2 below.

5.2 The Language of the Rational Number System using Index Form	
In this table let 'a', 'b', 'c' and 'd' represent integer variables and let 'x', 'y' and 'z' represent index-form variables.	
1. **Arithmetic Terms**	An individual rational number or sum and/or product of rational numbers using indices (each product is called a **factor** of the term), constructed from integer numbers, is referred to as an **arithmetic term** or simply a **term**. In this way, '2', '3^{-1}', '5×1^{-1}', '5×2^{-1}' or '$5^{-1} \times 3$' or '$5^{-1} \times -2 \times 6$' are all finite valid strings of symbols called **arithmetic terms**.
2. **Arithmetic Equations**	When we have an equation with Arithmetic terms on either side of an '=' or '≠' sign, we call this equation an **arithmetic equation**. For example, '$2^{-1} \times 2^{-1} = 4^{-1}$' is a finite valid string of symbols called an **arithmetic equation**.
3. **Algebraic Terms and Equations**	If any terms or equations contain one or more variables, then we refer to them as **algebraic** terms or equations. For example, '$a \times b^{-1}$' is a finite valid string of symbols called an **algebraic term** and '$x = 2$', '$3^{-1} = y$', '$5 = x^{-1}$', '$a \times b^{-1} = c \times d^{-1}$', '$x = 7 \times a$' and '$x \times y \times z = 1$' are finite valid strings of symbols called **algebraic equations**.

Table 5.2

To express the set of integers $\{ \ldots , -2, -1, 0, 1, 2, \ldots \}$ in index form, we would write them as $\{ \ldots , -2 \times 1^{-1}, -1 \times 1^{-1}, 0 \times 1^{-1}, 1 \times 1^{-1}, 2 \times 1^{-1}, \ldots \}$. However, as we did with fractions, when referring to integers as indices, it is traditional practice in Mathematics to leave out the multiplication by '1^{-1}' to make the text more readable. This allows us to write indices and integers alongside each other as follows: $\{ \ldots , -3, 8 \times (-3)^{-1}, -2, -1, 1 \times (-4)^{-1}, 0, 2 \times 3^{-1}, 1, 7 \times 6^{-1}, 2, 3, \ldots \}$ and still treat all the numbers in this set as numbers in index form.

Once again, we start our investigation by assuming that the variables 'x', 'y' and 'z' can be assigned any rational number (in index form) and can be combined with the operators '$+$' and/or '$\times$' to form terms and also '$=$' and '$\neq$' to form equations.

Order of Operations

Although we have introduced indices as a newly-defined mathematical operation, the Order of Operations table that applied to the Rational Numbers using Fractions also applies without change to the Rational Numbers using Index Form. These are outlined in Table 5.3 below.

5.3 Order of Operations for the Rational Number System using Index Form
The process used to evaluate arithmetic and algebraic terms is determined by the order of operations and proceeds according to the following steps:

1.	The **parentheses** have highest priority and control the order of evaluation. Parentheses may be removed by evaluating inside the parentheses **or** by using the Distributive Axiom.
2.	Next evaluate **exponents** (i.e. an integer index to which a base number is raised).
3.	Where there are no parentheses, then we evaluate **multiplication** '$\times$' and **division** '$\div$' from left to right.
4.	Finally, evaluate **addition** '$+$' and **subtraction** '$-$' from left to right.

Table 5.3

The Order of Operations for the Rational Number System using Index Form can be remembered by using the acronym '**PEMDAS**' which stands for:

- Parentheses
- Exponents
- Multiplication
- Division
- Addition
- Subtraction.

The acronym is dependent on the number system being used and has been extended in this chapter to include operations with the Rational Numbers using Index Form.

Order of Operations – PEMDAS	
Evaluate the following term using Order of Operations:	
$Term$	$= (3 \times 5) + 2^3 \times 7 \times 2^{-1} \div 2 - 12$
Parentheses	$= (3 \times 5) + 2^3 \times 7 \times 2^{-1} \div 2 - 12$
Exponents	$= 15 + 2^3 \times 7 \times 2^{-1} \div 2 - 12$
Multiplication	$= 15 + 8 \times 7 \times 2^{-1} \div 2 - 12$
Division	$= 15 + 28 \div 2 - 12$
Addition	$= 15 + 14 - 12$
Subtraction	$= 29 - 12$
Answer	$= 17$

When counting with rational numbers using indices, we retain the ten elementary properties of the Rational Number System and replace the Inverse Property of Multiplication we learnt about Chapter Four with the alternative symbol '$\square^{-1}$'.

The Tenth Elementary Property of the Rational Number System using Index Form

Inverse Property of Multiplication

This property guarantees us that for every non-zero integer we assign to the variable 'a', there is another number called its *inverse*, such that when we multiply this number by its inverse we get the number '1'. This is the inverse number with respect to multiplication.

This property is modelled in our Rational Number System using Index Form by multiplying a variable 'a' by its multiplicative inverse 'a^{-1}' to get the integer number '1'. In symbols, we model this property using the following equation:

$$a \times a^{-1} = 1$$

In this Chapter, we exchange the **Inverse Axiom for Multiplication** used for fractions with the equivalent axiom using index form. The statement of the axioms required for the Rational Numbers using Index Form are shown in Table 5.4.

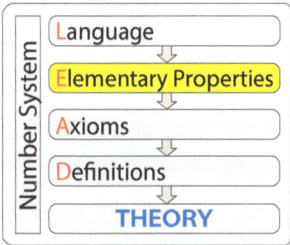

The Axioms of the Rational Number System using Index Form

By starting with the Axioms of the Integer Number System and adding this new property, we have extended the set of axioms available for us to use. This new set of axioms now applies to the broader set of numbers called the Rational Numbers using Index Form.

If we assume (as we did in Chapter Four) that 'x', 'y', and 'z' are variables that can be assigned any rational number, then the Axioms of the Integer Number System are now extended in scope and apply to the Rational Numbers using Index Form. Also, we assume that 'a' is a variable that can be assigned an integer number which is non-zero if 'a' is being used in the form 'a^{-1}'.

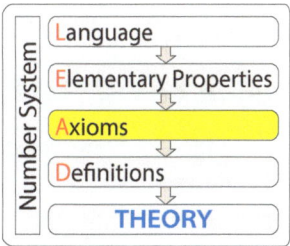

The axioms for use in the Rational Number System using Index Form are set out in Table 5.4 on the following page.

5.4 The **Axioms** of the Rational Number System using Index Form		
Axiom Type	**Addition '+'**	**Multiplication '×'**
Closure	1. Let 'a' be a non-zero integer variable and 'x', 'y' and 'z' be rational number variables. This Closure Axiom states that the following operations: '$x+y$' and '$x \times y$' result in unique rational numbers.	
Commutative	2. $\qquad x+y=y+x$	3. $\qquad x \times y = y \times x$
Associative	4. $(x+y)+z=x+(y+z)$	5. $(x \times y) \times z = x \times (y \times z)$
Identity	6. $\qquad 1+0=1$	7. $\qquad 1 \times 1 = 1$
Distributive	8. $\qquad\qquad x \times (y+z) = x \times y + x \times z$	
Inverse	9. $\quad 1 + -1 = 0$	10. $\quad a \times a^{-1} = 1$

Table 5.4

According to this new axiom – the Inverse Axiom for Multiplication – every non-zero integer number variable 'a' has a multiplicative inverse denoted by 'a^{-1}' that is always related to it.

The Axioms of Logic

As in Chapter Four, before we can reason correctly from the Axioms of the Rational Number System using Index Form, we again need to use the Axioms of Logic. The Axioms of Logic have not changed in structure from those in Chapter Four, however, we assign rational numbers using index form to the variables in this chapter.

Definitions of the Rational Number System using Index Form

Now that we have assigned meaning to the symbol '$\square^{-1}$' in the Axioms of the Rational Number System using Index Form, we will use this symbol to create alternate definitions for the rational numbers that we introduced in Chapter Four.

In the Theory Section we will explore the relationship between this new symbol, the other symbols in our alphabet, the axioms and definitions.

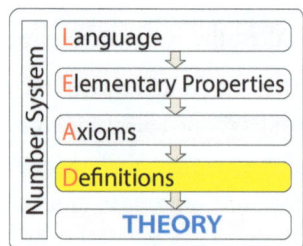

Number System
Language
↓
Elementary Properties
↓
Axioms
↓
Definitions
↓
THEORY

After creating the index form of rational numbers, we can now expand the set of integers to become the set of Rational Numbers using Index Form.

Definition 5.1   Rational Numbers using Index Form

A **rational number** '$a \times b^{-1}$' is defined as any arithmetic term formed by using any number in the first set below multiplied by any number from the second set below:

1.   The set of **integers**: $\{..., -3, -2, -1, 0, 1, 2, 3, ...\}$

2.   The set of **multiplicative inverses** of the integers (excluding '0') in index form: $\{..., (-3)^{-1}, (-2)^{-1}, (-1)^{-1}, 1^{-1}, 2^{-1}, 3^{-1}, ...\}$.

If 'a' is assigned an integer from the first set and 'b^{-1}' is assigned a multiplicative inverse from the second set, a rational number using index form is again written as '$a \times b^{-1}$'.

It should be noted that by this definition, the integer number '0' does not have a multiplicative inverse in the set of multiplicative inverses.

To demonstrate Definition 5.1, the product derived from multiplying a number from the set of integers with a number from the set of multiplicative inverses (such as: '-3×2^{-1}') results in a rational number using index form, which is equivalent to the fraction '$\frac{-3}{2}$'. Therefore, just as we wrote the general form of the fraction as '$\frac{a}{b}$', we write the general index form as '$a \times b^{-1}$', where 'a' and 'b' are integer variables and 'b' is not assigned '0'. As we did in Chapter Four for fractions, we will show that every rational number can be written in the index form '$a \times b^{-1}$'.

In Chapter Four we defined the operation of finding the reciprocal of a fraction (inverting the fraction) and creating the language for this operation. We create the same language for index form in our next definition.

Definition 5.2   Reciprocal of an Index Form

If 'a' and 'b' are integer variables (where 'a' and 'b' are not assigned '0'), then we define the **reciprocal** of a rational number in index form, '$a \times b^{-1}$', as the rational number '$b \times a^{-1}$'.

We now set out to define the concept of division of one rational number by another rational number in index form using this definition of the reciprocal.

Definition 5.3   Division of an Index Form

If 'a', 'b', 'c' and 'd' are integer variables (where 'b', 'c' and 'd' are not assigned '0'), then we define the rational number '$a \times b^{-1}$' **divided** by the rational number '$c \times d^{-1}$' as follows:

$$(a \times b^{-1}) \div (c \times d^{-1}) = (a \times b^{-1}) \times (d \times c^{-1})$$

This equation tells us that: 'To divide a rational number by another rational number, multiply the first rational number by the **reciprocal** of the second rational number'.

Similar to multiplicative inverses for integers, it is possible to define a multiplicative inverse for **rational numbers** in index form. We create this concept and its language in our next definition.

Definition 5.4   Multiplicative Inverse of a Rational Number

If 'a', 'b', 'c' and 'd' are integer variables (where 'b', 'c' and 'd' cannot be assigned '0'), then we define the multiplicative inverse of '$a \times b^{-1}$' as the rational number '$c \times d^{-1}$' that makes the following equation true:

$$(a \times b^{-1}) \times (c \times d^{-1}) = 1$$

Note:

The multiplicative inverse of a rational number is defined so that when you multiply a rational number by its multiplicative inverse you get the number '1'.

In Chapter Four we were able to create new symbols for negative fractions when they were expressed in compressed notation. As compressed notation is not available for the Rational Numbers using Index Form, it is not possible to define the unique negative of a rational number using index form. However, we can still define the unary operator applied to a Rational Number using Index Form.

Definition 5.5   Unary Operator '−' Applied to Index Form

For integer variables 'a' and 'b' (where 'b' is not assigned '0') the action of the negative symbol '−' on a rational number using index form is defined as:

$$-(a \times b^{-1}) = -1 \times a \times b^{-1}$$

So far in this chapter we have used an index of '−1' to represent rational numbers using index form. In Chapter Two we used natural number indices to abbreviate multiple products of a number that we called the base. For example, '$2 \times 2 \times 2$' was abbreviated to '2^3'.

We now wish to extend the use of index form so that the index can be any number symbol from the integer alphabet. However, at this point, the only index symbols we have been using are $\{-1, 0, 1, 2, 3, ...\}$. In the following definition we extend the range of the index to include the negative integers symbols $\{..., -4, -3, -2\}$ and define the application of all indices to base rational numbers of the form '$a \times b^{-1}$'.

Definition 5.6   General Indices for Rational Numbers

Let a rational number be given by '$a \times b^{-1}$' where 'a' and 'b' are integer variables and **neither** 'a' **nor** 'b' are assigned '0'. Let 'n' and 'm' be index variables that are assigned numbers from the set of symbols $\{..., -3, -2, -1, 0, 1, 2, 3, ...\}$ and can be combined with the set of symbols $\{'+', '\times', '=', '\neq', '(', ')'\}$.

The definitions of **indices**, using the index form '$a \times b^{-1}$', are given by the following equations:

1. $(a \times b^{-1})^0 = 1$

2. $(a \times b^{-1})^1 = (a \times b^{-1}) = a \times b^{-1}$

3. $(a \times b^{-1})^{-1} \times (a \times b^{-1})^1 = 1$ ... definition of multiplicative inverse

4. $(a \times b^{-1})^{m+n} = (a \times b^{-1})^m \times (a \times b^{-1})^n$

5. $(a \times b^{-1})^{m \times n} = ((a \times b^{-1})^m)^n$

6. '$m = n$' is true if and only if '$(a \times b^{-1})^m = (a \times b^{-1})^n$' is true and both '$a$' **and** '$b$' are not assigned integers from the set $\{1, -1\}$. '$m \neq n$' is also true if and only if '$(a \times b^{-1})^m \neq (a \times b^{-1})^n$' is true and both '$a$' **and** '$b$' are not assigned the integers from the set $\{1, -1\}$.

7. The parentheses symbols as applied to indices have their usual role as grouping symbols.

Notes:

1. Definition 5.6(1) provides the definition of the index '0'. Here we have to assume that both 'a' and 'b' are non-zero variables to avoid the occurrence of the term '0^0' which can't be defined unambiguously.

2. Definition 5.6(2) provides the definition of the index '1'. This index is used in the same way as it was in Chapter 2 for Natural Numbers, namely '$a^1 = a$'.

3. Definition 5.6(3) provides the definition of the index '-1'. In the Theory Section below we will derive an explicit equation for this multiplicative inverse.

4. Definition 5.6(4) provides the definition of the addition operator for indices. That is: 'when multiplying two rational numbers with the same base, you simply add their indices'.

5. Definition 5.6(5) provides the definition of the multiplication operator for indices. That is: 'when raising a rational number with an index of 'm' to an additional index 'n', we simply multiply the two indices'.

6. Definition 5.6(6) provides the definition of the equal operator and the not-equal operator for indices. In the exception cases, where 'a' and 'b' are assigned values from the set $\{1, -1\}$, we have to treat these cases individually to determine if '$m = n$'.

7. Definition 5.6(7) describes the use of the parentheses symbols. For example, we can use parentheses using index form as follows:

$$(a \times b^{-1})^{(m+n)+p} = (a \times b^{-1})^{m+n} \times (a \times b^{-1})^{p}$$

At this point it is worth including a similar observation we made for fractions – namely, we can consistently represent rational numbers using both positive and negative indices, so that instead of writing rational numbers as '$a \times b^{-1}$' we could write them as '$a^1 \times b^{-1}$'. Once again, traditional practice in Mathematics is to leave out this positive index to avoid any confusion.

For the index numbers from the symbols '... , -3, -2, -1, 0, 1, 2, 3, ...', we assume the usual definitions apply to these symbols. For example, '$2 = 1 + 1$', '$3 = 2 + 1$', '$-2 = -1 \times 2$', '$-3 = -1 \times 3$', etc.

Having established the above basic symbols of the alphabet, axioms and definitions, we can complete the alphabet for the Rational Number System using Index Form. This alphabet is summarised in Table 5.5 below.

5.5 The Alphabet of the Rational Number System using Index Form	
Symbols	Meaning from:
'0', '1', '+', '×', '(', ')'	The Axioms of the Natural Number System
'=', '≠'	The Axioms of Logic
'2', '3', '4', '5', '6', '7', '8', '9'	Definition 2.1
','	Definition 2.3
'...'	Definition 2.8
'a', 'b', 'c', ... , 'a_1', 'a_2', 'a_3', ...	Common variables to which we assign Integer Numbers
'-1'	The Axioms of the Integer Number System

5.5 The **Alphabet** of the Rational Number System using Index Form	
'—'	Definitions 3.4 and 3.5
$\dfrac{1}{\Box}$	The Axioms of the Rational Number System using Fractions
'x', 'y', 'z', ...	Common variables to which we assign rational numbers using fractions or index form
$\dfrac{a}{b}$	Definition 4.2
'÷'	Definition 4.4
'$\Box^{-1}$'	The Axioms of the Rational Number System using Index Form
'$(a \times b^{-1})^m$'	Definition 5.6

Table 5.5

We have now completed our approach to developing the Rational Numbers using Index Form as an alternative to the set of rational numbers using fraction form presented in Chapter Four. In the next section of this chapter we will explore the theory that follows from this extension to our number system axioms.

First, we will demonstrate that doing arithmetic with rational numbers using index form is equivalent to doing arithmetic with rational numbers using fraction form.

Later, in the Theory Section we will also show how to extend the use of the index '$\Box^{-1}$' to include the full range of integer indices.

Theory of the Rational Number System Using Index Form

We now have sufficient language, axioms and definitions to proceed with the proof of some basic theorems. To develop the theory of the Rational Number System using Index Form we will investigate the simple consequences of applying the Axioms of Logic to manipulate the Axioms of the Rational Number System using Index Form in order to prove all the basic theorems involving the multiplicative inverse of a non-zero integer.

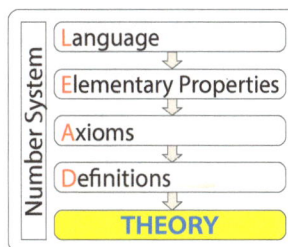

The extended set of numbers using indices has an expanded set of theorems that builds on the theorems from previous chapters. The first part of the Theory Section follows the same proofs as those in Chapter Four using index notation. However, we have extended the notation in the latter half of the Theory Section to show how all integers can be used as indices.

Theorem 5.1 Reflexive Theorem for Rational Numbers using Index Form

If 'x' is a rational number variable, where we write '$x = a \times b^{-1}$' with 'a' and 'b' as integer variables and 'b' is not assigned '0', then the equation '$x = x$' is a true equation.

Proof

We have already proved that '$a = a$' for an integer variable, so we begin by showing that '$b^{-1} = b^{-1}$' for multiplicative inverses.

We begin this process with the Reflexive Theorem for Integers as follows:

$$b = b \qquad \text{... by Theorem 3.5}$$
Assume: $$b^{-1} = b^{-1} \qquad \text{... by assumption}$$
∴ $$b \times b^{-1} = b \times b^{-1} \qquad \text{... by Deduction Axiom}$$
∴ $$1 = 1 \qquad \text{... by Inverse Axiom '×'}$$

Since this last equation is true, our original assumption that '$b^{-1} = b^{-1}$' must also be true. Therefore, when we multiply the equation '$a = a$' and '$b^{-1} = b^{-1}$' we arrive at:

$$a \times b^{-1} = a \times b^{-1} \qquad \text{... by Deduction Axiom}$$

Hence for every assignment of integers to the variables 'a' and 'b' (where '0' cannot be assigned to 'b') we have shown that '$x = x$' is a true equation for index form variables. Therefore, the theorem has been proved.

We can now extend the Reflexive Theorem outlined above to apply to terms as well as simple Rational Numbers using Index Form. For example, the following theorem will assure us that: '$(2 \times 3^{-1}) + 7 \times (5 \times 6^{-1}) = (2 \times 3^{-1}) + 7 \times (5 \times 6^{-1})$' is a true arithmetic equation.

Theorem 5.2 Reflexive Theorem for Terms

If 't' is a general term, then this term also has the reflexive property so that:

$$t = t$$

Proof

A term 't' is a finite string of symbols constructed from rational numbers in index form and variables using the operators '$+$' and '$\times$' (supported by parentheses).

Following the proof for integer number terms in Theorem 3.6, the equation '$t = t$' is also true for terms in the Rational Number System using Index Form because Theorem 5.1 assures us that '$x = x$' for every variable and every rational number using index form assigned to a variable. Therefore, this theorem is also true for the Rational Numbers using Index Form.

Now that we have established the Reflexive Theorem for Terms, '$t = t$', we will develop several theorems that will help us operate with terms. First, we need to show how some of the axioms also apply to terms and not just rational numbers using index form and variables.

Theorem 5.3   Properties of Terms Theorem

If three general terms (i.e. constructed from rational numbers using index form and/or variables) are given by 't_1', 't_2' and 't_3', then these terms also obey the Axioms of the Rational Number System using Index Form.

Proof

Following the proof for integer number terms in Theorem 3.7, this proof follows exactly the same logic and, hence, this theorem is also true for the Rational Numbers using Index Form.

In Chapter Three, we observed that it was tedious using the Transitive Axiom in its restrictive form to carry out number and variable substitution. Consequently, we will again state the Substitution Theorem as applied to integer numbers, variables and terms.

Theorem 5.4   Substitution Theorem

Assume we are given a general term in a rational number variable 'x' with coefficients 'c_0', 'c_1', 'c_2' up to 'c_n' which are terms. This general term can be written as:

$$c_0 \times x^0 + c_1 \times x^1 + ... + c_n \times x^n.$$

If '$x = y$', then 'y' can be substituted for 'x' everywhere in this term without changing the value of the term so that:

$$c_0 \times x^0 + c_1 \times x^1 + ... + c_n \times x^n = c_0 \times y^0 + c_1 \times y^1 + ... + c_n \times y^n$$

Proof

From Theorem 5.2 we can conclude that all the coefficients which are terms have the reflexive property so that: '$c_0 = c_0$', '$c_1 = c_1$', ... , '$c_n = c_n$'.

The remainder of this proof follows directly on from Theorem 3.8 whereby if '$x = y$', we can effectively substitute 'y' for 'x' in the above algebraic term with coefficients in the variable 'x' and not change the value of that term. In this way, the theorem is also true for the Rational Numbers using Index Form.

Now that we have the Reflexive Theorem and the Substitution Theorem, we will use these theorems to simplify the following proofs.

The next result we are required to prove follows directly on from the Inverse Axiom for Multiplication using index form which gives us the simplest of results, namely that: '$1^{-1} = 1$'.

Theorem 5.5   Numbers Equal to their Multiplicative Inverses Theorem

The integers '1' and '−1' are equal to their own multiplicative inverses so that:

$$1 = 1^{-1}$$
$$-1 = (-1)^{-1}$$

Proof

We follow proof by deduction starting with the Inverse Axiom for '×' to get:

$1 \times 1^{-1} = 1$ ... by Inverse Axiom '×'

∴ $1 \times 1^{-1} = 1 \times 1$ ... by Identity Axiom '×'

But: $1 = 1$ ... by Theorem 5.1

∴ $1^{-1} = 1$ ... by Deduction Axiom

Similarly, for '−1' we have:

$-1 \times (-1)^{-1} = 1$ ... by Inverse Axiom '×'

$-1 \times (-1)^{-1} = -1 \times -1$ ... by Theorem 3.3(3)

But: $-1 = -1$ ... by Theorem 5.1

∴ $(-1)^{-1} = -1$ ... by Deduction Axiom

Therefore, this simplest of theorems is proved.

In the Approach Section of this chapter, we sought to write integers as rational numbers using index form. The next theorem provides the justification for this approach. For example, the following arithmetic equation is assured by the theorem: '$3 = 3 \times 1^{-1}$'.

Theorem 5.6   Integers as Rational Numbers using Index Form Theorem

If 'a' is an integer number variable, then 'a' can be written as the rational number variable '$a \times 1^{-1}$', that is:

$$a = a \times 1^{-1}$$

Proof

The proof of this theorem follows easily from the previous theorem as shown below:

$$a = a \times 1 \qquad \text{... by Theorem 3.10}$$
$$= a \times 1^{-1} \qquad \text{... by Theorem 5.5}$$
$$\therefore \quad a = a \times 1^{-1}$$

Therefore, an integer variable 'a' can be expressed as the rational number variable using index form: '$a \times 1^{-1}$'.

As an example, the number '5' can be turned into a Rational Number using Index Form as '5×1^{-1}'. We should also be aware that the number '0' can be written as a Rational Number using Index Form as '0×1^{-1}'.

The next theorem we prove is an extension of the result '$a \times 0 = 0$', for the integer variable 'a', so that it applies to a term constructed from rational numbers using index form. For example, we are assured that arithmetic equations of the form '$(2 \times 3^{-1}) \times 0 = 0$' are always true.

Theorem 5.7   Null Multiplication of Index Forms Theorem

If 'x' is a rational number (using an index form) variable, then '0' is the null multiplier of 'x' so that:

$$x \times 0 = 0$$

Proof

The rational number variable 'x' can be written in the form '$x = a \times b^{-1}$', where 'a' and 'b' are integer variables and 'b' cannot be assigned '0'. We have to prove that '$a \times b^{-1} \times 0 = 0$'. We commence this process by showing that the multiplicative inverse 'b^{-1}' satisfies the equation '$b^{-1} \times 0 = 0$'.

$$0 = 0 \times 1 \qquad \text{... by Theorem 2.3}$$

$$= 0 \times b \times b^{-1} \qquad \text{... by Inverse Axiom '$\times$'}$$

$$= (0 \times b) \times b^{-1} \qquad \text{... by Associative Axiom '$\times$'}$$

$$= 0 \times b^{-1} \qquad \text{... by Theorem 3.9}$$

$$\therefore \qquad 0 \times b^{-1} = 0$$

It is now straightforward to show the more general result '$x \times 0 = 0$' is true as follows:

$$x \times 0 = (a \times b^{-1}) \times 0 \qquad \text{... by '$x = a \times b^{-1}$'}$$

$$= a \times (b^{-1} \times 0) \qquad \text{... by Associative Axiom '$\times$'}$$

$$= a \times 0 \qquad \text{... by first half of proof}$$

$$= 0 \qquad \text{... by Theorem 3.9}$$

$$\therefore \qquad x \times 0 = 0$$

Therefore, the equation '$x \times 0 = 0$' gives a true equation when 'x' is assigned any rational number using index form. Therefore, the theorem has been proved.

It is worth noting that from Theorem 5.3 we can also declare that for a general term 't' the equation '$t \times 0 = 0$' must also be a true equation.

The next theorem we set out to prove is an extension of the result '$a \times 1 = a$' for Integers from Chapter Three, as it applies to the Rational Numbers using Index Form. For example, we are assured that arithmetic equations of the form '$(2 \times 3^{-1}) \times 1 = 1$' are always true.

Theorem 5.8   Identity Theorem for Multiplication of Index Forms

Let 'x' represent a rational number variable. The number '1' is the multiplicative identity for 'x', so that:

$$x \times 1 = x$$

Proof

First, let 'a' and 'b' be integer variables where 'b' is not assigned '0'. Next, we have to prove: '$a \times b^{-1} \times 1 = a \times b^{-1}$'. We start with the left-hand side of this equation as follows:

$$a \times b^{-1} \times 1 = (a \times 1) \times b^{-1} \qquad \text{... by Commutative Axiom '}\times\text{'}$$

$$= a \times b^{-1} \qquad \text{... by Theorem 3.10}$$

$$\therefore \qquad x \times 1 = x \qquad \text{... since '}x = a \times b^{-1}\text{'}$$

Therefore, the equation '$x \times 1 = x$' gives a true equation when 'x' is assigned any rational number using index form. Therefore, the theorem has been proved.

Note:

From Theorem 5.3 we can also declare that for a general term 't' the equation '$t \times 1 = t$' must also be a true equation.

The next theorem we set out to prove is the equivalent of the result '$x + 0 = x$' for fractions from Chapter Four, so that it now applies to all rational numbers using index form. For example, we are assured that arithmetic equations of the form '$(2 \times 3^{-1}) + 0 = 2 \times 3^{-1}$' are always true.

Theorem 5.9 Identity Theorem for Addition of Index Forms

If 'x' is a rational number variable, then the additive identity for 'x' is '0' so that:

$$x + 0 = x$$

Proof

This proof follows identical reasoning to the equivalent proof in Chapter Four (Theorem 4.10). However, in this case we use Theorems 5.7 and 5.8 as opposed to Theorems 4.8 and 4.9.

Once again, the next theorem we set out to prove is equivalent to the Inverse Theorem for Addition of Fractions from Chapter Four, so that it applies to all rational numbers using index form. For example, we are assured that arithmetic equations of the form '$(2 \times 3^{-1}) - 2 \times 3^{-1} = 0$' are always true.

Theorem 5.10   Inverse Theorem for Addition of Index Forms

If 'x' is a rational number variable, then '$-x$' is the additive inverse of 'x' so that:

$$x + -x = 0$$

Proof

This proof follows identical reasoning to the equivalent proof in Chapter Four (Theorem 4.11). However, in this case we use Theorems 5.7 and 5.8 as opposed to Theorems 4.8 and 4.9.

The next theorem we set out to prove is the equivalent of the result '$-(-(x)) = x$' for fractions from Chapter Four, so that it applies to all rational numbers using index form. For example, the following theorem will assure us that: '$-(-(2 \times 3^{-1})) = 2 \times 3^{-1}$' is a true equation.

Theorem 5.11   Double Negatives

If 'x' is a rational number variable then the double negative of 'x' is simply 'x' itself so that:

$$-(-(x)) = x$$

Proof

This proof follows identical reasoning to the equivalent proof in Chapter Four (Theorem 4.12).

The next theorem provides us with the simple rule for combining two rational numbers using index forms which do have the same multiplicative inverses. For example, the following theorem will assure us that arithmetic equations of the following form '$(2 \times 3^{-1}) + (5 \times 3^{-1}) = (2 + 5) \times 3^{-1}$' are always true.

Theorem 5.12   Addition in Index Form for Multiplicative Inverses Theorem

If 'a', 'b' and 'c' are integer variables (where 'b' is not assigned '0'), then we can add the two rational numbers using index form '$a \times b^{-1}$' and '$c \times b^{-1}$' as follows:

$$a \times b^{-1} + c \times b^{-1} = (a + c) \times b^{-1}$$

Proof

Given 'a', 'b' and 'c' are integer variables (where 'b' cannot be assigned '0'), then:

$$a \times b^{-1} + c \times b^{-1} = (a + c) \times b^{-1} \qquad \text{... by Distributive Axiom}$$

$$\therefore \qquad a \times b^{-1} + c \times b^{-1} = (a + c) \times b^{-1}$$

This proves the theorem and the same proof holds for the subtraction of two rational numbers using index form which do have the same multiplicative inverses.

Now that we know how to add and subtract rational numbers expressed in their index form, the next step is to prove the simple multiplication operations for these numbers. For example, the following arithmetic equation must be true as a result of the theorem below: $(2 \times 3^{-1}) \times (5 \times 7^{-1}) = (2 \times 5) \times (3^{-1} \times 7^{-1})$.

Theorem 5.13 Theorem for Multiplication of Index Forms

If 'a', 'b', 'c' and 'd' are integer variables (where 'b' and 'd' are not assigned '0'), then we can multiply the index forms '$a \times b^{-1}$' and '$c \times d^{-1}$' as follows:

$$(a \times b^{-1}) \times (c \times d^{-1}) = (a \times c) \times (b \times d)^{-1}$$

Proof

First, we have to prove the simpler result: '$b^{-1} \times d^{-1} = (b \times d)^{-1}$'. We commence this process by applying mathematical deduction to this equation as follows:

$$b^{-1} \times d^{-1} = (b \times d)^{-1} \qquad \text{... by assumption}$$

$$(b \times d) \times b^{-1} \times d^{-1} = (b \times d) \times (b \times d)^{-1} \qquad \text{... by multiplying by '}b \times d\text{'}$$

$$(b \times d) \times b^{-1} \times d^{-1} = 1 \qquad \text{... by Inverse Axiom '}\times\text{'}$$

$$b \times b^{-1} \times d \times d^{-1} = 1 \qquad \text{... by Commutative Axiom '}\times\text{'}$$

$$1 \times 1 = 1 \qquad \text{... by Inverse Axiom '}\times\text{'}$$

Hence, we have proved that '$b^{-1} \times d^{-1} = (b \times d)^{-1}$' is a true equation. We now wish to prove it is also true for general rational numbers using index form and not just multiplicative inverses. We start with the left-hand side of the equation stated in the theorem as follows:

$$(a \times b^{-1}) \times (c \times d^{-1}) = a \times b^{-1} \times c \times d^{-1} \qquad \text{... by removing parentheses}$$

$$= a \times c \times b^{-1} \times d^{-1} \qquad \text{... by Commutative Axiom '}\times\text{'}$$

$$= a \times c \times (b \times d)^{-1} \quad \text{... by first half of this proof}$$

$$\therefore \quad (a \times b^{-1}) \times (c \times d^{-1}) = (a \times c) \times (b \times d)^{-1}$$

This proves the theorem.

———————————

A situation which often occurs is when a rational number using index form has the same factor (called a 'common' factor) in both its integer part and its multiplicative inverse part, as we saw with fractions in Chapter Four. The following theorem helps us to simplify these products using index form. For example, the following theorem will assure us that arithmetic equations of the following form '$(2 \times 3^{-1}) \times (5 \times 3^{-1})^{-1} = 2 \times 5^{-1}$' are always true.

Theorem 5.14   Cancelling a Factor Theorem

If 'a', 'b' and 'c' are integer variables (where 'b' and 'c' are not assigned '0'), then:

$$(a \times c) \times (b \times c)^{-1} = a \times b^{-1}$$

Proof

We begin with the left-hand side of the above equation and reason by deduction:

$$(a \times c) \times (b \times c)^{-1} = a \times c \times b^{-1} \times c^{-1} \quad \text{... by Theorem 5.13}$$

$$= a \times b^{-1} \times c \times c^{-1} \quad \text{... by Commutative Axiom '}\times\text{'}$$

$$= a \times b^{-1} \times 1 \quad \text{... by Inverse Axiom '}\times\text{'}$$

$$\therefore \quad (a \times c) \times (b \times c)^{-1} = a \times b^{-1} \quad \text{... by Theorem 5.8}$$

Consequently, this proves the theorem.

———————————

The theorem outlined above gives rise to the necessity for additional language to describe the state of play. The terms on either side of the equation '$(a \times c) \times (b \times c)^{-1} = a \times b^{-1}$' in Theorem 5.14 are called **equivalent terms**. For the term '$(a \times c) \times (b \times c)^{-1}$', we say that we have cancelled the common factor 'c' from the terms of this equation to arrive at '$a \times b^{-1}$'.

We will now show how to add two rational numbers using index form which have different multiplicative inverses. Once again, in this theorem we will create the equivalent of a common denominator as we did with fractions to enable us to carry out additions of different multiplicative inverses. For example, the following theorem will assure us that arithmetic equations of the following form:

'$(2 \times 3^{-1}) + (5 \times 7^{-1}) = (2 \times 7 + 5 \times 3) \times (3 \times 7)^{-1}$' are always true.

Theorem 5.15   Theorem for Addition of Rational Numbers using Index Form

If 'a', 'b', 'c' and 'd' are integer variables (where 'b' and 'd' are not assigned '0'), then we can add the rational numbers using index form '$a \times b^{-1}$' and '$c \times d^{-1}$' as follows:

$$a \times b^{-1} + c \times d^{-1} = (a \times d + c \times b) \times (b \times d)^{-1}$$

Proof

In Theorem 5.12 we were able to add two index forms with like multiplicative inverses. To add two index forms with unlike multiplicative inverses, we must first create a common multiplicative inverse of '$b^{-1} \times d^{-1}$'. We start this process with the left-hand side of the above equation and argue by deduction:

$$a \times b^{-1} + c \times d^{-1} = a \times b^{-1} \times 1 + c \times d^{-1} \times 1 \qquad \text{... by Theorem 5.8}$$

$$= a \times b^{-1} \times d \times d^{-1} + c \times d^{-1} \times b \times b^{-1}$$

$$\text{... by Inverse Axiom '×'}$$

$$= (a \times d) \times (b \times d)^{-1} + (c \times b) \times (b \times d)^{-1}$$

$$\text{... by Theorem 5.13}$$

$$= (a \times d + c \times b) \times (b \times d)^{-1} \qquad \text{... by Distributive Axiom}$$

$$\therefore \quad a \times b^{-1} + c \times d^{-1} = (a \times d + c \times b) \times (b \times d)^{-1}$$

Therefore, this proves the theorem.

––––––––––––––––––

Note:

When deriving an equation for **subtracting** two index forms with unlike multiplicative inverses, the proof follows the same steps as this theorem.

In Theorem 5.6 we showed that all integers could be written in index form as '$a \times 1^{-1}$', where 'a' is an integer variable. In Theorem 5.13 when we multiply any two rational numbers in index form the outcome is always a rational number using index form. In Theorem 5.15 (above), when we add any two rational numbers in index form the outcome is always a rational number using index form. These three theorems together prove that any rational number can be written in index form.

Now that we have proved theorems using addition, subtraction and multiplication of rational numbers, we need to prove a basic theorem which links the multiplicative inverse to **division** of rational numbers. First, we must work out an expression for the multiplicative inverse and then show how it relates to the division operation. For example, the following theorem will assure us that arithmetic equations of the following form '$(2 \times 3^{-1})^{-1} = 3 \times 2^{-1}$' are always true.

Theorem 5.16   Multiplicative Inverse Theorem using Index Form

If both 'a' and 'b' are non-zero integer variables, then we can write the term for the multiplicative inverse as a Rational Number using Index Form as:

$$(a \times b^{-1})^{-1} = b \times a^{-1}$$

Proof

Given both 'a' and 'b' are non-zero integer variables, we start with the specific notation for the multiplicative inverse and prove the above equation by deduction as follows:

$$(a \times b^{-1}) \times (a \times b^{-1})^{-1} = 1 \qquad \text{... by Definition 5.6(3)}$$

$$a^{-1} \times (a \times b^{-1}) \times (a \times b^{-1})^{-1} = 1 \times a^{-1} \qquad \text{... by Deduction Axiom}$$

$$(a^{-1} \times a) \times b^{-1} \times (a \times b^{-1})^{-1} = 1 \times a^{-1} \qquad \text{... by Associative Axiom '×'}$$

$$1 \times b^{-1} \times (a \times b^{-1})^{-1} = 1 \times a^{-1} \qquad \text{... by Inverse Axiom '×'}$$

$$b^{-1} \times (a \times b^{-1})^{-1} = a^{-1} \qquad \text{... by Theorem 5.8}$$

$$b \times b^{-1} \times (a \times b^{-1})^{-1} = a^{-1} \times b \qquad \text{... by Deduction Axiom}$$

$$1 \times (a \times b^{-1})^{-1} = a^{-1} \times b \qquad \text{... by Inverse Axiom '×'}$$

$$(a \times b^{-1})^{-1} = a^{-1} \times b \qquad \text{... by Theorem 5.8}$$

$$\therefore \quad (a \times b^{-1})^{-1} = b \times a^{-1} \qquad \text{... by Commutative Axiom '×'}$$

This proves the theorem.

Note:

Only now are we able to declare that the multiplicative inverse of a rational number using index form is simply the reciprocal of that number.

Now that we have an expression for the multiplicative inverse, we can prove that **division** using index form is the same as multiplying by its multiplicative inverse.

For example, the following theorem will assure us that arithmetic equations of the following form '$(2 \times 3^{-1}) \div (5 \times 7^{-1}) = (2 \times 3^{-1}) \times (5 \times 7^{-1})^{-1}$' are always true.

Theorem 5.17   Division using Index Form as the Multiplicative Inverse Theorem

If 'a', 'b', 'c' and 'd' are integer variables (where 'b', 'c' and 'd' are not assigned '0'), then:

$$(a \times b^{-1}) \div (c \times d^{-1}) = (a \times b^{-1}) \times (c \times d^{-1})^{-1}$$

Proof

We start with the left-hand side of the above equation and use deductive reasoning as follows:

$$
\begin{aligned}
a \times b^{-1} \div (c \times d^{-1}) \quad &= a \times b^{-1} \times c^{-1} \times d && \text{... by Definition 5.3} \\
&= a \times b^{-1} \times d \times c^{-1} && \text{... by Commutative Axiom '$\times$'} \\
&= a \times b^{-1} \times (c \times d^{-1})^{-1} && \text{... by Theorem 5.16}
\end{aligned}
$$

$\therefore \quad a \times b^{-1} \div (c \times d^{-1}) = a \times b^{-1} \times (c \times d^{-1})^{-1}$

Consequently, this proves the theorem.

In Chapter Three on the Integer Number System, we created an additive inverse for each integer and then proved that applying the inverse operation twice, left the original number unchanged. A natural consequence of Theorem 5.17 is that when we apply the inverse of an index form twice (inverse of an inverse), we get the index form back again. For example, the following theorem will assure us that arithmetic equations of the following form '$((2 \times 3^{-1})^{-1})^{-1} = (2 \times 3^{-1})$' are always true.

Theorem 5.18   Inverse of an Inverse Theorem

If both 'a' and 'b' are non-zero integer variables, then the inverse of the inverse operation applied to an index form, results in the original number in index form, as follows:

$$((a \times b^{-1})^{-1})^{-1} = a \times b^{-1}$$

Proof

We start with the above equation and use deductive reasoning as follows:

$$((a \times b^{-1})^{-1})^{-1} = a \times b^{-1} \quad \text{... by assumption}$$

$$((a \times b^{-1})^{-1})^{-1} \times (a \times b^{-1})^{-1} = a \times b^{-1} \times (a \times b^{-1})^{-1}$$

$$\text{... by multiplying by } `(a \times b^{-1})^{-1}"$$

$$\therefore \qquad 1 = 1 \qquad \text{... by Definition 5.6(3)}$$

From the Closure Axiom, each side of this equation is simply a rational number in index form multiplied by its inverse. In this way, the theorem is proved.

———————————————

Next, we prove a theorem that illustrates concisely how the negative symbol '−' interacts with rational numbers expressed in index form, in the same way that we did for fractions. For example, the following theorem will assure us that arithmetic equations of the following form '$-(2 \times 3^{-1}) = -2 \times 3^{-1} = 2 \times (-3)^{-1}$' are always true.

Theorem 5.19   Applying the Unary Operator '−' to Index Forms Theorem

If 'a' and 'b' are integer variables (where 'b' is not assigned '0'), then we prove the following equations:

$$-(a \times b^{-1}) = -a \times b^{-1} = a \times (-b)^{-1}$$

Proof

We will start with the left-hand-side of this equation and use deductive reasoning:

$$-(a \times b^{-1}) = -1 \times (a \times b^{-1}) \qquad \text{... by Definition 5.5}$$

$$= (-1 \times a) \times b^{-1} \qquad \text{... by Associative Axiom '}\times\text{'}$$

$$\therefore \quad -(a \times b^{-1}) = -a \times b^{-1} \qquad \text{... by Definition 3.1 of '}-a\text{'}$$

To prove the second result, use the result above as follows:

$$-(a \times b^{-1}) = -a \times b^{-1} \qquad \text{... by first half of Theorem}$$

$$= -a \times b^{-1} \times 1 \qquad \text{... by Theorem 5.8}$$

$$= -a \times b^{-1} \times (-1 \times (-1)^{-1}) \quad \text{... by Inverse Axiom '}\times\text{'}$$

$$= -a \times -1 \times (b^{-1} \times (-1)^{-1}) \quad \text{... by Commutat. Axiom '}\times\text{'}$$

$$= a \times (b \times -1)^{-1} \qquad \text{... by Theorem 5.13}$$

$$= a \times (-b)^{-1} \qquad \text{... by Definition 3.1 of '}-b\text{'}$$

$$\therefore \quad -(a \times b^{-1}) = a \times (-b)^{-1}$$

Consequently, we have completed the proof of the theorem.

Note:

It is worth pointing out that there is a simple conclusion that follows from the above theorem when we assign the variable 'a' the value of '1'; namely '$-(b^{-1}) = (-b)^{-1}$'. That is, the operation of negation (i.e. making a term negative by multiplying by '-1') commutes with the operation of taking the reciprocal.

So far, in this Theory Section we have shown that the Rational Numbers using Index Form written as '$a \times b^{-1}$' are equivalent to the Rational Numbers using Fraction Form written as '$\frac{a}{b}$'. We have also shown that the four operations of '$+$', '$-$', '$\times$' and '$\div$' produce identical results in either representation of these rational numbers.

We now wish to extend the above use of the index form to represent rational numbers using more general indices other than the index '-1'. By doing this we will be able to create a convenient notation in which to represent decimal numbers in later chapters of this textbook.

Rational Numbers using Integer Indices

You will recall in Definition 2.5 in Chapter Two that we defined the index for natural numbers. In the section above we made extensive use of the index of '-1' as part of an alternative notation for a rational number. In this section we will now extend the index of a rational number to cover all indices that can belong to the set of integer index symbols, i.e. $\{... , -3, -2, -1, 0, 1, 2, 3, ...\}$.

At this point in the chapter we do not know whether these index symbols behave like integers. However, in the theorems below, we will develop all the properties for manipulating these index symbols with the operations of addition and multiplication, and establish their relationship to integers. We will call these number symbols and the other symbols (namely: '$+$', '$\times$', '$=$', '$\neq$', '(' and ')') used to operate on them, the Index Number System.

Before we can perform operations on the Index Number System, we must first prove that the index symbols also possess the properties of the Axioms of Logic due to their relationship to their base rational numbers and Definition 5.6.

Theorem 5.20   Properties of Logic for the Index Number System Theorem

If 'm', 'n', 'p' and 'q' are index variables, '-1', '0' and '1' are index constant symbols and '$+$' and '$\times$' are the index operator symbols, then the index symbols have the same properties as the properties expressed in the Axioms of Logic as follows:

1. Declaration Property
2. Symmetry Property
3. Transitive Property
4. Not-Equal-To Property
5. Deduction Property
6. Mathematical Induction Property.

Proof

We commence this proof by showing that each of these properties follows from Definition 5.6.

The **declaration property** can be written in the following way: this basic property of the Index Number System follows from declaring the equations in Definition 5.6 to be true equations for indices.

The **symmetry property** can be written in the following way: given '$m = n$', then '$n = m$'. This property is easily demonstrated by using the definition of '=' for the index variable equation '$m = n$' as follows:

$$(a \times b^{-1})^m = (a \times b^{-1})^n \qquad \text{... by Definition 5.6(6)}$$

As the equal sign for rational numbers is symmetric, we can rewrite this equation as:

$$(a \times b^{-1})^n = (a \times b^{-1})^m \qquad \text{... by Symmetry Axiom}$$

$$\therefore \qquad n = m \qquad \text{... by Definition 5.6(6)}$$

Hence, we can declare that the indices also possess the Symmetry Property.

The **transitive property** can be written in the following way: given '$m = n$' and '$n = p$' are true equations when we assign indices to these variables, then the equation '$m = p$' is true for the Index Number System. This property is easily proved using the definition of the '=' symbol for indices that follows:

$$(a \times b^{-1})^m = (a \times b^{-1})^n \qquad \text{... by Definition 5.6(6)}$$

$$(a \times b^{-1})^n = (a \times b^{-1})^p \qquad \text{... by Definition 5.6(6)}$$

Now, applying the Transitive Axiom for rational numbers using index variables we have:

$$(a \times b^{-1})^m = (a \times b^{-1})^p \qquad \text{... by Transitive Axiom}$$

$$\therefore \qquad m = p \qquad \text{... by Definition 5.6(6)}$$

Consequently, we can declare that when we assign indices to index variables, these general indices also possess the transitive property.

The **not-equal-to property** follows directly from Definition 5.6(6).

The **deduction property** can be written in the following way: if '$m = n$' and '$p = q$' for index variables, then the equations '$m + p = n + q$' and '$m \times p = n \times q$' are true.

The first two of these four equations can be written as:

$$(a \times b^{-1})^m = (a \times b^{-1})^n \qquad \text{... by Definition 5.6(6)}$$

And:

$$(a \times b^{-1})^p = (a \times b^{-1})^q \qquad \text{... by Definition 5.6(6)}$$

Multiplying these two equations together as rational numbers gives:

$$(a \times b^{-1})^m \times (a \times b^{-1})^p = (a \times b^{-1})^n \times (a \times b^{-1})^q \quad \text{... by Deduction Axiom}$$

$$\therefore \quad (a \times b^{-1})^{m+p} = (a \times b^{-1})^{n+q} \qquad \text{... by Definition 5.6(4)}$$

$$\therefore \quad m + p = n + q \qquad \text{... by Definition 5.6(6)}$$

Similarly, given '$(a \times b^{-1})^m = (a \times b^{-1})^n$' and '$p = q$' we can write:

$$((a \times b^{-1})^m)^p = ((a \times b^{-1})^n)^q \qquad \text{... by Definition 5.6(6)}$$

$$\therefore \quad (a \times b^{-1})^{m \times p} = (a \times b^{-1})^{n \times q} \qquad \text{... by Definition 5.6(5)}$$

$$\therefore \quad m \times p = n \times q \qquad \text{... by Definition 5.6(6)}$$

Consequently, the deduction property holds for general indices by Definition 5.6 and the corresponding deduction property of their rational number bases.

The **mathematical induction property** applies to the truth of a numbered sequence of mathematical statements. If these statements contain general indices, then this mathematical induction property will apply to these general indices as well. Therefore, this property also holds true for general indices.

In summary, it is apparent that Definition 5.6 defines general indices in relation to their rational number bases. Therefore, Definition 5.6 with rational number bases guarantees us that the properties modelled in the Axioms of Logic are also properties of these general indices.

Consequently, this theorem is proved.

Our next step is to use Definition 5.6 and the properties of logic in Theorem 5.20 to determine the properties of these general indices using the addition operator for indices in the Index Number System. This is the purpose of our next theorem.

Theorem 5.21   Additive Properties for Indices Theorem

Given 'm', 'n' and 'p' are the index variables and the index operator symbol for addition is given by '$+$', then these index symbols have the following properties:

1. $\qquad m + n = n + m$

2. $\qquad (m + n) + p = m + (n + p)$

3. $\qquad 1 + 0 = 1$

4. $\qquad 1 + -1 = 0$

Proof

Proof of Equation 1:

We commence this process by showing that each of these properties follows from Definition 5.6. The first result that we prove is that the index addition operation is commutative.

$$(a \times b^{-1})^{m+n} = (a \times b^{-1})^m \times (a \times b^{-1})^n \text{ ... by Definition 5.6(4)}$$
$$= (a \times b^{-1})^n \times (a \times b^{-1})^m \text{ ... by Commutative Axiom '}\times\text{'}$$
$$\therefore \quad (a \times b^{-1})^{m+n} = (a \times b^{-1})^{m+n} \qquad \text{ ... by Definition 5.6(4)}$$

By Definition 5.6(6), we can now write '$m + n = n + m$' for integer indices. Hence, the index addition operation is commutative under addition because the rational number bases are commutative under multiplication.

Proof of Equation 2:

Next, we show that the index addition operation is associative by using Definition 5.6.

$$(a \times b^{-1})^{m+(n+p)} = (a \times b^{-1})^m \times (a \times b^{-1})^{n+p}$$
$$\text{ ... by Definition 5.6(4, 7)}$$
$$= (a \times b^{-1})^m \times ((a \times b^{-1})^n \times (a \times b^{-1})^p)$$
$$\text{ ... by Definition 5.6(4)}$$
$$= ((a \times b^{-1})^m \times (a \times b^{-1})^n) \times (a \times b^{-1})^p$$
$$\text{ ... by Assoc. Axiom '}\times\text{'}$$

$$= (a \times b^{-1})^{m+n} \times (a \times b^{-1})^p$$

$$\text{... by Definition 5.6(4)}$$

$$= (a \times b^{-1})^{(m+n)+p} \qquad \text{... by Definition 5.6(4, 7)}$$

$$\therefore \quad (a \times b^{-1})^{m+(n+p)} = (a \times b^{-1})^{(m+n)+p}$$

By Definition 5.6(6), we can write '$m + (n + p) = (m + n) + p$'. In this way, the indices have the associative property under index addition because the rational number bases are associative under multiplication.

Proof of Equation 3:

Next, we show that the index symbol '0' is the additive identity by using Definition 5.6.

$$(a \times b^{-1})^{1+0} = (a \times b^{-1})^1 \times (a \times b^{-1})^0 \quad \text{... by Definition 5.6(4)}$$

$$= (a \times b^{-1}) \times 1 \qquad \text{... by Definition 5.6(1, 2)}$$

$$= (a \times b^{-1}) \qquad \text{... by Identity Axiom '$\times$'}$$

$$\therefore \quad (a \times b^{-1})^{1+0} = (a \times b^{-1})^1 \qquad \text{... by Definition 5.6(2)}$$

By Definition 5.6(6), we can write '$1 + 0 = 1$'. Therefore, the index symbol '0' is the additive identity for indices.

Proof of Equation 4:

Finally, we show that the index symbol '−1' is the additive inverse of the index symbol '1' by using Definition 5.6.

$$(a \times b^{-1})^{1+-1} = (a \times b^{-1})^1 \times (a \times b^{-1})^{-1} \qquad \text{... by Definition 5.6(4)}$$

$$= 1 \qquad \text{... by Definition 5.6(3)}$$

$$= (a \times b^{-1})^0 \qquad \text{... by Definition 5.6(1)}$$

$$\therefore \quad (a \times b^{-1})^{1+-1} = (a \times b^{-1})^0$$

By Definition 5.6(6), we can write '$1 + -1 = 0$'. In this way, the index symbol '−1' is the additive inverse for the index symbol '1'.

Consequently, all four equations of this theorem have been proved.

With the properties of addition of indices proven above, we can now derive general equations that express '$(a \times b^{-1})^n$' and '$(a \times b^{-1})^{-n}$' as a product of factors.

Theorem 5.22   General Index Form Theorem

If 'a' and 'b' are integer variables (where 'b' cannot be assigned '0' and 'n' is a positive index), then we can express '$(a \times b^{-1})^n$' and '$(a \times b^{-1})^{-n}$' as products of rational numbers as follows:

1. $(a \times b^{-1})^n = (a \times b^{-1}) \times ... \times (a \times b^{-1})$ ... with 'n' factors

2. $(a \times b^{-1})^{-n} = (a \times b^{-1})^{-1 \times n} = (a \times b^{-1})^{n \times -1}$

$\qquad\qquad = (a \times b^{-1})^{-1} \times ... \times (a \times b^{-1})^{-1}$ ... with 'n' factors

Proof

Proof of Equation 1:

We start with the left-hand side of Equation 1 above using the same definitions for the positive indices as we used for natural numbers introduced in Chapter Two. Therefore, the index '2' is defined by '$2 = 1 + 1$' and using deductive reasoning as follows:

$$(a \times b^{-1})^2 = (a \times b^{-1})^{1+1} \qquad \text{... by Definition 5.6(6)}$$

$$= (a \times b^{-1})^1 \times (a \times b^{-1})^1 \qquad \text{... by Definition 5.6(4)}$$

$$= (a \times b^{-1}) \times (a \times b^{-1}) \qquad \text{... by Definition 5.6(2)}$$

$$\therefore \quad (a \times b^{-1})^2 = (a \times b^{-1}) \times (a \times b^{-1})$$

Once again, we can start with the left-hand side of Equation 1 using the definition of the index '3' defined by '$3 = 2 + 1$' to get:

$$(a \times b^{-1})^3 = (a \times b^{-1})^{2+1} \qquad \text{... by Definition 5.6(6)}$$

$$= (a \times b^{-1})^2 \times (a \times b^{-1})^1 \qquad \text{... by Definition 5.6(4)}$$

$$= (a \times b^{-1}) \times (a \times b^{-1}) \times (a \times b^{-1})$$

$$\text{... by first part of proof}$$

$$\therefore \quad (a \times b^{-1})^3 = (a \times b^{-1}) \times (a \times b^{-1}) \times (a \times b^{-1})$$

Finally, we can generalise this equation to an equation with an index of 'n', so that '$(a \times b^{-1})^n = (a \times b^{-1}) \times ... \times (a \times b^{-1})$' is the equation with the right-hand side containing a base '$(a \times b^{-1})$' with 'n' factors. Here the term '$(a \times b^{-1})^n$' is a rational number by the Closure Axiom for Rational Numbers because it is just a product of rational numbers. Hence, Definition 5.6 yields the expected result for raising a rational number to a positive index.

Proof of Equation 2:

We start the proof of Equation 2 by proving '$(a \times b^{-1})^{-1 \times n} = (a \times b^{-1})^{n \times -1}$' which claims that the indices '-1' and 'n' commute. We commence this process by using Definition 5.6(3) to evaluate '$(a \times b^{-1})^{n \times -1}$' as follows:

$$(a \times b^{-1})^{n \times -1} = ((a \times b^{-1})^n)^{-1} \qquad \text{... by Definition 5.6(5)}$$

However:

$$(a \times b^{-1})^n \times ((a \times b^{-1})^n)^{-1} = 1 \qquad \text{... by Definition 5.6(3)}$$

This last equation is true by Definition 5.6(3) as we are multiplying the rational number '$(a \times b^{-1})^n$' by its multiplicative inverse. Next, we use the expression for '$(a \times b^{-1})^n$' from the first proof in this theorem to derive an expression for '$((a \times b^{-1})^n)^{-1}$'.

$$((a \times b^{-1})^n)^{-1} \times (a \times b^{-1})^n = 1$$
$$((a \times b^{-1})^n)^{-1} \times (a \times b^{-1}) \times ... \times (a \times b^{-1}) = 1 \qquad \text{... by 'n' factors}$$
$$\therefore \qquad ((a \times b^{-1})^n)^{-1} = (a \times b^{-1})^{-1} \times ... \times (a \times b^{-1})^{-1}$$
$$\text{... by 'n' factors}$$

In this last equation we have isolated '$((a \times b^{-1})^n)^{-1}$' on the left-hand side of the equal sign by multiplying both sides by '$(a \times b^{-1})^{-1} \times ... \times (a \times b^{-1})^{-1}$' containing '$n$' factors and using the Deduction Axiom. These 'n' factors '$(a \times b^{-1})^{-1}$', can be written as '$((a \times b^{-1})^{-1})^n$' by the first proof in this theorem. Hence, we have the result:

$$\therefore \qquad ((a \times b^{-1})^n)^{-1} = ((a \times b^{-1})^{-1})^n \qquad \text{... by results above}$$

or,

$$(a \times b^{-1})^{n \times -1} = (a \times b^{-1})^{-1 \times n} \qquad \text{... by Definition 5.6(5)}$$

This commutative property of the positive index 'n' with '-1' entitles us to use our definition of negative numbers in Chapter 3, namely: '$-n = -1 \times n$' for indices. So finally, we can write:

$$(a \times b^{-1})^{-n} = (a \times b^{-1})^{-1 \times n} = (a \times b^{-1})^{n \times -1} \text{ ... by results above}$$
$$\therefore \qquad (a \times b^{-1})^{-n} = (a \times b^{-1})^{-1} \times ... \times (a \times b^{-1})^{-1} \text{ ... with 'n' factors}$$

Consequently, both parts of this theorem have been proved.

_____

With the general expressions for '$(a \times b^{-1})^n$' and '$(a \times b^{-1})^{-n}$' now available for us to use, we are ready to prove the equivalent theorem to Theorem 5.21 for multiplication.

Theorem 5.23 Multiplicative Properties for Indices Theorem

Given 'm', 'n' and 'p' are the index variables and the index operator symbol for multiplication is given by '$\times$', then these index symbols have the following properties:

1. $m \times n = n \times m$
2. $(m \times n) \times p = m \times (n \times p)$
3. $1 \times 1 = 1$

Proof

Proof of Equation 1:

We start by showing that each of these properties follows from Definition 5.6. The first result that we prove is that the index multiplication operation is commutative.

First, we consider the case where 'm' and 'n' are positive indices:

$$(a \times b^{-1})^{m \times n} = ((a \times b^{-1})^m)^n \qquad \text{... by Definition 5.6(5)}$$

$$= (a \times b^{-1})^m \times ... \times (a \times b^{-1})^m$$

$$\text{... '}n\text{' factors of '}(a \times b^{-1})^m\text{'}$$

$$= (a \times b^{-1}) \times ... \times (a \times b^{-1})$$

$$\text{... '}n \times m\text{' factors of '}a \times b^{-1}\text{'}$$

$$= (a \times b^{-1})^n \times ... \times (a \times b^{-1})^n$$

$$\text{... '}m\text{' factors of '}(a \times b^{-1})^n\text{'}$$

$$= ((a \times b^{-1})^n)^m \qquad \text{... by Theorem 5.22}$$

$$\therefore \quad (a \times b^{-1})^{m \times n} = (a \times b^{-1})^{n \times m} \qquad \text{... by Definition 5.6(5)}$$

By Definition 5.6(6), we can now write '$m \times n = n \times m$' is true for positive indices. Next, we carry out the same proof when one of the indices is negative. We assume the second index is negative and write '$-n$' in place of 'n' as follows:

$$(a \times b^{-1})^{m \times -n} = ((a \times b^{-1})^m)^{-n} \qquad \text{... by Definition 5.6(5)}$$

$$= ((a \times b^{-1})^m)^{-1 \times n} \qquad \text{... by Theorem 5.22}$$

$$= (((a \times b^{-1})^m)^{-1})^n \qquad \text{... by Definition 5.6(5)}$$

$$= ((a \times b^{-1})^m)^{-1} \times ... \times ((a \times b^{-1})^m)^{-1}$$

$$\text{... '}n\text{' factors '}((a \times b^{-1})^m)^{-1}\text{'}$$

$$= ((a \times b^{-1})^{-1})^m \times ... \times ((a \times b^{-1})^{-1})^m$$

$$\text{... by Theorem 5.22}$$

$$= (a \times b^{-1})^{-1} \times ... \times (a \times b^{-1})^{-1}$$

$$\text{... by '}n \times m\text{' factors}$$

$$= ((a \times b^{-1})^{-1})^n \times ... \times ((a \times b^{-1})^{-1})^n$$

$$\text{... by '}m\text{' factors}$$

$$= (a \times b^{-1})^{-1 \times n} \times ... \times (a \times b^{-1})^{-1 \times n}$$

$$\text{... by Definition 5.6(5)}$$

$$= (a \times b^{-1})^{-n} \times ... \times (a \times b^{-1})^{-n}$$

$$\text{... by Theorem 5.22}$$

$$= ((a \times b^{-1})^{-n})^m \qquad \text{... by Theorem 5.22}$$

$$\therefore \quad (a \times b^{-1})^{m \times -n} = (a \times b^{-1})^{-n \times m} \qquad \text{... by Definition 5.6(5)}$$

Using the same style of argument, the case for assigning a negative number to each index – namely '$(a \times b^{-1})^{-m \times -n}$' – can be easily proved. Consequently, we can write '$m \times n = n \times m$' is true for all indices.

Proof of Equation 2:

Next, we show that we can prove that the index multiplication operation is associative.

First, we consider the case where 'm', 'n' and 'p' are positive indices.

$$(a \times b^{-1})^{m \times (n \times p)} = ((a \times b^{-1})^m)^{n \times p} \qquad \text{... by Definition 5.6(5)}$$

$$= (a \times b^{-1})^m \times ... \times (a \times b^{-1})^m \quad \text{... by '}n \times p\text{' factors}$$

$$= (a \times b^{-1}) \times ... \times (a \times b^{-1}) \qquad \text{... by '}m \times n \times p\text{' factors}$$

$$= (a \times b^{-1})^{m \times n} \times ... \times (a \times b^{-1})^{m \times n}$$

$$\text{... by '}p\text{' factors}$$

$$= ((a \times b^{-1})^{m \times n})^p \qquad \text{... by Theorem 5.22}$$

$$\therefore \quad (a \times b^{-1})^{m \times (n \times p)} = (a \times b^{-1})^{(m \times n) \times p}$$

For the case where one, two or all 'm', 'n' and 'p' are negative indices, we can use the same style of proof used in the first part of this theorem. Therefore, the index multiplication operation is associative under multiplication because the rational number bases are associative under multiplication.

Proof of Equation 3:

Next we show that the index symbol '1' is the multiplicative identity by using Definition 5.6.

$$(a \times b^{-1})^{1 \times 1} = ((a \times b^{-1})^1)^1 \qquad \text{... by Definition 5.6(5)}$$

$$= (a \times b^{-1})^1 \qquad \text{... by Definition 5.6(2)}$$

$$\therefore \quad (a \times b^{-1})^{1 \times 1} = (a \times b^{-1})^1$$

By Definition 5.6(6), we can write '$1 \times 1 = 1$'. Hence, the index symbol '1' has the required property.

In this way, all three equations of this theorem have been proved.

With the proof of this theorem we have completed our demonstration that the set of indices, $\{..., -3, -2, -1, 0, 1, 2, 3, ...\}$, to which we raise a rational number possess all the same properties as the integers. Therefore, we can declare that the Index Number System is the same as the Integer Number System. This is a major simplification since we now only have to remember the definition of these indices in order to be able to easily work with rational numbers written using index form.

The following extension of Theorem 5.22, which calculates the product of two factors with the same indices but different bases will be very useful in later chapters of this textbook.

Theorem 5.24   Multiplying Different Bases with the Same Index Theorem

Given 'a', 'b', 'c' and 'd' are integer variables (where 'b' and 'd' are not assigned '0' and 'n' is an integer index), then we can write:

$$(a \times b^{-1})^n \times (c \times d^{-1})^n = ((a \times b^{-1}) \times (c \times d^{-1}))^n \qquad \text{... with 'n' factors}$$

Proof

This proof is a direct application of Theorem 5.22. If the integer variable 'n' is first assigned a positive value, then from Theorem 5.22 we have expressions for the two products above, as:

$$(a \times b^{-1})^n = (a \times b^{-1}) \times ... \times (a \times b^{-1}) \qquad ...\text{ by `}n\text{' factors}$$

$$(c \times d^{-1})^n = (c \times d^{-1}) \times ... \times (c \times d^{-1}) \qquad ...\text{ by `}n\text{' factors}$$

$$\therefore \quad (a \times b^{-1})^n \times (c \times d^{-1})^n =$$

$$(a \times b^{-1}) \times ... \times (a \times b^{-1}) \times (c \times d^{-1}) \times ... \times (c \times d^{-1})$$
$$...\text{ by `}2n\text{' factors}$$

$$= ((a \times b^{-1})(c \times d^{-1})) \times \ ... \times ((a \times b^{-1})(c \times d^{-1}))$$
$$...\text{ by Assoc. Axiom `}\times\text{'}$$

$$\therefore \quad (a \times b^{-1})^n \times (c \times d^{-1})^n = ((a \times b^{-1})(c \times d^{-1}))^n \quad ...\text{ by Theorem 5.22}$$

If the integer index is next assigned a negative value, then from the second part of Theorem 5.22 we can use the equation for '$(a \times b^{-1})^{-n}$' and prove the equivalent result that '$(a \times b^{-1})^{-n} \times (c \times d^{-1})^{-n} = ((a \times b^{-1})(c \times d^{-1}))^{-n}$'.

Consequently, both parts of this theorem have been proved.

We have now completed the basic theory of the Rational Number System using Index Form. This theory provides a very useful tool to manipulate rational numbers in later chapters of this textbook, particularly the chapters on representing numbers in decimal format.

Applications of the Rational Number System Using Index Form

Not only do numbers raised to an index have uses in many branches of Mathematics, they also have uses in Science and Engineering where either very small numbers or very large numbers are required to describe some aspect of the physical world.

In fact, the sciences use a standard form of notation called scientific notation, where the first number is expressed as a decimal with one digit to the left of the decimal point followed by the remaining digits and then multiplied by '10' raised to an integer index depending on the size of the decimal number. For example, the decimal number '123.456' is expressed in scientific notation as '1.23456×10^2' using index notation.

In our first application we prove the equivalent result that we proved for fractions in Application 4.1: '$-\frac{2}{3} + \frac{4}{5} = \frac{2}{15}$'.

Application 5.1

Evaluate the arithmetic term: '$-2 \times 3^{-1} + 4 \times 5^{-1}$'.

We evaluate the term '$-2 \times 3^{-1} + 4 \times 5^{-1}$' by using our theory as follows:

$$-2 \times 3^{-1} + 4 \times 5^{-1} = -2 \times 3^{-1} \times 1 + 4 \times 5^{-1} \times 1 \quad \text{... by Theorem 5.8}$$

$$= -2 \times 3^{-1} \times 5 \times 5^{-1} + 4 \times 5^{-1} \times 3 \times 3^{-1}$$
$$\text{... by Inverse Axiom '}\times\text{'}$$

$$= -2 \times 5 \times 3^{-1} \times 5^{-1} + 4 \times 3 \times 5^{-1} \times 3^{-1}$$
$$\text{... by Assoc. Axiom '}\times\text{'}$$

$$= (-2 \times 5 + 4 \times 3) \times 3^{-1} \times 5^{-1} \quad \text{... by Distributive Axiom}$$

$$= (-10 + 12) \times (3 \times 5)^{-1} \quad \text{... by Theorem 5.13}$$

$$= 2 \times 15^{-1} \quad \text{... by simplifying terms}$$

$\therefore \quad -2 \times 3^{-1} + 4 \times 5^{-1} = 2 \times 15^{-1}$

Hence, when we evaluate the term '$-2 \times 3^{-1} + 4 \times 5^{-1}$' in index form, we get the rational number in index form: '2×15^{-1}'.

In our second application, we prove the equivalent result that we proved for fractions in Application 4.2: '$-5\frac{2}{3} + 3\frac{4}{5} = -1\frac{13}{15}$'.

Application 5.2

Evaluate the arithmetic term in mixed index form: '$-5 - 2 \times 3^{-1} + 3 + 4 \times 5^{-1}$'.

We can evaluate this term by using our theory as follows:

$-5 - 2 \times 3^{-1} + 3 + 4 \times 5^{-1}$

$$= -(5 \times 1^{-1} + 2 \times 3^{-1}) + 3 \times 1^{-1} + 4 \times 5^{-1}$$
$$\text{... by Theorem 5.6}$$

$$= -(5 \times 3 + 2 \times 1) \times (1 \times 3)^{-1} + (3 \times 5 + 4 \times 1) \times (1 \times 5)^{-1}$$
$$\text{... by Theorem 5.15}$$

$$= -17 \times 3^{-1} + 19 \times 5^{-1} \quad \text{... by simplifying terms}$$

$$= (-17 \times 5 + 19 \times 3) \times (3 \times 5)^{-1} \text{... by Theorem 5.15}$$

$$= -28 \times 15^{-1} \quad \text{... by simplifying terms}$$

$$= (-15 + -13) \times 15^{-1} \quad \text{... by rewriting '}-28\text{'}$$

$$= -15 \times 15^{-1} + -13 \times 15^{-1} \qquad \text{... by Distributive Axiom}$$

$$= -1 + -13 \times 15^{-1} \qquad \text{... by Inverse Axiom 'x'}$$

$$= -1 \times 1^{-1} + -13 \times 15^{-1} \qquad \text{... by Theorem 5.6}$$

$$\therefore -5 - 2 \times 3^{-1} + 3 + 4 \times 5^{-1} = -1 \times 1^{-1} + -13 \times 15^{-1}$$

In this way, when we evaluate the term '$-5 - 2 \times 3^{-1} + 3 + 4 \times 5^{-1}$' we get the index form: '$-1 \times 1^{-1} + -13 \times 15^{-1}$'.

The next application will illustrate the use of the Distributive Axiom in practical situations with numbers in index form.

Application 5.3

Evaluate the arithmetic term: '$3 \times 4^{-1} \times (-5 - 2 \times 3^{-1} + 4)$' using the Distributive Axiom.

We can evaluate this term by using our theory as follows:

$$3 \times 4^{-1} \times (-5 - 2 \times 3^{-1} + 4)$$

$$= 3 \times 4^{-1} \times ((-5 \times 1^{-1} + -2 \times 3^{-1}) + 4)$$

$$\text{... by Theorem 5.6}$$

$$= 3 \times 4^{-1} \times ((-5 \times 3 + -2 \times 1) \times (1 \times 3)^{-1} + 4)$$

$$\text{... by Theorem 5.15}$$

$$= 3 \times 4^{-1} \times (-17 \times 3^{-1} + 4) \qquad \text{... by simplifying terms}$$

$$= 3 \times 4^{-1} \times (-17 \times 3^{-1} + 4 \times 1^{-1}) \qquad \text{... by Theorem 5.6}$$

$$= 3 \times 4^{-1} \times -17 \times 3^{-1} + 3 \times 4^{-1} \times 4 \times 1^{-1}$$

$$\text{... by Distributive Axiom}$$

$$= -17 \times 4^{-1} + 3 \times 1^{-1} \qquad \text{... by Inverse Axiom 'x'}$$

$$= (-17 \times 1 + 3 \times 4) \times (4^{-1} \times 1^{-1})$$

$$\text{... by Theorem 5.15}$$

$$= -5 \times 4^{-1} \qquad \text{... by Theorem 5.13}$$

$$= -(4 + 1) \times 4^{-1} \qquad \text{... by rewriting '-5'}$$

$$= -4 \times 4^{-1} + -1 \times 4^{-1} \qquad \text{... by Distributive Axiom}$$

$$= -1 + -1 \times 4^{-1} \qquad \text{... by Inverse Axiom 'x'}$$

$$= -1 \times 1^{-1} + -1 \times 4^{-1} \qquad \text{... by Theorem 5.6}$$

$$\therefore 3 \times 4^{-1} \times (-5 - 2 \times 3^{-1} + 4)$$

$$= -1 \times 1^{-1} - 1 \times 4^{-1}$$

In this way, when we evaluate the term '$3 \times 4^{-1} \times (-5 - 2 \times 3^{-1} + 4)$', we get the term in index form: '$-1 \times 1^{-1} - 1 \times 4^{-1}$'. In this proof we have successfully used the Distributive Axiom to demonstrate the use of the definitions and theorems of this chapter.

_____

Next we give an application that demonstrates the use of the division operation with index forms. This is equivalent to Application 4.4 in Chapter Four.

Application 5.4

Evaluate the arithmetic term: '$(3 \times 4^{-1}) \div (5 \times 8^{-1}) \times (-5 - 2 \times 3^{-1}) + 4$'.

We evaluate this term using our theory as follows:

$$(3 \times 4^{-1}) \div (5 \times 8^{-1}) \times (-5 - 2 \times 3^{-1}) + 4$$

$$= (3 \times 4^{-1}) \times (8 \times 5^{-1}) \times (-5 - 2 \times 3^{-1}) + 4$$
$$\text{... by Definition 5.3}$$

$$= (3 \times 8) \times (4^{-1} \times 5^{-1}) \times (-5 - 2 \times 3^{-1}) + 4$$
$$\text{... by Theorem 5.13}$$

$$= (3 \times 2 \times 4) \times (4^{-1} \times 5^{-1}) \times (-5 - 2 \times 3^{-1}) + 4$$
$$\text{... by rewriting '8'}$$

$$= 6 \times 5^{-1} \times (-5 - 2 \times 3^{-1}) + 4$$
$$\text{... by Inverse Axiom '}\times\text{'}$$

$$= (6 \times 5^{-1} \times -5) + (6 \times 5^{-1} \times -2 \times 3^{-1}) + 4$$
$$\text{... by Distributive Axiom}$$

$$= -1 \times (6 \times 1) + (6 \times -2 \times 5^{-1} \times 3^{-1}) + 4$$
$$\text{... by Inverse Axiom '}\times\text{'}$$

$$= -6 + 4 + (3 \times 2 \times -2 \times 5^{-1} \times 3^{-1})$$
$$\text{... by rearranging terms}$$

$$= -6 + 4 + (2 \times -2 \times 5^{-1}) \qquad \text{... by Inverse Axiom '}\times\text{'}$$

$$= -2 + -4 \times 5^{-1} \qquad \text{... by simplifying terms}$$

$\therefore (3 \times 4^{-1}) \div (5 \times 8^{-1}) \times (-5 - 2 \times 3^{-1}) + 4 = -2 + -4 \times 5^{-1}$

Hence, when we evaluate the arithmetic term: '$(3 \times 4^{-1}) \div (5 \times 8^{-1}) \times (-5 - 2 \times 3^{-1}) + 4$' we get the arithmetic term using index form as: '$-2 \times 1^{-1} + -4 \times 5^{-1}$'.

The final application illustrates the application of some of the definitions for the index form where the indices can be negative and positive integers.

Application 5.5

Evaluate the arithmetic term: '$(3 \times 4^{-1})^3 \div (5 \times 8^{-1}) \times (8 \times 5^{-1})^{-2} \times (3 \times 4^{-1})^{-4}$'.

We evaluate this term using our theory as follows:

$(3 \times 4^{-1})^3 \div (5 \times 8^{-1}) \times (8 \times 5^{-1})^{-2} \times (3 \times 4^{-1})^{-4}$

$\qquad = (3 \times 4^{-1})^3 \times (3 \times 4^{-1})^{-4} \div (5 \times 8^{-1}) \times (8 \times 5^{-1})^{-2}$

$\qquad\qquad\qquad\qquad$ ... by rearranging terms

$\qquad = (3 \times 4^{-1})^{(3-4)} \div (5 \times 8^{-1}) \times (8 \times 5^{-1})^{-2}$

$\qquad\qquad\qquad\qquad$ ... by Definition 5.6(4)

$\qquad = (3 \times 4^{-1})^{-1} \times (8 \times 5^{-1})^1 \times (8 \times 5^{-1})^{-2}$

$\qquad\qquad\qquad\qquad$ ... by Theorem 5.16

$\qquad = (3 \times 4^{-1})^{-1} \times (8 \times 5^{-1})^{-1}$ $\qquad$ ... by Definition 5.6(3)

$\qquad = 4 \times 3^{-1} \times 5 \times 8^{-1}$ $\qquad$ ... by Theorem 5.16

$\qquad = 4 \times 5 \times 3^{-1} \times 8^{-1}$ $\qquad$ ... by rearranging terms

$\qquad = 4 \times 5 \times 3^{-1} \times 2^{-1} \times 4^{-1}$ $\qquad$ ... by Theorem 5.13

$\qquad = 5 \times 3^{-1} \times 2^{-1}$ $\qquad$ ... by Inverse Axiom '$\times$'

$\qquad = 5 \times 6^{-1}$ $\qquad$ ... by Theorem 5.13

$\therefore (3 \times 4^{-1})^3 \div (5 \times 8^{-1}) \times (8 \times 5^{-1})^{-2} \times (3 \times 4^{-1})^{-4} = 5 \times 6^{-1}$

In this way, when we evaluate the arithmetic term '$(3 \times 4^{-1})^3 \div (5 \times 8^{-1}) \times (8 \times 5^{-1})^{-2} \times (3 \times 4^{-1})^{-4}$' we get the arithmetic term using index forms as: '5×6^{-1}'.

In most of these applications it is easier to calculate the result using fraction notation or a combination of fraction notation and index notation. This explains why fractions play a dominant role in our manipulation of rational numbers.

However, the Rational Numbers using Index Form have many practical applications such as: expressing decimal numbers, expressing prefixes for units of measure, memory sizing and computer speeds and many more. In this textbook, the use of the index form of a rational number will be indispensable for expressing rational numbers and terms in sequences and series in later chapters.

Summary of the Rational Number System Using Index Form

At this point, we have covered the key objectives of this chapter and so can summarise our findings up to now. In terms of the Rational Number System using Index Form, the most important definitions that have been introduced so far are:

1. Define the **multiplicative inverse** of the non–zero integer variable 'a' as 'a^{-1}' by the relationship '$a \times a^{-1} = 1$'.

2. Define a **rational number using index** form as an arithmetic term formed from the integers $\{... , -3, -2, -1, 0, 1, 2, 3, ...\}$ and their multiplicative inverses $\{... , (-3)^{-1}, (-2)^{-1}, (-1)^{-1}, 1^{-1}, 2^{-1}, 3^{-1}, ...\}$ in index form.

3. Define the **division** of the rational number '$a \times b^{-1}$' by the rational number '$c \times d^{-1}$' (where 'a', 'b', 'c' and 'd' are integer variables and 'b', 'c' and 'd' cannot be assigned '0') by:

$$(a \times b^{-1}) \div (c \times d^{-1}) = (a \times b^{-1}) \times (d \times c^{-1})$$

4. Define the **index of '-1'** using the equation:

$$(a \times b^{-1}) \times (a \times b^{-1})^{-1} = 1$$

5. The definitions of **indices** using the rational number base '$a \times b^{-1}$' and indices from the set $\{... , -2, -1, 0, 1, 2, ...\}$ are given by the following equations:

    5.1 $(a \times b^{-1})^0 = 1$

    5.2 $(a \times b^{-1})^1 = (a \times b^{-1}) = a \times b^{-1}$

    5.3 $(a \times b^{-1})^{-1} (a \times b^{-1}) = 1$

    5.4 $(a \times b^{-1})^{m+n} = (a \times b^{-1})^m \times (a \times b^{-1})^n$

    5.5 $(a \times b^{-1})^{m \times n} = ((a \times b^{-1})^m)^n$

    5.6 '$m = n$' is true if and only if '$(a \times b^{-1})^m = (a \times b^{-1})^n$' is true and both '$a$ **and** 'b' are not assigned integers from the set $\{1, -1\}$. '$m \neq n$' is also true if and only if $(a \times b^{-1})^m \neq (a \times b^{-1})^n$' is true and both '$a$' **and** '$b$' are not assigned the integers from the set $\{1, -1\}$.

5.7   The parentheses symbols as they apply to indices have their usual role as grouping symbols.

The Axioms of the Rational Number System using Index Form that model the partitioning of a number line into equal portions are set out in Table 5.6 below. In this table, rational number variables and an integer variable are used to describe the multiplicative inverse.

5.6 The **Axioms** of the Rational Number System using Index Form		
Axiom Type	**Addition '+'**	**Multiplication '×'**
Closure	1. In this table, the rational number variables are 'x', 'y' and 'z' and the non–zero integer variable is 'a'. The Closure Axiom states that the operations with the following numbers: '$x + y$' and '$x \times y$' result in unique rational numbers.	
Commutative	2. $x + y = y + x$	3. $x \times y = y \times x$
Associative	4. $(x + y) + z = x + (y + z)$	5. $(x \times y) \times z = x \times (y \times z)$
Identity	6. $1 + 0 = 1$	7. $1 \times 1 = 1$
Distributive	8. $x \times (y + z) = x \times y + x \times z$	
Inverse	9. $1 + -1 = 0$	10. $a \times a^{-1} = 1$

Table 5.6

Once again, the Axioms of Logic are used to manipulate the axioms in the table above have not changed in form since Table 2.5. However, there has been a small change as they can now be applied to the broader set of numbers – the Rational Numbers using Index Form.

Similarly, the change in notation from fraction form to index form in progressing from Chapter Four to this chapter has been a small one; however, the benefit in terms of familiarity with index notation and its extension to integer indices is critical to our later treatment of Rational Numbers as Repeating Decimals in Chapter Ten.

It is also worth noting that in both the treatments of the Rational Number System in Chapter Four and in this chapter, we chose to ensure every integer (except '0') had a multiplicative inverse. If we leave the unlimited number of variables aside, then in progressing from the Integer Number System to the Rational Number System, our base alphabet has become unlimited because of these new multiplicative inverses. This is in stark contrast to the progression from the Natural Number System to the Integer Number System where we only had to add the new symbol, the additive inverse '−1'. We will cover this observation in more detail in our discussion about Repeating Decimals in Chapter Ten.

The Language, Elementary Properties, Axioms and Definitions Sections of this chapter have given us another way of representing rational numbers. This chapter has also paved the way for writing rational numbers in another format (i.e. as decimals in a later chapter) which, in turn, has given us an extension of the representation of the Rational Number System and extended our mastery over this system.

In Chapter Six we will continue to extend the language we have used up to this point by using decimal notation to describe certain numbers in the elementary number systems. We will also use the same methods to explore a subset of the theory of rational numbers as it relates to the fundamentally different notation associated with the theory of the Finite Decimal System.

CHAPTER 6

THE FINITE DECIMAL SYSTEM

Overview of the Finite Decimal System

In Chapters Four and Five, we continued our study of the elementary number systems by first introducing the Rational Number System using Fractions and then using Index Form.

Now that we have mastered rational numbers using fractions and using index form, it is time to look at the 'intermediate' number system called the **Finite Decimal System** (which historically has **not** been referred to as an elementary number system). In this number system, every finite decimal (which is also known as a terminating decimal) can be expressed as a rational number; however, **not every** rational number can be expressed as a finite decimal. It is not until Chapter Ten, when we introduce 'repeating decimals', that we will be able to express every rational number as either a finite decimal or a repeating decimal.

Although we have already mastered fractions and index forms where every integer had a multiplicative inverse, in this chapter we will only be using the multiplicative inverse of the number '10' to create the extension to the Integer Number System called the Finite Decimal System. In the Finite Decimal System, we extend our model of the integer numbers to include those numbers **between** the integers that can be expressed using the multiplicative inverse of '10' which is denoted by '10^{-1}' (rather than '$\frac{1}{10}$' since finite decimals are more easily defined in terms of index notation).

As the name implies, finite decimals can only have digits positioned a **limited** number of places to the left and the right of the decimal point. In this way, when we add or multiply these decimals, they will still be allocated a limited number of places after the decimal point. This 'limited' requirement automatically eliminates repeating decimals (such as the decimal expansion of the fraction '$\frac{1}{3} = 0.333...$') from being members of the Finite Decimal System since they involve an unlimited number of digits to the right of the decimal point.

The symbols required for the Finite Decimal System consist of the symbols of the Integer Number System and the new symbols: '.' and '10^{-1}'. The symbol '.' is called the **decimal point** and, as mentioned above, the symbol '10^{-1}' is called the multiplicative inverse of the integer '10'. In this chapter, we will learn how to write finite decimal numbers in several formats.

Generally speaking, we know that a finite decimal is written in the format '123.456' and we interpret this number to mean: '$1 \times 100 + 2 \times 10 + 3 + 4 \times \frac{1}{10} + 5 \times \frac{1}{100} + 6 \times \frac{1}{1000}$'. However, these decimal components '$\frac{1}{10}$', '$\frac{1}{100}$' and '$\frac{1}{1000}$' can be built up from the single number '10^{-1}'. We will use this new number '10^{-1}' in place of the familiar fraction '$\frac{1}{10}$'. Therefore, the decimal number '123.456' can be written using index notation as: '$1 \times 10^2 + 2 \times 10^1 + 3 \times 10^0 + 4 \times 10^{-1} + 5 \times 10^{-2} + 6 \times 10^{-3}$'.

Since the Finite Decimal System is also a subsystem of the Rational Number System, some of the definitions and theorems for the Rational Number System will have equivalent definitions and theorems in the Finite Decimal System. We will use this concept to reduce the amount of replication in definitions and theorems within this chapter.

These new finite decimals will not only be very useful in their own right, but will be particularly advantageous in Chapter Ten (Repeating Decimals) where we extend the Finite Decimal System to allow us to express all Rational Numbers in decimal notation.

In this chapter, we will continue on our journey of mastering the number systems by building on our basic knowledge about the Integer Number System and the work on the Rational Number System using Fractions and Index Forms we acquired in Chapters Four and Five, respectively. This chapter will reinforce most of our learning about the elementary number systems we have developed so far.

Pictorial Representation of Finite Decimals

As in previous chapters, we will provide a pictorial representation of the finite decimals, but now we will include labels for some integer values and several points between integer values along the number line as finite decimals. This concept is illustrated in Figure 6.1 below.

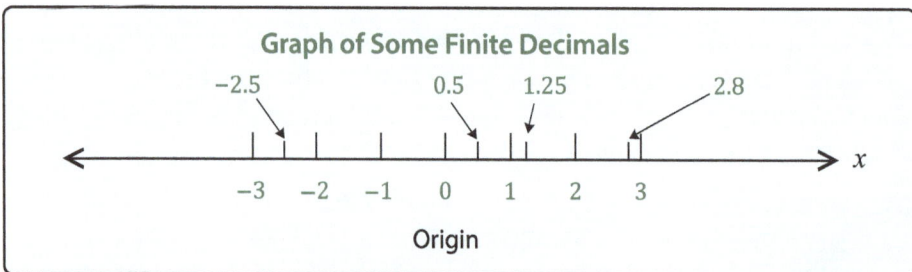

Figure 6.1

Sometimes these integer coordinates are written in decimal notation whereby the integer '3' could be written as '3.0'. However, we will define finite decimals to terminate in a non-zero digit in order to avoid an infinite string of zeroes after the decimal point. As mentioned above, the point along the number line corresponding to the rational number '$\frac{1}{3}$', for example, cannot be labelled with a finite decimal coordinate and so its label does not belong to the Finite Decimal System.

Background and Context of the Finite Decimal System

No doubt, everyone is familiar with decimal notation since we use it in our daily lives to carry out many detailed calculations and financial transactions. As decimal numbers expressed in general decimal notation are important in later chapters of this textbook, it is helpful to have an understanding of their historical origins.

Most historians agree the Decimal System for representing numbers was generally adopted around 600 A.D. in India, where people started to use a 'place value' numbering system. This involved writing a large integer as a string of digits from the set of ten digits {0, 1, 2, 3, 4, 5, 6, 7, 8, 9} where the position of each digit in the string contributed an amount to the total value of the integer based on its position in this string.

This decimal system was quickly adopted by Arabic people and became known as Arabic numerals in Europe where it is believed it was in general use about 1300 A.D. However, in modern times it is often referred to as the Hindu-Arabic Number System.

In Chapter Two we learnt about decimal notation as a way of writing natural numbers larger than the number '9'. Decimal notation allows us to write numbers as multiples of '10' so that the first digit on the right-hand end of the string represents the number of **units**, the second digit from the right-hand end represents the number of **tens**, the third digit represents the number of **hundreds**, and so on. This concept is the base-10 positional notation for integer numbers.

Later, in the Definition Section of this chapter, we will expand this decimal notation to represent numbers which have a finite integer part and a finite fraction part. The finite fraction part of a finite decimal number is called a 'decimal fraction'. For example, the number '532.7' can be written as the integer part: '$532 = 5 \times 100 + 3 \times 10 + 2 \times 1$' and the decimal fraction part: '$0.7 = 7 \times 10^{-1}$'. The dot symbol '.' is called the decimal point and is inserted between the finite integer part and the finite decimal fraction part of the number '532.7' to separate the finite integer part from the finite decimal fraction part. The decimal point will be defined in Definition 6.4 of this chapter.

The introduction of finite decimal notation gives us the tools for applying the operations of addition, subtraction and multiplication. However, undertaking the full operation of division for all decimal numbers will not be covered until Chapter Ten where we use repeating decimals to represent certain multiplicative inverses such as '$\frac{1}{3}$'. Some divisions result in finite decimals and so there is an opportunity for limited division operation (for example, '6.6' divided by '2.2' equals '3'). Yet, as the fraction '$\frac{1}{3}$' demonstrates, the operation of division is not closed for the Finite Decimal System as it requires an infinite number of decimal places for its description.

Now that we have the set of symbols for the Finite Decimal System, it only remains for us to assign meaning to these new symbols using the same method we used in Chapter Four. This approach is outlined in the next section.

Approach to the Finite Decimal System

In this section our aim will be to assign meaning to the alphabetic symbols of the language of the Finite Decimal System. This language is an extension of the language we used for the Integer Number System.

Our approach in this chapter is the same as in previous chapters as, once again, we will be again using our knowledge of the Integer Number System to develop the notation for finite decimals. Although we will assign meaning to adding, subtracting and multiplying finite decimals in this chapter, it will only be possible to fully define division of decimals by extending the Finite Decimal System to include '**repeating decimals**' that have an infinite repeating pattern which will be covered in Chapter Ten.

As mentioned above, some simple fractions and index forms such as '$\frac{1}{3}$' and '3^{-1}' cannot be expressed as a finite decimal number – that is, with a finite number of digits after the decimal point. Repeating decimals such as '0.333...' have a repeating pattern that goes on indefinitely.

In this chapter, we establish that the alphabet for the Finite Decimal System (excluding variables) consists of the symbols $\{0, 1, -1, 10^{-1}, +, \times, (,), =, \neq\}$ and only the integer '10' is required to have a multiplicative inverse.

Language of the Finite Decimal System

At this point, we again extend the alphabet that we developed for the Integer Number System to also include finite decimals by using the multiplicative inverse '10^{-1}'. In this section, we demonstrate how to build up the familiar decimal notation for finite

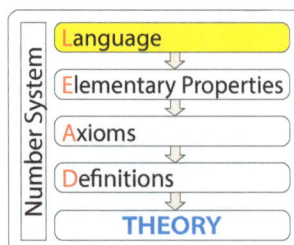

Number System

Language → Elementary Properties → Axioms → Definitions → THEORY decimals (such as '123.456', '0.432', '−2.0', etc.) from the basic symbols of the alphabet of the Finite Decimal System.

The starting alphabet of the Finite Decimal System is summarised in Table 6.1 below.

6.1 The Alphabet of the Finite Decimal System	
Symbols	**Meaning from:**
'0', '1', '+', '×', '(', ')'	The Axioms of the Natural Number System
'=', '≠'	The Axioms of Logic
'2', '3', '4', '5', '6', '7', '8', '9'	Definition 2.1
'a', 'b', 'c', ...	The Axioms of the Integer Number System
','	Definition 2.3
'−1'	The Axioms of the Integer Number System
'−'	Definitions 3.4 and 3.5
'10^{-1}'	The Axioms of the Finite Decimal System

Table 6.1

In this starting language of the Finite Decimal System, we again use variables to represent finite decimal numbers. In this way, a variable such as 'x', 'y' or 'z' is a symbol that can be assigned any finite decimal number.

The use of finite decimals and variables is summarised in Table 6.2 below.

6.2 The Language of the Finite Decimal System	
In this table, 'x', 'y' and 'z' represent finite decimal variables and 'a', 'b', and 'c' represent integer variables.	
1. **Arithmetic Terms**	An individual finite decimal or sum and/or product of finite decimals constructed from the finite decimal alphabet is referred to as an **arithmetic term**, or simply a **term**. Consequently: '$5 \times 21 + 10^{-1}$', '$545 \times 10^{-1} \times 2 \times 10^{-1}$', '$7 + 5 - 6 \times 10^{-1}$', '$-5 + 5 \times 10^{-1}$', '$-3 \times 10^{-1}$', '$(2 + 10^{-1}) \times 10^{-1}$', '5', '$-10^{-1} + -1 \times 4$' are all finite valid strings of symbols called **arithmetic terms**.

6.2 The Language of the Finite Decimal System	
2. **Arithmetic Equations**	When we have an equation with arithmetic terms on either side of an '$=$' or '$\neq$' sign, we call this equation an **arithmetic equation**. A comprehensive example is: '$4 \times (-2 + 5 \times 10^{-1} + 125 \times 10^{-1} \times 10^{-1}) = -1$', is a valid string of symbols called an **arithmetic equation**.
3. **Algebraic Terms and Equations**	If any of the above terms or equations contains one or more variables, then we refer to them as **algebraic** terms or equations. For example, '$3 \times 10^{-1} \times x$' is a finite valid string of symbols called an **algebraic term** and '$y \times 15 \times 10^{-1} = z \times 3$', '$x \times y \times z \times 10^{-1} = 5 \times x$' and '$x = a \times 10^{-1} + b$' are finite valid strings of symbols called **algebraic equations**.

Table 6.2

Note:

Table 6.2 does not include decimals in 'standard decimal notation' as this method of representing finite decimals will not be discussed until later in Definition 6.4.

In summary, we start our investigation of the Finite Decimal System by assuming that the variables 'x', 'y' and 'z' can be assigned to any finite decimal number and can be combined with the operators '$+$' and/or '$\times$' to form terms and equations.

Order of Operations

As we identified in previous chapters, there is an inherent ambiguity in the order in which we could apply operators if no parentheses are used to determine this order. The same Order of Operations convention that we applied to the Integer Number System is extended here to include the finite decimal numbers.

To evaluate a term containing finite decimal numbers we use the Order of Operations shown in Table 6.3 on the following page.

6.3 Order of Operations for the Finite Decimal System

The process used to evaluate arithmetic and algebraic terms is determined by the **order of operations** and proceeds according to the following steps:

1.	The **parentheses** have highest priority and control the order of evaluation. Parentheses may be removed by evaluating inside the parentheses **or** by using the Distributive Axiom.
2.	Next evaluate **exponents** (i.e. an integer index to which a base number is raised).
3.	Where there are no parentheses, then we evaluate **multiplication** '×' from left to right.
4.	Finally, evaluate **addition** '+' and **subtraction** '−' from left to right.

Table 6.3

The Order of Operations for the Finite Decimal System can be remembered by using the same acronym '**PEMAS**' which stands for:

- Parentheses
- Exponents
- Multiplication
- Addition
- Subtraction.

The acronym is dependent on the number system being used and has been amended in this chapter to include operations used in the Finite Decimal System.

Order of Operations – PEMAS

Evaluate the following term using Order of Operations:

Term	$= (3 \times 5) + 2^3 \times 3.5 \times 0.5 - 12$
Parentheses	$= (3 \times 5) + 2^3 \times 3.5 \times 0.5 - 12$
Exponents	$= 15 + 2^3 \times 3.5 \times 0.5 - 12$
Multiplication	$= 15 + 8 \times 3.5 \times 0.5 - 12$
Addition	$= 15 + 14 - 12$
Subtraction	$= 29 - 12$
Answer	$= 17$

The Tenth Elementary Property of the Finite Decimal System

Inverse Property of Multiplication for '10'

This property describes how **only** the integer '10' is required to have a *multiplicative inverse* which is denoted by '10^{-1}' in index form. That is, we only need to know that the equation '$10 \times 10^{-1} = 1$' is true and no other finite inverses are required.

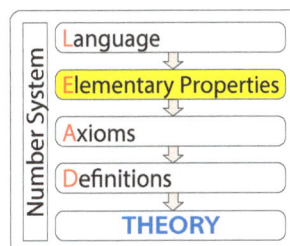

Number System:
- Language
- Elementary Properties
- Axioms
- Definitions
- THEORY

In our Finite Decimal System, this property is modelled by the equation:

$$10 \times 10^{-1} = 1$$

This property gives us the multiplicative inverse for '10', and will also give us the multiplicative inverse for '100' and so on by using the Axioms of the Finite Decimal System.

In this chapter, we have simplified the **Inverse Axiom for Multiplication** used for Fractions and Index Form so that only the number '10' has a multiplicative inverse. The axioms required for the Finite Decimal System are shown in Table 6.4 in the next section.

The Axioms of the Finite Decimal System

By starting with the Axioms of the Integer Number System and adding this new property, we have extended the set of axioms available for us to use. This new set of axioms now applies to the broader set of numbers called the Finite Decimal Numbers and is referred to as the Axioms of the Finite Decimal System.

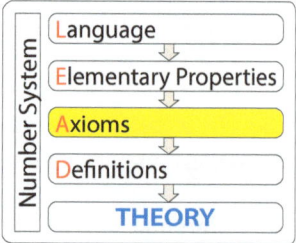

Number System
- Language
- Elementary Properties
- Axioms
- Definitions
- THEORY

The Axioms of the Finite Decimal System are shown in Table 6.4 below.

6.4 The Axioms of the Finite Decimal System		
Axiom Type	**Addition '+'**	**Multiplication '×'**
Closure	1. In this table we use the symbols 'x', 'y' and 'z' as finite decimal variables. The Closure Axiom states that the following operations with any finite decimals assigned to the variables 'x' and 'y' (namely '$x + y$' and '$x \times y$') result in unique finite decimals.	
Commutative	2. $\quad x + y = y + x$	3. $\quad x \times y = y \times x$
Associative	4. $(x + y) + z = x + (y + z)$	5. $(x \times y) \times z = x \times (y \times z)$
Identity	6. $\quad 1 + 0 = 1$	7. $\quad 1 \times 1 = 1$
Distributive	8. $\quad x \times (y + z) = x \times y + x \times z$	
Inverse	9. $\quad 1 + -1 = 0$	10. $\quad 10 \times 10^{-1} = 1$

Table 6.4

These axioms only express the relationships between the numbers, the operators '+' and '×', and the grouping symbols '(' and ')'. To establish all of the relationships between the symbols (i.e. including '=' and '≠') will require us to use the Axioms of Logic.

The Axioms of Logic

As in Chapter Three, before we can deduce correctly from these axioms for the Finite Decimal System, we need to use the Axioms of Logic. Although the Axioms of Logic have not changed in structure from those used in Chapter Two, in this chapter the variables we need to use to state the axioms must be assigned finite decimal numbers.

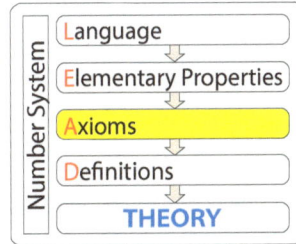

Definitions of the Finite Decimal System

In this section we create new symbols and words that describe concepts we use to model properties of the Finite Decimal System. We will explore the relationship between these concepts using the symbols in our basic alphabet and their relationship to each other in the Theory Section below.

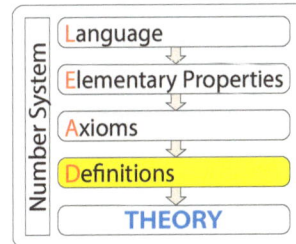

In our first definition, we must formalise what we mean by a 'finite decimal' in the Finite Decimal System.

Definition 6.1   Finite Decimals

We define a **finite decimal** (also called a **terminating decimal**) as any arithmetic term (i.e. finite valid string of symbols) formed from the alphabet; that is, those terms that do not contain the '=' and '≠' symbol.

Examples of finite decimals include: '-3×10^{-1}', '$(2 + 10^{-1}) \times 10^{-1}$', '5' and '$-10^{-1} + -1 \times 4$'.

Notes:

1. We treat the symbol '10^{-1}' in the same way we would treat any integer number.

2. In Definition 6.4 below we will give the familiar format of a finite decimal that we first learnt about in primary school.

As in Chapter Three, we will define the minus operator (unary operator) and the subtraction operator for finite decimals in order to simplify notation.

Definition 6.2   Minus Operator and Subtraction Operator

Let 'x' and 'y' be finite decimal variables. The **minus operator** is defined by:

$$-(x) = -1 \times x$$

The **subtraction operator** is defined by:

$$x - y = x + -y$$

On the right-hand side of this last equation, the combination of symbols '$-y$' represents the minus operator applied to this variable as defined in the first part of this definition.

––––––––––––––––

In order to obtain the index form of a finite decimal number, we must begin by building on our work in Chapter Five and define index notation for the base of '10' in relation to the Finite Decimal System.

Definition 6.3   Base-10 Indices for Finite Decimals

Let 'n' and 'm' be index variables that can be assigned numbers from the set of symbols $\{..., -3, -2, -1, 0, 1, 2, 3, ...\}$ that can be combined with the set of symbols $\{'+', '\times', '=', '\neq', '(', ')'\}$. The definitions for these **indices** (using a base of '10') are given by the following equations:

1. $10^0 = 1$

2. $10^1 = 10$

3. $10^{-1} \times 10^1 = 1$ ... by Axiom 10 above

4. $10^{m+n} = 10^m \times 10^n$

5. $10^{m \times n} = (10^m)^n$

6. '$m = n$' if and only if '$10^m = 10^n$'

    '$m \neq n$' if and only if '$10^m \neq 10^n$'

7. The parentheses symbols (in the same way they apply to indices) have their usual role as grouping symbols.

––––––––––––––––

Using this definition of indices for a base of '10', most of the equivalent theorems in Chapter Five (Rational Number System using Index Form) can be proved using equivalent steps with the restricted base of '10'. In this way, it is easy to show that the set of indices given by the set {... , −3, −2, −1, 0, 1, 2, 3, ...}, to which we raise the base integer '10', possess all the same properties as the integers. This again provides us with a major simplification whereby from this point forward we will treat index manipulation as integer operations.

In the same way we showed that rational numbers can always be written in fractional form, a general finite decimal can be written in decimal notation. Therefore, we are finally in a position to provide the definition of the decimal notation of a finite decimal.

Definition 6.4   Decimal Point Notation for Finite Decimals

Decimal notation requires a marker to signal that the string of digits to the left of that marker represents an integer and that the string of digits to the right of that marker is a proper fraction. This marker is the dot symbol '.' and is called the **decimal point**. For example, the decimal number '5.76' is spoken as 'five point seven six'.

Definition 6.5   Decimal Notation for Finite Decimals

Let the variables 'n' and 'm' be assigned positive integers that are subscripts identifying the '$n + 1$' digits to the left of the decimal point and the 'm' digits to the right of the decimal point, respectively. Also, let the symbols '$a_n, ... , a_0$' and '$d_1, ... , d_m$' be variables, each of which can be assigned any one of the ten digits {0, 1, 2, 3, 4, 5, 6, 7, 8, 9}. The final outcome is that a finite **decimal number** can be either a positive string or negative string of variables:

$$\text{`}a_n ... a_1 a_0 . d_1 ... d_m\text{' or `}-(a_n ... a_1 a_0 . d_1 ... d_m)\text{'.}$$

The contribution of each variable to the value of the decimal number is determined by its position, as follows:

- Variable 'a_n' contributes '$a_n \times 10^n$' to the value of the decimal number
  $\vdots$
- Variable 'a_1' contributes '$a_1 \times 10^1$' to the value of the decimal number
- Variable 'a_0' contributes '$a_0 \times 10^0$' to the value of the decimal number
- Decimal point '.' separates the integer string from the fraction string
- Variable 'd_1' contributes '$d_1 \times 10^{-1}$' to the value of the decimal number

$\vdots$

- Variable 'd_m' contributes '$d_m \times 10^{-m}$' to the value of the decimal number.

Therefore, in index form we can write this finite decimal number as:

$$a_n \dots a_1 a_0 . d_1 \dots d_m$$
$$= a_n \times 10^n + \dots + a_1 \times 10^1 + a_0 \times 10^0 + d_1 \times 10^{-1} + \dots + d_m \times 10^{-m}$$

For example: $123.45 = 1 \times 10^2 + 2 \times 10^1 + 3 \times 10^0 + 4 \times 10^{-1} + 5 \times 10^{-2}$

In the case of a **negative finite decimal number**, the negative symbol '$-$' is distributed throughout the sequence of digits that make up the finite decimal number as follows:

$$-a_n \dots a_1 a_0 . d_1 \dots d_m$$
$$= -a_n \times 10^n - \dots - a_1 \times 10^1 - a_0 \times 10^0 - d_1 \times 10^{-1} - \dots - d_m \times 10^{-m}$$

The string of symbols to the left of the decimal point (i.e. '$a_n \dots a_1 a_0$') is called the integer number part of the finite decimal number, while the string of symbols to the right of the decimal point (i.e. '$d_1 \dots d_m$') is called the decimal fraction part of the finite decimal number.

This description is rather long as we have provided the definition of the base-10 positional notation system as part of the definition of this decimal notation for a finite decimal.

Finally, we can give a definition of those fractions which can be expressed as finite decimals. Since '$\frac{1}{10} = 10^{-1}$' derives its meaning from Axiom 10 of the Axioms of the Finite Decimal System, the only fractions that are finite are those that have a denominator made up of factors of '2' and/or '5', e.g. '$\frac{1}{2}$', '$\frac{1}{4}$' or '$\frac{1}{20}$'. All other fractions such as '$\frac{1}{3}$', '$\frac{1}{6}$', '$\frac{1}{7}$', '$\frac{1}{9}$', etc. will lead to an unlimited number of digits in their decimal fraction expression and are therefore not finite decimals.

This definition provides a clear link between a subset of fractions and the finite decimals.

Definition 6.6   Finite Decimal Fractions

The fractions '$\frac{1}{2}$' and '$\frac{1}{5}$' are defined in terms of the fraction '$\frac{1}{10}$' as follows:

$$\frac{1}{2} = \frac{5}{10} \left(= 5 \times \frac{1}{10} \right)$$

$$\tfrac{1}{5} = \tfrac{2}{10} \left(= 2 \times \tfrac{1}{10}\right)$$

If 'p', 'i' and 'j' are integer variables, then from the two fractions '$\tfrac{1}{2}$' and '$\tfrac{1}{5}$' we can define the **finite decimal fractions** as '$p \times (\tfrac{1}{2})^i \times (\tfrac{1}{5})^j$'.

Note:

This definition of finite decimal fractions identifies all those fractions that will be shown to have finite decimal expansions. From Chapter Five, we see that we can write a finite decimal fraction in index notation as '$p \times 2^{-i} \times 5^{-j}$'. Likewise, from Chapter Four, we observe we could also write it in fraction form as '$\tfrac{p}{2^i \times 5^j}$'. It is important to also note that this decimal fraction must be equivalent to '$\tfrac{q}{10^k}$' for some values assigned to the integer variables 'q' and 'k' so we can always make the denominator a power of '10'.

Some of the simplest examples of finite decimal fractions include the following set: $\{\tfrac{1}{2}, \tfrac{1}{4}, \tfrac{3}{4}, \tfrac{1}{5}, \tfrac{2}{5}, \tfrac{3}{5}, \tfrac{4}{5}, \tfrac{1}{8}, \tfrac{3}{8}, \tfrac{5}{8}, \tfrac{7}{8}, \tfrac{1}{10}, \tfrac{3}{10}, \tfrac{7}{10}, \tfrac{9}{10}, \tfrac{1}{16}, \tfrac{2}{16}, ...\}$ which can be written in the form '$\tfrac{p}{2^i \times 5^j}$' as: $\{\tfrac{1}{2^1 \times 5^0}, \tfrac{1}{2^2 \times 5^0}, \tfrac{3}{2^2 \times 5^0}, \tfrac{1}{2^0 \times 5^1}, \tfrac{2}{2^0 \times 5^1}, \tfrac{3}{2^0 \times 5^1}, \tfrac{4}{2^0 \times 5^1}, \tfrac{1}{2^3 \times 5^0}, \tfrac{3}{2^3 \times 5^0}, \tfrac{5}{2^3 \times 5^0}, \tfrac{7}{2^3 \times 5^0}, \tfrac{1}{2^1 \times 5^1}, \tfrac{3}{2^1 \times 5^1}, \tfrac{7}{2^1 \times 5^1}, \tfrac{9}{2^1 \times 5^1}, \tfrac{1}{2^4 \times 5^0}, \tfrac{1}{2^4 \times 5^0}, ...\}$.

In the Theory Section below we will demonstrate that every finite decimal can be written as a finite decimal fraction and that every finite decimal fraction can be written as a finite decimal. This result provides the justification that the Finite Decimal System is a subsystem of the Rational Number System using Fractions.

The new symbols we have defined in this section can now be appended to our alphabet of the Integer Number System in order to form the new symbols required for the Finite Decimal System (see Table 6.5 below).

6.5 The **Alphabet** of the **Finite Decimal System**	
Symbols	**Meaning from:**
'0', '1', '+', '×', '(', ')'	The Axioms of the Natural Number System
'=', '≠'	The Axioms of Logic
'2', '3', '4', '5', '6', '7', '8', '9'	Definition 2.1
'a', 'b', 'c', ...	The Axioms of the Natural Number System
','	Definition 2.3

6.5 The **Alphabet** of the Finite Decimal System	
'−1'	The Axioms of the Integer Number System
'x', 'y', 'z', ...	The Axioms of the Finite Decimal System
'10^{-1} (= 0.1)'	The Axioms of the Finite Decimal System
'_'	Definition 6.2
'10^m'	Definition 6.3
'.'	Definition 6.4
'$a_n ... a_1 a_0 . d_1 ... d_m$'	Definition 6.5
'$\frac{1}{2}$', '$\frac{1}{5}$'	Definition 6.6

Table 6.5

In this chapter we have introduced the concept of a finite decimal number which also emerges as a natural generalisation of our normal counting system. In the next section we will explore how to carry out mathematical operations with finite decimals.

Theory of the Finite Decimal System

In this section we will derive the general operations for addition, subtraction and multiplication of finite decimal numbers and then illustrate these operations in the Application Section below.

Number System

- Language
- Elementary Properties
- Axioms
- Definitions
- **THEORY**

The first theorem we will prove is a property of decimals that demonstrates the ease with which finite decimals can be multiplied by '10' and '10^{-1}'.

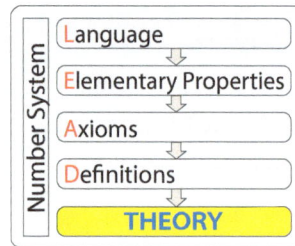

Theorem 6.1   Finite Decimals by Products of '10' and '10^{-1}'

Let a variable 'x' be assigned a value so that: '$x = a_n ... a_1 a_0 . d_1 d_2 ... d_m$'. The product of 'x' by '10' and '10^{-1}' is given by:

1. When multiplying a finite decimal number assigned to 'x' by '10', the decimal point moves one place to the right of its initial position, i.e. '$x = a_n ... a_1 a_0 d_1 . d_2 ... d_m$'

2. When multiplying a finite decimal number assigned to 'x' by '10^{-1}', the decimal point moves one place to the left of its initial position, i.e. '$x = a_n ... a_1 . a_0 d_1 d_2 ... d_m$'.

Proof

Using Definition 6.5, this result follows easily by expressing 'x' in index form to obtain the following result:

$$x = a_n \times 10^n + \ldots + a_1 \times 10^1 + a_0 \times 10^0 + d_1 \times 10^{-1} + d_2 \times 10^{-2} + \ldots + d_m \times 10^{-m}$$

$$x \times 10 = a_n \times 10^n \times 10^1 + \ldots + a_1 \times 10^1 \times 10^1 + a_0 \times 10^0 \times 10^1 + d_1 \times 10^{-1} \times 10^1 + d_2 \times 10^{-2} \times 10^1 + \ldots + d_m \times 10^{-m} \times 10^1$$

$$\ldots \text{by Distributive Axiom}$$

$$= a_n \times 10^{n+1} + \ldots + a_1 \times 10^{1+1} + a_0 \times 10^{0+1} + d_1 \times 10^{-1+1} + d_2 \times 10^{-2+1} + \ldots + d_m \times 10^{-m+1}$$

$$\therefore \quad x \times 10 = a_n \ldots a_1 a_0 d_1 . d_2 \ldots d_m$$

Therefore, we have demonstrated that multiplying the decimal 'x' by '10' has moved the decimal point one place to the right of its initial position; in this case, it was initially between 'a_0' and 'd_1' and now has moved to between 'd_1' and 'd_2'.

To multiply 'x' by '10^{-1}' we proceed as follows:

$$x \times 10^{-1} = a_n \times 10^{n-1} + \ldots + a_1 \times 10^{1-1} + a_0 \times 10^{0-1} + d_1 \times 10^{-1-1} + d_2 \times 10^{-2-1} + \ldots + d_m \times 10^{-m-1}$$

$$\ldots \text{by Distributive Axiom}$$

$$\therefore \quad x \times 10^{-1} = a_n \ldots a_1 . a_0 \, d_1 \, d_2 \ldots d_m$$

Therefore, we have demonstrated that multiplying the decimal 'x' by '10^{-1}' has moved the decimal point one place to the left of its initial position; in this second case, it was initially between 'a_0' and 'd_1' and now has moved to between 'a_1' and 'a_0'.

Further demonstrations of multiples by '10' or '10^{-1}' are just iterations of the process outlined above. In this way, we have completed the proof.

———————————

To demonstrate the value of Theorem 6.1, we will look at an example of multiplying the number '3.1' by the number '2.3'. We use Theorem 6.1 to convert each decimal back to an integer product with a power of '10'. From Theorem 6.1 we have:

$$2.3 = 23 \times 10^{-1} \qquad \ldots \text{by Theorem 6.1}$$

$$3.1 = 31 \times 10^{-1} \qquad \text{... by Theorem 6.1}$$

$$\therefore \quad 2.3 \times 3.1 = 23 \times 10^{-1} \times 31 \times 10^{-1} \qquad \text{... by substitution}$$

$$= (23 \times 31) \times 10^{-1} \times 10^{-1} \qquad \text{... by Associative Axiom}$$

$$= 713 \times 10^{-2} \qquad \text{... by '}23 \times 31 = 713\text{'}$$

$$\therefore \quad 2.3 \times 3.1 = 7.13 \qquad \text{... by Theorem 6.1}$$

The next theorem will provide the process for adding two finite decimal numbers in decimal notation.

Theorem 6.2   Addition of Finite Decimal Numbers

Let two finite decimal number variables 'x' and 'y' be assigned decimal values where 'n' and 'p' are positive integers indicating there are '$n+1$' and '$p+1$' digits to the left of the decimal point in the numbers 'x' and 'y', respectively. The positive integers 'm' and 'q' indicate there are 'm' and 'q' digits to the right of the decimal point of 'x' and 'y', respectively, so that:

$$\text{'}x = a_n \dots a_1 a_0.d_1 \dots d_m\text{' and '}y = b_p \dots b_1 b_0.e_1 \dots e_q\text{'}$$

For convenience, we will assume the decimal number 'y' has more digits to the left and right of the decimal point than 'x', that is, 'p' is greater than 'n' and 'q' is great than 'm'. The addition (or sum) of two finite decimals '$x + y$' in index form is as follows:

$$x + y = b_p \times 10^p + \dots + b_{n+1} \times 10^{n+1} + (a_n + b_n) \times 10^n + \dots$$

$$+ (a_1 + b_1) \times 10^1 + (a_0 + b_0) \times 10^0 + (d_1 + e_1) \times 10^{-1}$$

$$\dots + (d_m + e_m) \times 10^{-m} + e_{m+1} \times 10^{-m-1} + \dots + e_q \times 10^{-q}$$

Proof

We begin to demonstrate the proof by writing out the finite decimals 'x' and 'y' in their expanded form as given in Definition 6.5:

$$a_n \dots a_1 a_0.d_1 \dots d_m$$

$$= a_n \times 10^n + \dots + a_1 \times 10^1 + a_0 \times 10^0 + d_1 \times 10^{-1} + \dots + d_m \times 10^{-m}$$

$$b_p \dots b_1 b_0.e_1 \dots e_q$$

$$= b_p \times 10^p + \dots + b_1 \times 10^1 + b_0 \times 10^0 + e_1 \times 10^{-1} + \dots + e_q \times 10^{-q}$$

We now add these two finite decimals by adding their corresponding coefficients with the same power of '10'. You will recall that we have already assumed the decimal number 'y' contains more digits to the left and right of the decimal point than 'x'. Consequently, we can write the sum 'x + y' as:

$$a_n \ldots a_1 a_0.d_1 \ldots d_m + b_p \ldots b_1 b_0.e_1 \ldots e_q$$
$$= b_p \times 10^p + \ldots + b_{n+1} \times 10^{n+1} + (a_n + b_n) \times 10^n + \ldots + (a_1 + b_1) \times$$
$$10^1 + (a_0 + b_0) + (d_1 + e_1) \times 10^{-1} + \ldots + (d_m + e_m) \times 10^{-m} + e_{m+1} \times$$
$$10^{-m-1} + \ldots + e_q \times 10^{-q}$$

This complicated sum is just the simple addition algorithm for decimal numbers that we used in primary school. For example: $12.3 + 456.78 = (1 \times 10^1 + 2 \times 10^0 + 3 \times 10^{-1}) + (4 \times 10^2 + 5 \times 10^1 + 6 \times 10^0 + 7 \times 10^{-1} + 8 \times 10^{-2})$. If the general coefficient '$a_i + b_i$' or '$d_j + e_j$' (where subscripts 'i' and 'j' take on the values from '0' to 'n' and '1' to 'm', respectively) add up to more than '9', then we carry a '1' over to the term to its left. For the example above, we see that we will have to carry a '1' from the first digits to the right of the decimal point (i.e. '3' and '7'). This completes the proof.

For the case of subtraction, we derive the subtraction process for finite decimals in decimal notation by using the definition of the subtraction operator. To subtract two finite decimals in decimal notation, we add the opposite of the finite decimal being subtracted. For example: $12.3 - 456.78 = (1 \times 10^1 + 2 \times 10^0 + 3 \times 10^{-1}) + (-4 \times 10^2 + -5 \times 10^1 + -6 \times 10^0 + -7 \times 10^{-1} + -8 \times 10^{-2})$.

This last theorem demonstrates the multiplication operation for finite decimals in decimal notation. This operation provides the derivation of the long multiplication algorithm that we learnt in primary school.

Theorem 6.3 Multiplication of Finite Decimals

Let two finite decimal variables 'x' and 'y' be assigned decimal values, where 'n' and 'p' are positive integers indicating there are '$n + 1$' and '$p + 1$' digits to the left of the decimal point in the numbers 'x' and 'y', respectively. The positive integers 'm' and 'q' indicate that there are 'm' and 'q' digits are to the right of the decimal point of 'x' and 'y' respectively, so that:

$$\text{'}x = a_n \ldots a_1 a_0.d_1 \ldots d_m\text{' and '}y = b_p \ldots b_1 b_0.e_1 \ldots e_q\text{'}$$

The product '$x \times y$' is given by:

$$x \times y = a_n \times 10^n \times (b_p \ldots b_1 b_0 . e_1 \ldots e_q)$$
$$+$$
$$\vdots$$
$$+ a_1 \times 10^1 \times (b_p \ldots b_1 b_0 . e_1 \ldots e_q)$$
$$+ a_0 \times 10^0 \times (b_p \ldots b_1 b_0 . e_1 \ldots e_q)$$
$$+ d_1 \times 10^{-1} \times (b_p \ldots b_1 b_0 . e_1 \ldots e_q)$$
$$+$$
$$\vdots$$
$$+ d_m \times 10^{-m} \times (b_p \ldots b_1 b_0 . e_1 \ldots e_q)$$

Proof

Using Definition 6.5, this result follows on easily by expressing 'x' in index form to get:

$$x = a_n \times 10^n + \ldots + a_1 \times 10^1 + a_0 \times 10^0 + d_1 \times 10^{-1} + \ldots$$
$$+ d_m \times 10^{-m}$$

If we leave 'y' in its compact form, we can write the product as follows:

$$x \times y = (a_n \times 10^n + \ldots + a_1 \times 10^1 + a_0 \times 10^0 + d_1 \times 10^{-1} +$$
$$\ldots + d_m \times 10^{-m}) \times (b_p \ldots b_1 b_0 . e_1 \ldots e_q)$$

If we now apply the Distributive Axiom to this last product, we obtain the result stated in the theorem (and the simple multiplication algorithm for decimal numbers that we used in primary school). For example: $2.3 \times 6.7 = (2 \times 10^0 + 3 \times 10^{-1})$ $\times (6 \times 10^0 + 7 \times 10^{-1})$. Consequently, the result has been proved.

––––––––––––––––

We will now set out to prove the equivalence of finite decimals and finite decimal fractions as our last theorem of this chapter.

Theorem 6.4   Finite Decimal Fractions are Equivalent to Finite Decimals

Let a finite decimal be given by '$a_n \ldots a_1 a_0 . d_1 \ldots d_m$', where '$n$' and '$m$' are positive integers and the symbols '$a_n \ldots a_0$' and '$d_1 \ldots d_m$' are variables that are assigned digits from the set of digits $\{0, 1, 2, 3, 4, 5, 6, 7, 8, 9\}$ and 'd_m' is a non-zero digit.

We can now assign values to the integer variables 'p', 'i' and 'j' in the finite decimal fraction '$\dfrac{p}{2^i \times 5^j}$' so that:

$$a_n \ldots a_1 a_0 . d_1 \ldots d_m = \frac{p}{2^i \times 5^j}$$

Proof

This result will be shown to be straightforward and will follow from the simple arithmetic equation: '$10 = 2 \times 5$'.

Firstly, we show that every finite decimal can be written as a finite decimal fraction. To do this we start with the following simple relationship:

$$a_n \ldots a_1 a_0 . d_1 \ldots d_m = \frac{a_n \ldots a_1 a_0 d_1 \ldots d_m}{10^m} \qquad \text{... by Theorem 6.1}$$
$$= \frac{a_n \ldots a_1 a_0 d_1 \ldots d_m}{2^m \, 5^m} \qquad \text{... by Theorem 5.25}$$

If we now let 'p' be assigned '$a_n \ldots a_1 a_0 \, d_1 \ldots d_m$' and '$i$' and '$j$' be assigned '$m$', then we have translated the general finite decimal into a finite decimal fraction.

Secondly, we can show that the reverse argument can be applied. Given integer assignments to the integer variables 'p', 'i' and 'j' and assuming that 'i' is less than 'j' so there exists an integer 'k' such that '$j = i + k$', we can write the finite decimal fraction '$\frac{p}{2^i \times 5^j}$' as:

$$\frac{p}{2^i \times 5^j} = \frac{p}{2^i \times 5^j} \times \frac{2^k}{2^k} \qquad \text{... by Identity Axiom '×'}$$
$$= \frac{p \times 2^k}{2^{i+k} \times 5^j} \qquad \text{... by Theorem 5.25}$$
$$= \frac{p \times 2^k}{2^j \times 5^j} \qquad \text{... by '$i + k = j$'}$$
$$\therefore \quad \frac{p}{2^i \times 5^j} = \frac{p \times 2^k}{10^j} \qquad \text{... by '$2 \times 5 = 10$'}$$

However, the term '$p \times 2^k$' is simply an integer which can be written in decimal form as '$a_n \ldots a_1 a_0 \, d_1 \ldots d_m$'. By now dividing by '$10^j$' we convert it to a standard finite decimal. In this way, the equivalence between finite decimals and finite decimal fractions has been proved.

In conclusion, the Theory Section above has shown the basic properties of the Finite Decimal System and demonstrated its relationship as a subsystem of the Rational Number System using Fractions.

Applications of the Finite Decimal System

In this section we again explore some simple applications of the theory developed in this chapter that are relevant as high school syllabus questions for 'Numbers and Algebra'. These simple applications highlight the manipulation of arithmetic terms and equations which contain finite decimals. The first application illustrates the origin of the common algorithm used to add finite decimals.

Application 6.1   Addition of Finite Decimals

We can demonstrate this application by applying the operation of '+' to two decimals. In this case, we perform a simple addition of the numbers '1.2' and '34' in their decimal notation (using their definitions) to show how the Arithmetic works:

$$1.2 + 34 = (1 \times 10^0 + 2 \times 10^{-1}) + (3 \times 10^1 + 4 \times 10^0)$$

$$\qquad\qquad \text{... by Definition 6.5}$$

$$= (3 \times 10^1) + (4 + 1) \times 10^0 + (2 \times 10^{-1}) \qquad \text{... by Commutative}$$

$$\text{Axiom '+'}$$

$$= 3 \times 10^1 + 5 \times 10^0 + 2 \times 10^{-1} \qquad \text{... by simplifying terms}$$

$$= 35.2 \qquad\qquad \text{... by Definition 6.5}$$

$$\therefore\ 1.2 + 34 = 35.2$$

In the next application, we will again add two finite decimals; however, this time we will demonstrate the carry-over from one decimal place to another.

Application 6.2   Addition of Finite Decimals

Using Definition 6.1, we can add the two decimals '4.5' and '6.8' with decimal fraction parts as follows:

$$4.5 + 6.8 = (4 \times 10^0 + 5 \times 10^{-1}) + (6 \times 10^0 + 8 \times 10^{-1})$$

$$\qquad\qquad \text{... by Definition 6.5}$$

$$= (4 + 6) \times 10^0 + (5 + 8) \times 10^{-1} \qquad \text{... by Distributive Axiom}$$

$$= 10 \times 10^0 + 13 \times 10^{-1} \qquad \text{... by simplifying terms}$$

$$= 10^1 + (10 + 3) \times 10^{-1} \qquad \text{... by Definition 6.5}$$

$$= 10^1 + 10 \times 10^{-1} + 3 \times 10^{-1} \qquad \text{... by Distributive Axiom}$$

$$= 1 \times 10^1 + 1 \times 10^0 + 3 \times 10^{-1} \qquad \text{... by Definition 6.3}$$

$$\therefore 4.5 + 6.8 = 11.3 \qquad\qquad \text{... by Definition 6.5}$$

As we observed in the application above, this is a very tedious process. For this reason, we normally use a short-cut process to reach the answer more quickly using a process we call 'the Addition Algorithm'. When applied to the above application, this algorithm looks like:

$$4.5 + 6.8 = \quad \longrightarrow \quad \text{go to algorithm} \quad \longrightarrow \quad \begin{array}{r} 4.5 \\ +6.8 \\ \hline \end{array}$$

$$\therefore \quad 4.5 + 6.8 = 11.3 \quad \longleftarrow \quad \text{return to Arithmetic} \quad \longleftarrow \quad 11.3$$

By writing the digits corresponding to the same multiple of '10' under each other in this algorithm and then carrying numbers to the left when the numbers add up to more than '9', we can create an algorithm for an otherwise long and tedious (but logical) process.

Application 6.3 Subtraction of Finite Decimal

Using Definition 6.1, we can subtract two decimals (e.g. '6.7' and '4.3') with decimal fraction parts as follows:

$$6.7 - 4.3 = (6 \times 10^0 + 7 \times 10^{-1}) - (4 \times 10^0 + 3 \times 10^{-1})$$

$$\text{... by Definition 6.5}$$

$$= 6 \times 10^0 + 7 \times 10^{-1} - 4 \times 10^0 - 3 \times 10^{-1}$$

$$\text{... by Distributive Axiom}$$

$$= (6 - 4) \times 10^0 + (7 - 3) \times 10^{-1} \qquad \text{... by Distributive Axiom}$$

$$= 2 \times 10^0 + 4 \times 10^{-1} \qquad \text{... by simplifying terms}$$

$$\therefore 6.7 - 4.3 = 2.4 \qquad \text{... by Definition 6.5}$$

Where the number we are subtracting is smaller than the decimal we are subtracting from, we can use the subtraction of decimals algorithm as follows:

$$6.7 - 4.3 = \quad \longrightarrow \quad \text{go to algorithm} \quad \longrightarrow \quad \begin{array}{r} 6.7 \\ -4.3 \\ \hline \end{array}$$

$$\therefore \quad 6.7 - 4.3 = 2.4 \quad \longleftarrow \quad \text{return to Arithmetic} \quad \longleftarrow \quad 2.4$$

However, if the decimal we are subtracting is larger than the decimal we are subtracting from (e.g. '4.3 − 6.7'), we need to rewrite this decimal term as '−(6.7 − 4.3)', and then apply the algorithm above to the term in parentheses to get '−(2.4)'. Finally, we remove the parentheses to obtain the correct result of '−2.4'.

$$4.3 - 6.7 = \quad \longrightarrow \quad \text{change signs of decimals to '}6.7 - 4.3\text{'}$$

$$\longrightarrow \quad \text{go to algorithm} \quad \longrightarrow \quad \begin{array}{r} 6.7 \\ -4.3 \\ \hline \end{array}$$

$$2.4$$

$\longrightarrow$ change sign of result $\qquad -2.4$

$\therefore \quad 4.3 - 6.7 = -2.4 \longleftarrow$ return to Arithmetic

In the next example, we will illustrate the method of long multiplication using the basic axioms for manipulating numbers in decimal format.

Application 6.4 Multiplication of Finite Decimals

A similar process to the one described above is used to find the product of two finite decimals. For example, we can use the numbers '12' and '34' and multiply them. This multiplication can be done the long way using the theory of Arithmetic again as follows:

$$12 \times 34 \quad = (1 \times 10^1 + 2 \times 10^0) \times (3 \times 10^1 + 4 \times 10^0)$$

... by Definition 6.5

$$= (1 \times 10^1) \times (3 \times 10^1 + 4 \times 10^0) + (2 \times 10^0) \times (3 \times 10^1 + 4 \times 10^0)$$

... by Distributive Axiom

$$= (3 \times 10^2 + 4 \times 10^1) + (6 \times 10^1 + 8 \times 10^0)$$

... by Distributive Axiom

$$= 3 \times 10^2 + (4 + 6) \times 10^1 + 8 \times 10^0 \quad \text{... by Distributive Axiom}$$

$$= 3 \times 10^2 + 10 \times 10^1 + 8 \times 10^0 \quad \text{... by simplifying terms}$$

$$= 3 \times 10^2 + 1 \times 10^2 + 8 \times 10^0 \quad \text{... by simplifying terms}$$

$$= (3 + 1) \times 10^2 + 0 \times 10^1 + 8 \times 10^0 \quad \text{... by Distributive Axiom}$$

$$= 408 \quad \text{... by Definition 6.5}$$

$\therefore \quad 12 \times 34 = 408$

Fortunately, once again, this tedious process can be replaced by a simple algorithm for multiplying finite decimals. This algorithm is the familiar multiplication algorithm we learnt in primary school and can be carried out as follows:

$$12 \times 34 = \qquad \longrightarrow \quad \text{go to algorithm} \quad \longrightarrow \qquad 12$$

$$\underline{\times\ 34}$$

$$48$$

^{1360}

$$\therefore \quad 12 \times 34 = 408 \quad \longleftarrow \quad \text{return to Arithmetic} \quad \longleftarrow \quad 408$$

In some forms of the multiplication algorithm, the '1' in the prefix superscript of '^{1}3' above, indicates a carry-over operation from the decimal place to its right-hand side. Once again, we have carefully aligned the same multiples of '10' directly under each other.

Application 6.5 Multiplication of Finite Decimals with Decimal Fraction Parts

Similarly, we can multiply the two decimals '4.5' and '6.8' with decimal fraction parts as follows:

$$4.5 \times 6.8 = (4 \times 10^0 + 5 \times 10^{-1}) \times (6 \times 10^0 + 8 \times 10^{-1})$$
$$\text{... by Definition 6.5}$$

$$= (4 \times 10^0) \times (6 \times 10^0 + 8 \times 10^{-1}) + 5 \times 10^{-1} \times (6 \times 10^0 + 8 \times 10^{-1})$$
$$\text{... by Distributive Axiom}$$

$$= (4 \times 6 \times 10^0 \times 10^0) + (4 \times 8 \times 10^0 \times 10^{-1}) + (5 \times 6 \times 10^{-1} \times 10^0)$$
$$+ (5 \times 8 \times 10^{-1} \times 10^{-1})$$
$$\text{... by Distributive Axiom}$$

$$= 24 \times 10^0 + 32 \times 10^{-1} + 30 \times 10^{-1} + 40 \times 10^{-2}$$
$$\text{... by simplifying terms}$$

$$= (20 + 4) \times 10^0 + (60 + 2) \times 10^{-1} + 40 \times 10^{-2}$$
$$\text{... by breaking into '10's}$$

$$= 2 \times 10^1 + 4 \times 10^0 + 6 \times 10^0 + 2 \times 10^{-1} + 4 \times 10^{-1}$$
$$\text{... by Distributive Axiom}$$

$$= 2 \times 10^1 + 10 \times 10^0 + 6 \times 10^{-1} \quad \text{... by Distributive Axiom}$$

$$= 2 \times 10^1 + 1 \times 10^1 + 6 \times 10^{-1} \quad \text{... by simplifying terms}$$

$$= (2 + 1) \times 10^1 + 6 \times 10^{-1} \quad \text{... by Distributive Axiom}$$

$$= 3 \times 10^1 + 0 \times 10^0 + 6 \times 10^{-1} \quad \text{... by simplifying terms}$$

$$= 30.6$$

$$\therefore \ 4.5 \times 6.8 = 30.6$$

———————————

Again this tedious process of multiplication can be replaced by one of several simple algorithms for multiplying decimals. The algorithm below is one such example and can be performed in the following way:

$$4.5 \times 6.8 = \quad \longrightarrow \quad \text{go to algorithm} \quad \longrightarrow \quad
\begin{array}{r}
4.5 \\
\times\, 6.8 \\
\hline
360 \\
{}^{1}2700 \\
\hline
\end{array}$$

$$\therefore \quad 4.5 \times 6.8 = 30.6 \quad \longleftarrow \quad \text{return to Arithmetic} \quad \longleftarrow \quad 30.60$$

When we obtain the result from the algorithm, we need to insert the decimal point in the correct place. We do this by adding the number of digits to the right of the decimal point (called the number of decimal places) in the numbers being multiplied (i.e. '4.5' and '6.8') and then placing the decimal point this number of places to the left of the last digit in the algorithm. In this example, we have a combined total of two decimal places and so we move the decimal point two places to the left of the last digit.

We have now completed the Application Section of this chapter.

Summary of the Finite Decimal System

At this point, we have covered the objectives of this chapter and so can summarise our findings up to now. In terms of the Finite Decimal System, the most important definitions that have been introduced so far are:

1. Define **finite decimals** as arithmetic terms constructed from the alphabet and excluding the symbols from the set $\{'=', '\neq'\}$.

2. Define **indices** (which behave like integers) with a base of '10' to give another way of expressing finite decimals.

3. Define the **decimal notations for a positive and negative finite decimal** as: '$a_n \dots a_1 a_0.d_1 \dots d_m$' or '$-(a_n \dots a_1 a_0.d_1 \dots d_m)$', respectively.

4. For the case where '$d_1 = d_2 = d_3 = \dots = d_m = 0$', these decimals become equivalent to integers – as there is no decimal fraction part – and these integers make up the Integer Number System as a subsystem of the Finite Decimal System.

The Axioms of the Finite Decimal System that model the partitioning of a number line between the integers into portions of products of '10^{-1}' are shown in Table 6.6 below.

6.6 The Axioms of the Finite Decimal System		
Axiom Type	**Addition '+'**	**Multiplication '×'**
Closure	1. In this table the finite decimal variables are 'x', 'y' and 'z'. The Closure Axiom states that the operations with the following decimals (finite) '$x+y$' and '$x \times y$' result in unique finite decimals.	
Commutative	2. $\qquad x+y=y+x$	3. $\qquad x \times y = y \times x$
Associative	4. $(x+y)+z=x+(y+z)$	5. $(x \times y) \times z = x \times (y \times z)$
Identity	6. $\qquad 1+0=1$	7. $\qquad 1 \times 1 = 1$
Distributive	8. $\qquad\qquad x \times (y+z) = x \times y + x \times z$	
Inverse	9. $\quad 1 + -1 = 0$	10. $10 \times 10^{-1} = 1$

Table 6.6

Once again, the Axioms of Logic used to manipulate the Axioms in the table above have not changed in form since Table 2.5. However, they can now be applied to the set of numbers – the Finite Decimal Numbers.

The basic theorems in this chapter illustrate how the work we did with adding and multiplying finite decimal numbers in primary school fits into the elementary number system framework. Once again, we have applied the language, elementary properties, axioms and definitions (LEAD) approach to show how straightforward it is to master finite decimals.

Before we can proceed any further and show that the combination of Finite and Repeating Decimals is the third alternative way of representing rational numbers, we must first master some elementary topics. The first of these topics is Basic Algebra, which is a natural extension of the work we have already done up to this point with algebraic equations. In covering this topic in Chapter Seven, we will revert to using rational numbers so that we can obtain solutions to a broader range of algebraic equations.

CHAPTER 7

ALGEBRA

Overview of Algebra

In this chapter we will continue our journey of mastering the elementary number systems by incorporating 'Algebra' as a tool to find answers to a specific range of problems involving algebraic equations. The inclusion of Algebra into our 'toolkit' of mathematical concepts provides us with an important method of investigation and one that will help us to understand the remaining chapters of this textbook. The Algebra studied in secondary school Mathematics is normally called 'Elementary Algebra', but for the sake of brevity and to avoid any risk of confusion, we will continue using the single word 'Algebra'.

Algebra is a simple extension of Arithmetic that we started to learn about in our early years of schooling. However, by the time we have returned to it in later years in a slightly different form, it seems to be mysterious and too abstract to easily relate to. In this chapter we will aim to remove this mystery by demonstrating how Algebra relates to our previous mathematical knowledge and follows on naturally from it.

So how do we define Algebra? **Algebra is the process of identifying which values we must assign to variables in an algebraic equation in order to convert that equation into a true arithmetic equation.** For example, in the case of the algebraic equation '$x + y = 2x + 3$', when we assign 'x' the value '1' and 'y' the value '4', we obtain the true arithmetic equation '$1 + 4 = 2 \times 1 + 3$' (i.e. '$5 = 5$').

In previous chapters, we used variables in equations (i.e. algebraic equations) to express general properties of our number systems as axioms and theorems. We observed that for axioms which are algebraic equations, every assignment to the variables of those equations resulted in a true equation.

In this chapter, we will use a **single** variable (with one exception in Theorem 7.5) to write several different types of algebraic equations and then provide general procedures for finding the values of this variable which ensures these equations are 'true' equations.

Background and Context of Algebra

The history of Algebra stretches back to at least the time of Euclid in the 3rd century BC where some components of his text, *The Elements* reveal the beginnings of Algebra. The form of Algebra we identify with today has its origins around 800 AD where the Hindus appear to have been the first people to extend the basics of this subject.

Sometime during the 9th century AD, the name of the discipline we now call Algebra was altered due to an Arabic book in which part of the title was 'al-jabar'. This name is thought to have reached the English language around the middle of the 15th century AD.

A more revealing title for Algebra would have been the 'application of the logic of Arithmetic', since Algebra is concerned with algebraic equations and the values we can assign to their variables that make these equations into 'true' equations.

We first encounter Algebra in a disguised form in our first year of primary school. You will recall in primary school seeing an equation like:

$$\Box + 2 = 5$$

Your school teacher would have asked the class the following question: "What number must I put in the box above to make this a true equation?" You would have gleefully guessed the obvious answer, '3'. However, in secondary school, your teacher would have written this statement in an alternative form, such as:

$$\text{'}a + 2 = 5\text{' or '}x + 2 = 5\text{'}$$

Your teacher would have then asked the class the same question in a slightly different way: "What number must I **assign** to the variable 'a' (or 'x') above to make this a true equation?" You may have responded to this question with hesitation because guessing answers in secondary school Mathematics was not considered to be an acceptable approach to finding the answer to this equation.

As stated above, Algebra can be described as the process we use to determine which values we must assign to variables in (algebraic) equations in order to arrive at true arithmetic equations. Also, the manipulation of algebraic terms and equations is often referred to as 'doing' Algebra. Although this process may sound complex, the underlying principle is simple: the axioms, definitions and theorems in this chapter will demonstrate this simplicity for you.

Approach to Algebra

In this chapter, we will be applying Algebra to the Rational Number System as this is the most comprehensive system of numbers developed to this point. All the rules and processes of Algebra can be applied to any number system (for example, the Real Number System covered in Chapter Twelve).

In this section our aim will be to assign meaning to the process of finding numbers that make algebraic equations into true statements. This concept is referred to as the process of solving algebraic equations.

We have been using algebraic terms and equations in all of the previous chapters to write out axioms, definitions and theorems. In this chapter we will demonstrate how Algebra can be used to find the answer to everyday problems through solving simple algebraic equations.

Language of the Rational Number System for Algebra

At this point we will be using the alphabet we have developed so far to create algebraic terms and equations. We will not require any extra symbols in order to understand these algebraic terms and equations. The alphabet we will use in this chapter is outlined in Table 7.1 below.

7.1 The Alphabet of the Rational Number System for Algebra	
Symbols	**Meaning from:**
'0', '1', '+', '×', '(', ')'	The Axioms of the Natural Number System
'=', '≠'	The Axioms of Logic
'2', '3', '4', '5', '6', '7', '8', '9'	Definition 2.1
'a', 'b', 'c', ...	Rational number variables (these will no longer be reserved for integer variables)
','	Definition 2.3
'-1'	The Axioms of the Integer Number System
'–'	Definitions 3.4 and 3.5
'$\frac{1}{\Box}$'	The Axioms of the Rational Number System using Fractions
'x', 'y', 'z', ...	Rational number variables
'$\frac{a}{b}$'	Definition 4.2

7.1 The Alphabet of the Rational Number System for Algebra	
'$\div$'	Definition 4.4
'$\square^{-1}$'	The Axioms of the Rational Number System using Index Form
'$(a \times b^{-1})^m$'	Definition 5.6
'.'	Definition 6.4
'$a_n \dots a_1 a_0 . d_1 \dots d_m$'	Definition 6.5

Table 7.1

Once again, we use variables to represent numbers in algebraic terms and equations. As we will be focusing on the Rational Number System using both fractions and index form in this chapter, we will follow convention and use the letters 'x', 'y', 'z' etc. to denote these variables.

In Algebra when we multiply a number and a variable it is **conventional to drop the multiplication symbol**. For example, in the case of the term '$2 \times x$', we can rewrite this term as '$2x$' by dropping the multiplication symbol. Similarly, we write the product of two variables (e.g. '$y \times z$') as 'yz'. This is similar to writing mixed numbers in compact notation (e.g. '$4\frac{2}{3}$', as an abbreviation for '$4 + \frac{2}{3}$'). This simple convention helps to improve the readability of algebraic equations.

When leaving out the multiplication symbol, we must be sure to exercise care. If we assign a value to a variable, we have to revert to the unabbreviated form to avoid an incorrect answer. For example, in the term '$2x$' when we assign 'x' the value of '3', we must reinsert the multiplication sign so that this term becomes '2×3' and not the incorrect number '23'.

From the knowledge we have gained in previous chapters, we can now restate the language used to identify different parts of a string of symbols from the alphabet that contains at least one variable (see Table 7.2 below).

7.2 The Language of the Rational Number System for Algebra	
1. **Algebraic Term**	An individual variable, or sum and/or product of numbers with at least one variable, is referred to as an **algebraic term** or just a **term**. Consequently, 'x', '$5 + x$', '$5 \times y$', '$5x$', '$x \times z$', '$\frac{5}{1} \times \frac{x}{3}$' or '$\frac{z}{y} \times -x$' are all finite valid strings of symbols called **algebraic terms**.

7.2 The **Language** of the Rational Number System for Algebra	
2. **Algebraic Equation**	When we have an equation with at least one algebraic term on either side of an '=' or '≠' sign, we refer to this equation as an **algebraic equation**. For example, '$\frac{2 \times x}{3} \times -3 = -1$', '$x \times y \times z = 2x$' or '$x \neq 3$' are **algebraic equations**.

Table 7.2

Order of Operations

Having identified fractions and index forms of rational numbers, we must now go back and update our **Order of Operations** table to reflect the process for applying these operations to rational numbers in any notation. These concepts are set out in Table 7.3 below.

7.3 **Order of Operations** for the Rational Number System	
The process used to evaluate arithmetic and algebraic terms is established by the **order of operations** and proceeds according to the following steps:	
1.	The **parentheses** have highest priority and control the order of evaluation. Parentheses may be removed by evaluating inside the parentheses **or** by using the Distributive Axiom.
2.	Next evaluate **exponents** (i.e. an integer index to which a base number is raised).
3.	Where there are no parentheses, then we evaluate **multiplication** '×' and **division** '÷' from left to right.
4.	Finally, evaluate **addition** '+' and **subtraction** '−' from left to right.

Table 7.3

The Order of Operations for the Rational Number System can be remembered by using the acronym '**PEMDAS**' which stands for:

- Parentheses
- Exponents
- Multiplication
- Division
- Addition
- Subtraction.

Order of Operations – PEMDAS	
Evaluate the following term using Order of Operations:	
Term	$= 3(x + 1) + 2^3 \times 7 \times 2^{-1} + 2 - 12$
Parentheses	$= 3(x + 1) + 2^3 \times 7 \times 2^{-1} + 2 - 12$
Exponents	$= 3x + 3 + 2^3 \times 7 \times 2^{-1} + 2 - 12$
Multiplication	$= 3x + 3 + 8 \times 7 \times 2^{-1} + 2 - 12$
Division	$= 3x + 3 + 28 \div 2 - 12$
Addition	$= 3x + 3 + 14 - 12$
Subtraction	$= 3x + 17 - 12$
Answer	$= 3x + 5$

The acronym is dependent on the number system being used at that point and has its most general form (so far) when we use it to evaluate terms employing rational numbers in any format.

The Elementary Properties of the Rational Number System

There are no additional elementary properties of the Rational Number System required to evaluate algebraic terms and equations. We will continue to use the elementary properties developed for Rational Numbers as fractions and index forms.

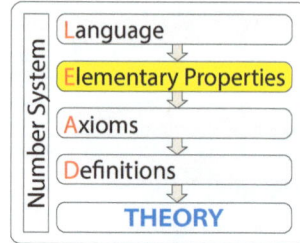

The Axioms of Logic

As in previous chapters, we need to use the Axioms of Logic to manipulate algebraic terms and equations. However, there are two theorems that we have proved in the past chapters describing the number systems that are also very useful for carrying out operations with algebraic terms and equations. These theorems are sometimes **incorrectly** referred to as the 'Axioms of Algebra'.

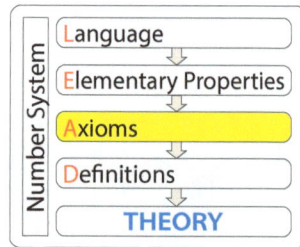

If the rational number variables are 'x' and 'y', then these two theorems can be restated as:

1. **The Reflexive Theorem '$x = x$'**

    If we assign a rational number to the variable 'x', then '$x = x$' is a true equation.

2. **The Substitution Theorem**

    If '$x = y$' is a true equation, then everywhere an 'x' appears in an algebraic term or equation, we can substitute 'y' for 'x' without altering the value of this term or the 'truth' value of this equation.

From this point forward, when solving algebraic equations we will be referring to the same Axioms of Logic we used in the previous chapters (i.e. Table 2.5, and the Reflexive and Substitution Theorems as derived from these axioms).

Definitions of the Rational Number System for Algebra

In this section we will introduce some of the common terminology used to describe operations with algebraic terms and equations.

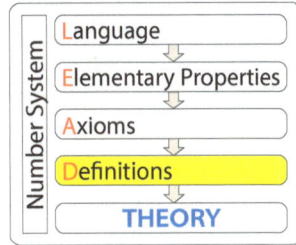

Number System:
Language ⇩
Elementary Properties ⇩
Axioms ⇩
Definitions ⇩
THEORY

Definition 7.1   Value of an Algebraic Term

After assigning specific values to all variables in an algebraic term, we obtain an arithmetic term which we can simplify to a specific rational number. This rational number is called the **value of the algebraic term**, and the process is called **evaluating a term**.

For example, in the case of a simple algebraic term such as '$x + 1$', we can evaluate this term when 'x' is assigned the value '2'. The value of this term is '$2 + 1$' which results in '3'. Consequently, the value of this term is '3'.

Although there may be several variables used in algebraic terms and equations, we may be interested in only one of these variables. This situation is highlighted in Definition 7.2 below.

Definition 7.2   The Phrase 'in the variable 'x''

A term or equation is said to be written '**in the variable 'x''** when the variable 'x' is the variable of interest when evaluating this term or equation.

For example, the algebraic equation '$ax + 2 = 0$' in the variable 'x' has two variables; however, we are only interested in eventually finding the value of 'x' in terms of the other variable 'a' and we call 'a' a constant with respect to 'x'.

The algebraic term '$2x + 1$' is a specific instance of a 'linear' term (see the definition below for general linear term). When we graph the term '$2x + 1$' relative to its variable 'x', we get a straight line. As the graph of this term results in a straight line, this term is called a **linear** algebraic term. This concept is illustrated in Figure 7.1 where we assign 'x' the values '0', '1' and '3' in the term '$2x + 1$' and we then graph the coordinates of the points given by $(0, 1)$, $(1, 3)$ and $(3, 7)$ to demonstrate the fact that these points lie along a straight line.

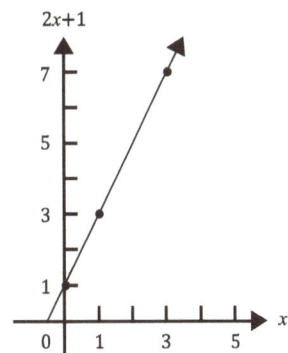

Definition 7.3 below formalises the general linear term.

Figure 7.1

Definition 7.3 General Linear Term

The **general linear term** in the variable 'x' (where 'x', 'a' and 'b' are rational number variables and 'a' (the coefficient of 'x') is not assigned '0') is:

$$ax + b$$

Note:

A 'linear' term in 'x' is an algebraic term where the highest power of its variable 'x' is '1' (i.e. we can write '$ax^1 + b$' as '$ax + b$'). Using our example of a linear term above, '$2x + 1$', '2' is the coefficient of 'x' and because '$x = x^1$', the highest power of 'x' in this term is '1'.

The next more complex algebraic term in 'x' is where we add the next highest power (namely '2') to the general linear term. Definition 7.4 below formalises this extension of the general linear term.

Definition 7.4 General Quadratic Term

The **general quadratic term** in the variable 'x' (where each of 'x', 'a', 'b' and 'c' is a rational number variable and 'a' (the coefficient of 'x^2') is not assigned '0') is given by the algebraic term:

$$ax^2 + bx + c$$

A specific quadratic term will have values assigned to the variables 'a', 'b' and 'c'.

Note:

When graphing the value of the general quadratic term against its variable 'x', we get a parabola.

In the example of the graph of a specific parabola, that is, we have assigned each of the variables 'a', 'b' and 'c', a value of '1' in the general quadratic term '$ax^2 + bx + c$'. On this graph we have plotted six values assigned to the variable 'x' against the corresponding six values of the algebraic term '$x^2 + x + 1$'. For example, when 'x' is assigned '-3', the term '$x^2 + x + 1$' evaluated as '$(-3)^2 + (-3) + 1 = 7$' which is the vertical coordinate on the graph in Figure 7.2. When all the other assignments to the variable 'x' are also graphed, we get the parabola in Figure 7.2.

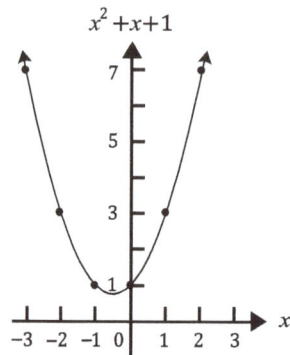

Figure 7.2

Comparing an algebraic term to either an algebraic term or arithmetic term using the equal operator '=' produces an algebraic equation. This leads us to our next definition (Definition 7.5) below.

Definition 7.5   Solving an Algebraic Equation

Solving an algebraic equation in a single variable (e.g. 'x') is the process of assigning numbers to this variable to make this algebraic equation true.

Notes:

1. The simplest example of solving an algebraic equation, i.e. '$x + 2 = 0$', is the process of finding the value of 'x' that will make this equation a true arithmetic equation. The result of this process assigns 'x' the value of '-2', so that the algebraic equation becomes the true arithmetic equation '$-2 + 2 = 0$'.

2. The method we use to solve an algebraic equation is to use the Axioms of Logic to isolate the variable so that we can easily see which values to assign to the variable to get a true arithmetic equation.

Now we use the definition of the general linear term to form the general linear equation (Definition 7.6).

Definition 7.6   General Linear Equation

The **general linear equation** in the variable 'x' (where each variable 'x', 'a' and 'b' is a rational number variable, and the highest power of 'x' is the index '1' and 'a' (the coefficient of 'x') is not assigned '0') is given by the algebraic equation:

$$ax + b = 0$$

A specific linear equation will have values assigned to the variables 'a' and 'b'.

As an example, if we assign 'a' the value of '2' and 'b' the value of '1' in the general linear equation, then we get the specific linear equation: '$2x + 1 = 0$'.

The next (more complex) algebraic equation in the variable 'x', occurs where we add the next highest power (namely '2') to the general linear equation. Definition 7.7 below formalises this extension of the general linear equation.

Definition 7.7 General Quadratic Equation

The **general quadratic equation** in the variable 'x' (where each variable 'x', 'a', 'b' and 'c' is a rational number variable where the highest power of 'x' is the index '2' and 'a' (the coefficient of 'x^2') is not assigned '0') is given by the algebraic equation:

$$ax^2 + bx + c = 0$$

A specific quadratic equation will have values assigned to the variables 'a', 'b' and 'c'.

Note:

In the definition of the general linear and quadratic equations, we have used the following convention for assigning coefficients: the coefficient of the highest power of 'x' is written as 'a', the coefficient of the second highest power of 'x' is written as 'b' and so on down the alphabet as the power 'n' in 'x^n' decreases. For example, in the linear equation '$2x + 1 = 2x^1 + 1x^0$' the coefficient of the highest power of 'x' is '2' and the coefficient of the second highest power of 'x' is '1'.

We can always write the general quadratic equation with a '0' on the right-hand side because if there were any arithmetic or algebraic terms on the right-hand side of the equal sign, we can always subtract these terms from both sides of the equation leaving a '0' on the right-hand side. For example, the quadratic equation '$3x^2 + 7x + 5 = x + 3$' can always be manipulated into the form '$3x^2 + 6x + 2 = 0$' by subtracting 'x' and '3' from both sides of this equation by using the Deduction Axiom.

This leads us to Definition 7.8 (the last definition of this chapter) which is a definition of a new symbol to extend our alphabet.

Definition 7.8 The Symbol '$\pm$'

The '$\pm$' symbol is used to abbreviate two options in the solution of an algebraic equation. The symbol '$\pm$', can be replaced by the plus symbol '+' and the negative symbol '−' to obtain the specific two options the '$\pm$' symbol is summarising.

The alphabet of the Rational Number System can now be extended by appending the new symbols defined in this section. This extension to the alphabet of the Rational Number System is summarised in Table 7.4 on the following page.

7.4 The **Alphabet** of the Rational Number System	
Symbols	**Meaning from:**
'0', '1', '+', '×', '(', ')'	The Axioms of the Natural Number System
'=', '≠'	The Axioms of Logic
'2', '3', '4', '5', '6', '7', '8', '9'	Definition 2.1
'a', 'b', 'c', ...	Rational number variables (or as specified)
','	Definition 2.3
'-1'	The Axioms of the Integer Number System
'$-$'	Definitions 3.4 and 3.5
'$\frac{1}{\Box}$'	The Axioms of the Rational Number System as Fractions
'x', 'y', 'z', ...	Rational number variables
'$\frac{a}{b}$'	Definition 4.2
'÷'	Definition 4.4
'$\Box^{-1}$'	The Axioms of the Rational Number System using Index Form
'$(a \times b^{-1})^m$'	Definition 5.6
'.'	Definition 6.4
'$a_n \ldots a_1 a_0 . d_1 \ldots d_m$'	Definition 6.5
'$\pm$'	Definition 7.8

Table 7.4

In the next section, we will explore the theory for solving these general algebraic equations.

Theory of Algebra

We now have adequate language, axioms and definitions in order to proceed with the proof of some basic theorems that solve the simplest algebraic equations using rational numbers. We begin the process by applying the Axioms of Logic to illustrate how we can solve the general linear equation.

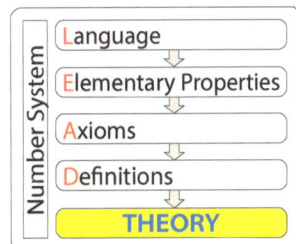

Number System
Language
⇩
Elementary Properties
⇩
Axioms
⇩
Definitions
⇩
THEORY

Theorem 7.1   Solving the General Linear Equation Theorem

Let 'a', 'b' and 'x' be rational number variables where 'a' is not assigned '0'. In this way, the solution of the general linear equation '$ax + b = 0$' is given in the variable 'x' as follows:

$$x = \frac{-b}{a}$$

Proof

We begin with the general linear equation '$ax + b = 0$' and apply the Axioms of Logic to isolate the variable 'x' as follows:

$$ax + b = 0 \qquad \text{... given}$$

$$ax + b - b = 0 - b \qquad \text{... by Deduction Axiom}$$

$$ax + 0 = -b \qquad \text{... by simplifying terms}$$

$$ax = -b \qquad \text{... by Identity Axiom '+'}$$

$$a \times x \times \tfrac{1}{a} = -b \times \tfrac{1}{a} \qquad \text{... by Deduction Axiom}$$

$$\therefore \qquad x = \frac{-b}{a} \qquad \text{... by simplifying terms}$$

Therefore, the general solution has been found and the theorem has been proved.

Note:

1. The general linear equation '$ax + b = 0$' does not always have a solution in the Natural Number System as some solutions require negative numbers in the first step of the proof above.

2. Likewise, the general linear equation '$ax + b = 0$' does not always have a solution in the Integer Number System as the solution requires division of numbers in the fifth step of the proof above.

The most general quadratic equation that can be solved using the Rational Number System is the algebraic equation '$ax^2 + bx + c = 0$', where '$b^2 - 4ac$' is a perfect square. Unlike the general linear algebraic equation, there does not exist a general solution to the general quadratic equation using the Rational Number System.

Next, we apply the Axioms of Logic to demonstrate how to solve the general quadratic equation using the Rational Number System. There are three key steps involved in finding the solution of the general quadratic equation that can be solved using rational numbers:

1. Write the given quadratic term as the product of two linear terms (Theorem 7.4).

2. Prove that if the product of two linear terms is equal to zero, then at least one of these terms must be equal to zero according to Theorem 7.5.

3. Find the solution to each linear equation (Theorem 7.1) and then solve the general quadratic equation where '$b^2 - 4ac$' is a perfect square using Theorem 7.6.

STEP 1 – The General Quadratic Term as the Product of Two Linear Terms

To prove that the general quadratic term can be written as the product of two linear terms, we will first require two lead-up theorems. The first lead-up theorem required for Step 1 is called the Squaring a Linear Term Theorem. This theorem involves taking a specific quadratic term '$x^2 + 2ax + a^2$' and demonstrating that it is the square of the linear term '$x + a$', i.e. '$(x + a)^2$'.

Theorem 7.2   Squaring a Linear Term Theorem

If 'x' and 'a' are rational number variables, then the quadratic term '$x^2 + 2ax + a^2$' can be written as the square of the linear term '$(x + a)$' as follows:

$$x^2 + 2ax + a^2 = (x + a)^2$$

Proof

Starting with the right-hand side of the above equation, we use proof by deduction as follows:

$$(x + a)^2 = (x + a)(x + a) \quad \text{... by Theorem 5.20}$$
$$= x(x + a) + a(x + a) \quad \text{... by Distributive Axiom}$$
$$= x^2 + xa + ax + a^2 \quad \text{... by Distributive Axiom again}$$
$$= x^2 + 2ax + a^2 \quad \text{... by collecting like terms}$$
$$\therefore \quad (x + a)^2 = x^2 + 2ax + a^2$$

This illustration concludes the proof of the squaring of a linear term.

Note:

The square of a linear term results in a quadratic term. When we manipulate a quadratic term into the form '$(x + a)^2$' we say: 'We are completing the square' of the quadratic term.

The second theorem of Step 1 is called the Difference of Two Squares Theorem. This theorem demonstrates how to write another specific quadratic term (namely '$x^2 - a^2$') as the product of two linear terms.

Theorem 7.3   Difference of Two Squares Theorem

If 'x' and 'a' are rational number variables, then the difference of the squares of these two variables can be written as follows:

$$x^2 - a^2 = (x + a)(x - a)$$

Proof

Starting with the right-hand side of the above equation, we use proof by deduction as follows:

$$(x + a)(x - a) = x(x - a) + a(x - a) \quad \text{... by Distributive Axiom}$$
$$= x^2 - xa + ax - a^2 \quad \text{... by Distributive Axiom again}$$
$$= x^2 - a^2 \quad \text{... by simplifying}$$
$$\therefore \quad (x + a)(x - a) = x^2 - a^2$$

This illustration concludes the proof of the factorisation of the difference of two variables squared.

We will now use the two theorems outlined above to demonstrate how we can write a general quadratic term as the product of two linear terms by completing the square.

Theorem 7.4   General Quadratic Term as Product of Two Linear Terms

If 'x', 'a', 'b', 'c' and 'd' are rational number variables where 'a' is not assigned '0' and where there exists a rational number 'd' such that '$d^2 = b^2 - 4ac$', then the general quadratic term in the variable 'x' (namely '$ax^2 + bx + c$') can be written as the product of two linear terms as follows:

$$ax^2 + bx + c = a(x + \tfrac{b + d}{2a})(x + \tfrac{b - d}{2a})$$

It is only possible to break this quadratic term into the product of two linear factors if 'a', 'b' and 'c' are related to each other by the algebraic term '$b^2 - 4ac$', which must be a perfect square. This will be required in the proof below where we will substitute 'd^2' for '$b^2 - 4ac$'.

Proof

We start the proof with the general quadratic term '$ax^2 + bx + c$' and then manipulate it into the product of two linear terms as follows:

$$ax^2 + bx + c = a(x^2 + \frac{b}{a}x + \frac{c}{a}) \qquad \text{... by factoring out '}a\text{'}$$

$$= a(x^2 + 2\frac{b}{2a}x + \frac{c}{a}) \qquad \text{... by the form of Theorem 7.2}$$

$$= a(x^2 + 2\frac{b}{2a}x + (\frac{b}{2a})^2 - (\frac{b}{2a})^2 + \frac{c}{a})$$

$$\text{... by adding '}(\frac{b}{2a})^2\text{' \& '}-(\frac{b}{2a})^2\text{'}$$

$$= a((x + \frac{b}{2a})^2 - (\frac{b}{2a})^2 + \frac{c}{a}) \qquad \text{... by completing the square}$$

$$= a((x + \frac{b}{2a})^2 - \frac{b^2}{4a^2} + \frac{4ac}{4a^2}) \qquad \text{... by expanding '}(\frac{b}{2a})^2\text{'}$$

$$= a((x + \frac{b}{2a})^2 - (\frac{b^2}{4a^2} - \frac{4ac}{4a^2})) \qquad \text{... by Distributive Axiom}$$

$$\therefore\ ax^2 + bx + c = a((x + \frac{b}{2a})^2 - (\frac{b^2 - 4ac}{4a^2})) \qquad \text{... by common denominator}$$

However, if '$b^2 - 4ac$' is the square of a rational number 'd', then we can write '$d^2 = b^2 - 4ac$' (and hence simplify the term on the right-hand side of our last equation) as follows:

$$ax^2 + bx + c = a((x + \frac{b}{2a})^2 - (\frac{d^2}{4a^2})) \qquad \text{... by substituting using '}d\text{'}$$

$$= a((x + \frac{b}{2a})^2 - (\frac{d}{2a})^2) \qquad \text{... by writing '}\frac{d^2}{4a^2}\text{' as a square}$$

$$= a(((x + \frac{b}{2a}) + (\frac{d}{2a}))((x + \frac{b}{2a}) - (\frac{d}{2a})))$$

$$\text{... by Theorem 7.3}$$

$$\therefore\ ax^2 + bx + c = a(x + \frac{b+d}{2a})(x + \frac{b-d}{2a})$$

This completes the proof that a general quadratic term can be written as the product of two linear terms (if we assume '$b^2 - 4ac$' can be written as the square of the rational number 'd').

Our objective is to eventually solve the general quadratic equation '$ax^2 + bx + c = 0$' where '$b^2 - 4ac$' can be written as a perfect square. Now that we are able to write the general quadratic term '$ax^2 + bx + c$' as the product of two linear terms (multiplied by 'a'), we need to know that making each one of these linear terms equal to zero gives a solution to this quadratic equation.

STEP 2 – The Product of Two Linear Terms Equal to Zero

If we can write the general quadratic equation as the product of two linear terms equal to zero, and we can demonstrate that each linear term must also be equal to zero, then we will have solved the general quadratic equation.

We now prove that the simple algebraic equation '$xy = 0$' has three solutions.

Theorem 7.5   Solving the Product of Two Linear Terms Theorem

If 'x' and 'y' are rational number variables and '$xy = 0$', then the solutions to this equation are:

1. 'x' is assigned '0' and 'y' is non-zero

2. 'y' is assigned '0' and 'x' is non-zero

3. both 'x' and 'y' are assigned '0'

Proof

To demonstrate this concept we use proof by deduction by assuming 'y' is not assigned '0'.

$$xy = 0 \qquad \text{... given}$$

$$xy \times \frac{1}{y} = 0 \times \frac{1}{y} \qquad \text{... by Deduction Axiom}$$

$$x \times 1 = 0 \times \frac{1}{y} \qquad \text{... by Theorem 4.17}$$

$$\therefore \qquad x = 0 \qquad \text{... by simplifying both sides}$$

In this way, if we assume 'y' is not assigned '0', then we are forced to assign 'x' the value '0' to make the equation '$xy = 0$' a true equation. If we assume 'x' is not assigned '0', then we can repeat this process to find the solution for 'y'. If both 'x' and 'y' are assigned '0', then from Theorem 2.4 the equation '$xy = 0$' is a true equation.

Consequently, we have found all three solutions to this equation and the theorem has been proved.

_____

All the elements of the proof of the general quadratic equation have now been established. Now we are ready for the final step – Step 3.

STEP 3 – Solving the general quadratic equation

The solution of the general quadratic equation relies on being able to split the general quadratic term into two linear terms and then use Theorem 7.5 above.

Theorem 7.6 Solution of the General Quadratic Equation Theorem

If 'x', 'a', 'b', 'c' and 'd' are rational number variables where 'a' is not assigned '0', then the general quadratic equation in the variable 'x' (namely '$ax^2 + bx + c = 0$') has the two solutions given by:

$$x = \frac{-b + d}{2a} \quad \text{or} \quad x = \frac{-b - d}{2a}$$

This is only possible if there exists a rational number 'd', such that '$d^2 = b^2 - 4ac$'. In this way, '$b^2 - 4ac$' can be written as the square of a rational number.

Proof

We start with the general quadratic equation '$ax^2 + bx + c = 0$' and write the general quadratic term as the product of two linear terms as follows:

$$0 = ax^2 + bx + c$$

$\therefore \qquad 0 = a(x + \frac{b + d}{2a})(x + \frac{b - d}{2a}) \qquad$ ... by Theorem 7.4

$\therefore \qquad \frac{1}{a} \times 0 = a \times \frac{1}{a}(x + \frac{b + d}{2a})(x + \frac{b - d}{2a}) \qquad$ ... by Deduction Axiom

$\therefore \qquad 0 = (x + \frac{b + d}{2a})(x + \frac{b - d}{2a}) \qquad$ ... by Inverse Axiom '$\times$'

According to Theorem 7.5, we know that the solution to this last equation is reached by setting each of these linear terms equal to '0'. In this way, beginning with the first equation, we solve the two linear equations '$x + \frac{b + d}{2a} = 0$' and '$x + \frac{b - d}{2a} = 0$' as follows:

$$x + \frac{b + d}{2a} = 0 \qquad \text{... given}$$

$$x + \frac{b + d}{2a} - \frac{b + d}{2a} = 0 - \frac{b + d}{2a} \qquad \text{... by Deduction Axiom}$$

$\therefore \qquad x = -\frac{b + d}{2a} \qquad$ ... by simplifying terms

Similarly, when we solve the other linear equation by the same method we obtain:

$$x \quad = -\frac{b-d}{2a}$$

Therefore, each of the solutions to these linear equations is also a solution to the general quadratic equation. Consequently, the theorem has been proved and we have discovered the solutions to the general quadratic equation – if '$b^2 - 4ac$' – can be written as the square of the rational number 'd'.

Note:

1. Using Definition 7.8, the two solutions of the general quadratic equation are often summarised into one compact mathematical equation as:

$$x = \frac{-b \pm d}{2a}$$

2. The solutions '$x = \frac{-b \pm d}{2a}$' are often called the **quadratic formula** for finding the solutions to the general quadratic equation '$ax^2 + bx + c = 0$'.

3. It doesn't matter whether 'd' is the positive number or negative number that squares to '$b^2 - 4ac$' since both 'd' and '$-d$' appear in the solution.

With the proof of this last theorem, we have demonstrated how to solve equations in a single variable 'x' which are raised to a power of no more than '2'. From this theorem, our key learning is that if we can reduce an algebraic term to a product of linear terms we can then find the solution of the corresponding algebraic equation.

We have now completed the introduction to the basic theory of Algebra. At this point we have access to a very powerful tool for solving algebraic equations in later chapters of this textbook.

Applications of Algebra

Algebra has extensive applications in our daily lives. The solution to almost any problem in scientific, engineering and commerce fields involves the use of Algebra. Our development of the elementary number systems (i.e. the development of the integers from the natural numbers, and the development of the rational numbers from the integers) can even be considered to be an exercise in broadening our ability to solve algebraic equations.

We will illustrate the basic use of the Axioms of Logic again to solve equations in the examples below by finding the value of the variable 'x' that makes these algebraic equations into 'true' arithmetic equations.

Application 7.1

Find the solutions of the equation '$x^2 - 1 = 0$'.

First, we write the equation '$x^2 - 1 = 0$' as the difference of two squares as:

$$'x^2 - 1^2 = 0' \text{ since } '1 = 1^2'.$$

We then apply Theorem 7.5 to find the solution.

$x^2 - 1^2 = 0$	... by substituting '1^2' for '1'
$(x + 1)(x - 1) = 0$	... by Theorem 7.3
∴ $(x + 1) = 0$ or $(x - 1) = 0$	... by Theorem 7.5 above
∴ $x = -1$ or $x = 1$	... by solving each equation

We check these solutions by substituting '-1' and '1' into the original equation '$x^2 - 1 = 0$'. In abbreviated notation, the two solutions shown above are sometimes written as: '$x = \pm 1$'. In this way, the solutions to the equation '$x^2 - 1 = 0$' are obtained by assigning the numbers '-1' and '1' to the variable 'x'.

Note:

In the example shown above, only one of the linear terms '$(x - 1)$' or '$(x + 1)$' can be '0' for any one assignment to the variable 'x'. In contrast, in Theorem 7.5 we had two variables to which we could assign values independently so they could both be zero at the same time.

Application 7.2

Find the solution to the algebraic equation '$\frac{(x + 2) \times 5}{3} + 7 = 22$'.

We apply the Axioms of Logic to isolate the variable 'x' as follows:

$\frac{(x+2) \times 5}{3} + 7 = 22$	... initial equation
$\frac{(x+2) \times 5}{3} + 7 - 7 = 22 - 7$	... by Deduction Axiom
$\frac{(x+2) \times 5}{3} + 0 = 15$	... by simplifying terms
$\frac{(x+2) \times 5}{3} = 15$	... by Additive Axiom '+'
$\frac{(x+2) \times 5}{3} \times \frac{3}{1} = 15 \times \frac{3}{1}$	... by Deduction Axiom
$\frac{(x+2) \times 5}{1} = 45$	... by Inverse Axiom '×'
$\frac{(x+2) \times 5}{1} \times \frac{1}{5} = 45 \times \frac{1}{5}$	... by Deduction Axiom

$$x + 2 = 9 \qquad \text{... by Inverse Axiom '}\times\text{'}$$

$$x + 2 - 2 = 9 - 2 \qquad \text{... by Deduction Axiom}$$

$$\therefore \qquad x = 7 \qquad \text{... by Additive Axiom '}+\text{'}$$

Using substitution, we must now check that when we assign 'x' the value '7' we obtain a 'true' equation:

$$\frac{(7+2) \times 5}{3} + 7 = 22$$

$$\therefore \qquad 22 = 22$$

This last equation is true by the Reflexive Theorem and so '7' is the solution to the algebraic equation '$\frac{(x+2) \times 5}{3} + 7 = 22$'.

_____

Application 7.3

Solve the quadratic equation: '$x^2 + x - 6 = 0$' using basic theorems. We start this process by stating the equation and manipulating it with the basic theorems that we outlined above, as follows:

$$x^2 + x - 6 = 0 \qquad \text{... given our equation to solve}$$

$$x^2 + x = 6 \qquad \text{... by adding Deduction Axiom}$$

$$x^2 + 2(\tfrac{1}{2})x + (\tfrac{1}{2})^2 = 6 + (\tfrac{1}{2})^2 \qquad \text{... by setting up as Theorem 7.2}$$

$$x^2 + 2(\tfrac{1}{2})x + (\tfrac{1}{2})^2 = \tfrac{24}{4} + \tfrac{1}{4} \qquad \text{... by setting up as Theorem 7.2}$$

$$(x + \tfrac{1}{2})^2 = \tfrac{25}{4} \qquad \text{... by using Theorem 7.2}$$

$$(x + \tfrac{1}{2})^2 = (\tfrac{5}{2})^2 \qquad \text{... by '}\tfrac{25}{4} = (\tfrac{5}{2})^2\text{'}$$

$$(x + \tfrac{1}{2})^2 - (\tfrac{5}{2})^2 = 0 \qquad \text{... by subtracting '}(\tfrac{5}{2})^2 = (\tfrac{5}{2})^2\text{'}$$

$$((x + \tfrac{1}{2}) - (\tfrac{5}{2}))\,((x + \tfrac{1}{2}) + (\tfrac{5}{2})) = 0 \qquad \text{... by Theorem 7.3}$$

$$\therefore \text{ either:} \qquad (x + \tfrac{1}{2}) - (\tfrac{5}{2}) = 0 \qquad \text{... by Theorem 7.5}$$

$$\text{or:} \qquad (x + \tfrac{1}{2}) + (\tfrac{5}{2}) = 0 \qquad \text{... by Theorem 7.5}$$

$$\therefore \text{ either:} \qquad x = +\tfrac{5}{2} - \tfrac{1}{2} = 2 \qquad \text{... by isolating '}x\text{'}$$

$$\text{or:} \qquad x = -\tfrac{5}{2} - \tfrac{1}{2} = -3 \qquad \text{... by isolating '}x\text{'}$$

We can check the solution by substituting '2' for 'x' in the original equation to obtain:

$$2^2 + 2 - 6 = 0$$

$$4 + 2 - 6 = 0$$

$$\therefore \qquad 0 = 0$$

This last equation is true by the Reflexive Theorem.

Now, substituting '-3' for 'x' in the original equation gives:

$$(-3)^2 + -3 - 6 = 0$$

$$9 - 3 - 6 = 0$$

$$\therefore \qquad 0 = 0$$

This last equation is true by the Reflexive Theorem.

Therefore, from our work above, we have demonstrated that the rational numbers that satisfy the quadratic equation '$x^2 + x - 6 = 0$' are the rational numbers '-3' and '2' simply because we are working in the Rational Number System.

———————————

In the next application we will solve a slightly more complex quadratic equation that has fractional solutions instead of whole number solutions. The mathematical steps required to find these fractional solutions are exactly the same as previous examples.

Application 7.4

In this application, we find the two rational numbers that are solutions of the equation '$12x^2 + 7x - 10 = 0$' by using basic theorems.

Once again, we follow the same steps we used in the previous example. We commence the process by stating the equation and manipulating it with the theorems above to isolate 'x', as follows:

$$12x^2 + 7x - 10 = 0 \qquad \text{... given our equation to solve}$$

$$12x^2 + 7x = 10 \qquad \text{... by adding '10 = 10'}$$

$$x^2 + \frac{7}{12}x = \frac{10}{12} \qquad \text{... by dividing '12 = 12'}$$

$$x^2 + 2(\tfrac{7}{24})x + (\tfrac{7}{24})^2 = \tfrac{10}{12} + (\tfrac{7}{24})^2 \qquad \text{... by adding '}(\tfrac{7}{24})^2 = (\tfrac{7}{24})^2\text{'}$$

$$x^2 + 2(\tfrac{7}{24})x + (\tfrac{7}{24})^2 = \tfrac{480}{(24)^2} + (\tfrac{7}{24})^2$$

... by common denominator

$$x^2 + 2(\tfrac{7}{24})x + (\tfrac{7}{24})^2 = \tfrac{529}{(24)^2}$$ ... by setting up to Theorem 7.2

$$x^2 + 2(\tfrac{7}{24})x + (\tfrac{7}{24})^2 = (\tfrac{23}{24})^2$$ ... by '529 $= (23)^2$'

$$(x + \tfrac{7}{24})^2 = (\tfrac{23}{24})^2$$ ... by using Theorem 7.2

$$(x + \tfrac{7}{24})^2 - (\tfrac{23}{24})^2 = 0$$ ... by subtracting '$(\tfrac{23}{24})^2$'

$$((x + \tfrac{7}{24}) - (\tfrac{23}{24}))\,((x + \tfrac{7}{24}) + (\tfrac{23}{24})) = 0$$ ... by using Theorem 7.3

∴ either: $$(x + \tfrac{7}{24}) - (\tfrac{23}{24}) = 0$$ ... by using Theorem 7.5

or: $$(x + \tfrac{7}{24}) + (\tfrac{23}{24}) = 0$$ ... by using Theorem 7.5

∴ either: $$x = +\tfrac{23}{24} - \tfrac{7}{24} = \tfrac{2}{3}$$ ... by isolating 'x'

or: $$x = -\tfrac{23}{24} - \tfrac{7}{24} = -\tfrac{5}{4}$$ ... by isolating 'x'

We can check the solution by substituting '$\tfrac{2}{3}$' for 'x' in the original equation to obtain:

$$12(\tfrac{2}{3})^2 + 7(\tfrac{2}{3}) - 10 = 0$$

$$\tfrac{48}{9} + \tfrac{42}{9} - \tfrac{90}{9} = 0$$

∴ $$0 = 0$$

This last equation is true by the Reflexive Theorem. Therefore, '$\tfrac{2}{3}$' is the rational number solution of the agebraic equation: '$12x^2 + 7x - 10 = 0$'.

Next, we substitute '$-\tfrac{5}{4}$' for 'x' in the original equation to obtain:

$$12(-\tfrac{5}{4})^2 + 7(-\tfrac{5}{4}) - 10 = 0$$

$$\tfrac{300}{16} - \tfrac{140}{16} - \tfrac{160}{16} = 0$$

∴ $$0 = 0$$

This last equation is a true statement by the Reflexive Theorem and so '$-\tfrac{5}{4}$' is also a rational number solution. In this way, we have found the two rational number solutions of this equation.

In the next application we will show how to find the solutions to the same quadratic equation by using the quadratic formula.

Application 7.5

In this application, we discover two rational numbers that are solutions of the equation '$12x^2 + 7x - 10 = 0$' by using the quadratic formula.

In this equation 'a', 'b' and 'c' have been assigned the values '12', '7' and '-10' respectively. The first step is to calculate the value of 'd' in the quadratic formula as follows:

$$d^2 = b^2 - 4ac \qquad \text{... by definition of '}d^2\text{'}$$

$$= 7^2 - 4 \times 12 \times -10 \quad \text{... by substituting values}$$

$$\therefore \qquad d^2 = 529 \qquad \text{... by simplifying terms}$$

From the above equation, we have either '$d = 23$' or '$d = -23$'. Using 'd' with the value of '23', we can write the first solution of this equation as:

$$x = -\frac{b+d}{2a}$$

$$= -\frac{7+23}{2 \times 12}$$

$$\therefore \qquad x = -\frac{5}{4}$$

We can write the second solution of this equation as:

$$x = -\frac{b-d}{2a}$$

$$= -\frac{7-23}{2 \times 12}$$

$$\therefore \qquad x = \frac{2}{3}$$

You will observe that substituting '-23' for 'd' would have given the same two solutions.

Consequently, we have arrived at the same two rational number solutions for the quadratic equation '$12x^2 + 7x - 10 = 0$' by using the quadratic formula as we did using basic theorems in Application 7.3 above.

––––––––––––––––––

We have now completed the process of applying the Axioms of Logic to solving some simple quadratic equations.

Summary of Algebra

At this point, we have covered the objectives of this chapter and so can summarise our findings up to now. In terms of Algebra, the most important definitions we have introduced in this chapter are:

1. **Solving** an algebraic equation is the process of assigning values to the variables (or equivalently, assigning specific numbers to variables) to obtain a true arithmetic equation.

2. Defining **the general linear term** in the variable 'x' as the algebraic term '$ax + b$'.

3. Defining **the general quadratic term** in the variable 'x' as the algebraic term '$ax^2 + bx + c$'.

4. Defining **the general linear equation** in the variable 'x' as the algebraic equation '$ax + b = 0$'.

5. Defining **the general quadratic equation** in the variable 'x' as the algebraic equation '$ax^2 + bx + c = 0$'.

The Axioms of Logic provide the rules for us to be able to manipulate algebraic equations in order to simplify these equations to assist us to isolate variables and determine their value in the Rational Number System. This process of isolating variables is what most people now understand by the term 'Algebra'.

This chapter on Algebra has provided another building block necessary in our journey to understanding the Real Number System (which is the last of the elementary number systems).

CHAPTER 8

RATIONAL NUMBERS USING ORDER

Overview of Rational Numbers using Order

In this chapter we will be focusing on modelling the concept of 'order' for the rational numbers in the same way in which we modelled the concepts of 'addition', 'multiplication' and 'equals' from Chapter Two onwards. This concept is a critical step on our journey of mastering the Rational Number System as it is essential for defining the concept of a 'limit' which, in turn, is necessary to fully understand rational numbers expressed as decimal numbers.

As in previous chapters, we will extend our language of the Rational Number System by introducing new symbols. In this chapter, these new symbols will allow us to place the rational numbers in 'order' along the number line.

This concept of order is distinct from the concept of the 'order of operations' for a number system which we used previously to describe the process of carrying out the operations of addition, multiplication and raising a number to an index in the absence of controlling parentheses.

Pictorial Representation of Rational Numbers using Order

In this chapter, we will formalise the order of rational numbers along a number line. The concept of ordering rational numbers describes the process of placing the rational numbers along a number line from 'smaller' to 'larger' as we move from left to right along the line. This type of order is also sometimes referred to by Mathematicians as 'linear order' as illustrated in Figure 8.1 below.

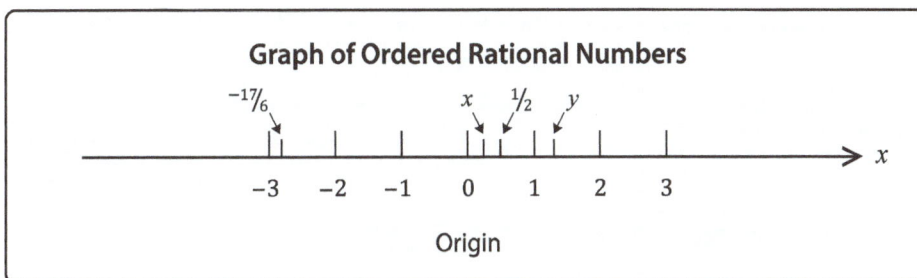

Figure 8.1

In this graph, the order between 'x' and 'y' is denoted by '$x < y$'. That is, we use the standard convention that says: "x represents a rational number less than y" because 'x' is to the left of 'y', or 'x' precedes 'y' along the number line. In this way, the larger rational numbers are always pictured further to the right along the number line. It is for this reason that the concept of ordering of numbers is also called 'linear' ordering.

Background and Context of Rational Numbers using Order

The process of ordering numbers has its origins in the early development of Mathematics. For example, we are aware that Archimedes ($278 - 212$ BC) identified the area of a circle by arguing that it couldn't be less than or greater than the area of a particular right-angle triangle with a height equal to the radius of the circle and a base length equal to the circumference. This case in point tells us Archimedes was using the order relation as it applied to the size of areas of geometrical figures in the plane.

So far, in the elementary number systems we have worked with we have used a set of axioms to manipulate equations in order to determine the 'true' or 'false' property of those equations. In this chapter, we will demonstrate how the ordering of numbers has a similar set of axioms that control how true 'inequalities' involving the 'order' symbols can be manipulated.

To continue our study of rational numbers, we now introduce four extra symbols for comparing rational numbers and placing them in their 'ordered' position along the number line.

The first of these order symbols '$<$' is called the '**less-than**' symbol. This symbol derives its meaning from a set of Axioms of Order and extends our alphabet of symbols. We will be explicitly defining the other three order symbols '$\leq$', '$>$' and '$\geq$' using the less-than symbol ('$<$') as a reference point. These four symbols are included in the alphabet in Table 8.1 below.

Once we have defined the order symbols, we can then define the process of determining the distance between two points along the number line. This operation of determining the distance between points along the number line is carried out by the '**absolute value**' operator and will be defined in the Definitions Section below.

Approach to Rational Numbers using Order

In this section, our aim is to take a similar approach to assign meaning to the symbol '$<$' by using axioms in the same way we have in previous chapters. After we have assigned meaning to the symbol '$<$', we will then be able to define new symbols in terms of this 'less-than' symbol.

Language of the Rational Number System using Order

The use of the less-than symbol allows us to now extend our mathematical language to include inequalities, as well as terms and equations (see Table 8.1 below). These inequalities are relationships that compare two terms using the less-than symbol ('$<$').

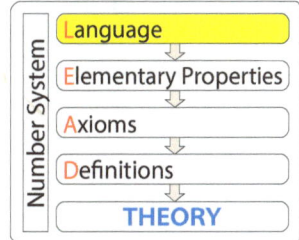

8.1 The Alphabet of the Rational Number System using Order	
Symbols	**Meaning from:**
'0', '1', '+', '×', '(', ')'	The Axioms of the Natural Number System
'=', '≠'	The Axioms of Logic
'2', '3', '4', '5', '6', '7', '8', '9'	Definition 2.1
'a', 'b', 'c', ... , 'x', 'y', 'z', ...	Rational number variables
','	Definition 2.3
'-1'	The Axioms of the Integer Number System
'$-$'	Definitions 3.4 and 3.5
'$\frac{1}{\square}$'	The Axioms of the Rational Number System as Fractions
'$\frac{a}{b}$'	Definition 4.2
'$\div$'	Definition 4.4
'$\square^{-1}$'	The Axioms of the Rational Number System using Index Form
'$(a \times b^{-1})^m$'	Definition 5.6
'.'	Definition 6.4
'$a_n ... a_1 a_0 . d_1 ... d_m$'	Definition 6.5
'$\pm$'	Definition 7.8
'$<$'	The Axioms of the Rational Number System using Order

Table 8.1

We again use variables to represent numbers in algebraic terms, equations and inequalities. We will be focusing on the Rational Number System using Order and will now use any letters of the alphabet 'a', 'b', 'c', ... , 'x', 'y', 'z' etc. to denote these variables.

We can now state the language we will be using to identify different parts of a string of symbols from the alphabet that contains the less-than symbol (see Table 8.2 below).

8.2 The Language of the Rational Number System using Order
In this table, let 'x', 'y' and 'z' represent rational number variables.

1. Arithmetic Inequalities	When we have an inequality with arithmetic terms on either side of a less-than symbol ('$<$'), we call this inequality an arithmetic inequality. For example, '$2+3<6$', '$-\frac{1}{2}<0$' and '$\frac{3}{4}\times 7<5$' are arithmetic inequalities.
2. Algebraic Inequalities	When we have an inequality with at least one algebraic term which can be on either side of a less-than symbol ('$<$'), we call this inequality an algebraic inequality. For example, '$3x+1<5y$', '$xyz<0$' and '$x^2+2x+1<1$' are algebraic inequalities.

Table 8.2

When ordering the rational numbers, we observe the simple property that these numbers can be placed sequentially from smaller to larger along the number line. We call this the 'order property' of rational numbers.

Order of Operations

With the introduction of ordering rational numbers, we can now apply the same Order of Operations to evaluate arithmetic and algebraic terms on either side of the less-than symbol.

The Order Property of the Rational Number System

This new property describes how every two **different** rational numbers assigned to the variables 'x' and 'y' must result in either:

1. '$x<y$' is true if the number assigned to 'x' is less than the number assigned to 'y', **or,**

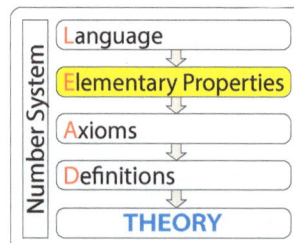

2. '$y < x$' is true if the number assigned to 'y' is less than the number assigned to 'x'.

This property is modelled in the Rational Number System by placing the variable that is assigned the smaller number to the left of the less-than symbol in the same way we do for the numbers along the number line (i.e. the smaller number is to the left of the larger number).

In symbols, we model this property by the inequality:

$$x < y$$

The axioms that model the less-than property '$<$' for Rational Numbers using Order are outlined in the section below.

The Axioms of Order for the Rational Number System

We will now define the '**Axioms of Order**' that give meaning to the less-than symbol '$<$' in the Rational Number System. These axioms model the property of order of the rational numbers by providing rules by which we can manipulate inequalities so that we preserve the truth of such inequalities.

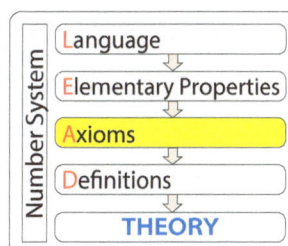

We observe the use of the equal sign in these axioms and hence these Axioms of Order rely on the Axioms of Logic to assign meaning to this new less-than symbol.

The Axioms of Order that model the concept of order for rational numbers and implicitly define the less-than symbol '$<$' are outlined in Table 8.3 below.

8.3 The Axioms of Order for the Rational Number System	
For the rational number variables 'x', 'y' and 'z', the order symbol '$<$' satisfies the following axioms:	
1. **Trichotomous Axiom for '$<$'**	Either '$x < y$', '$y < x$' or '$x = y$' (and only one of these three statements can be **true** for each assignment of a rational number to each variable and the other two statements are **false**).
2. **Transitive Axiom for '$<$'**	Given '$x < y$' and '$y < z$' are true inequalities, then '$x < z$' is also a **true** inequality.

8.3 The Axioms of Order for the Rational Number System	
3. **Additive Axiom for '$<$'**	Given '$z = z$' and '$x < y$' then '$x + z < y + z$'. If one of these two inequalities is **true**, then the other is **true**. If one is **false**, then the other is **false**.
4. **Multiplicative Axiom for '$<$'**	Given '$0 < x$' and '$0 < y$' then '$0 < xy$'. If any two of these inequalities are **true**, then the other inequality is also **true**. If one is **true** and another is **false** then the other is **false**.

Table 8.3

Note:

These axioms can be applied without alteration to the Integer Number System. Since the rational numbers are an extension of the integers, all of the theorems presented in this chapter that don't require a multiplicative inverse (division) can be proven for the Integer Number System as well.

The axioms described above encapsulate our intuitive understanding of this 'order' of rational numbers whereby we are able to place those numbers along the number line with smaller coordinates to the left of those numbers that have bigger coordinates.

The Axioms of Logic

The Axioms of Logic have not changed in structure from those in Table 2.5 in Chapter Two (with the inclusion of the notes in both Chapters Four and Five). The variables are again assigned rational numbers in this chapter. In the same way that the Axioms of Logic assign meaning to the equal and not-equal signs, these axioms are required to give meaning to the less-than symbol in the statement of the Trichotomous Axiom and the Additive Axiom in Table 8.3 above.

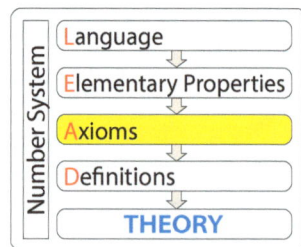

Number System
Language
⇓
Elementary Properties
⇓
Axioms
⇓
Definitions
⇓
THEORY

Definitions of the Rational Number System using Order

Having now assigned meaning to the symbol '$<$' in the Axioms of Order for the Rational Number System, we will be using this symbol to create definitions for other order symbols.

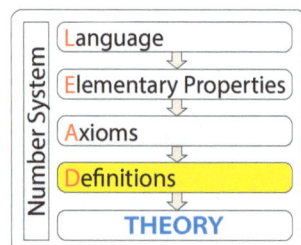

Number System
Language
⇓
Elementary Properties
⇓
Axioms
⇓
Definitions
⇓
THEORY

In the Theory Section below, we will explore the relationship between this new concept which uses the new order symbol, the other symbols in our alphabet, the axioms and definitions.

Definition 8.1 Inequality

An **inequality** is a finite valid string of symbols with a term on either side of an order symbol.

In Definition 8.2 below, we explicitly define the three order symbols '$\leq$', '$>$' and '$\geq$' in terms of the symbols '$<$' and '$=$'.

Definition 8.2 Order Symbols '$\leq$', '$>$' and '$\geq$'

The **order symbols** are used in inequalities to describe the order of two rational numbers. Given any two rational number variables 'x' and 'y', the following symbols are used to represent the inequalities between these two variables when we assign rational numbers to these variables. The definitions for these three symbols are:

1. '$x \leq y$' is true if and only if **either** '$x < y$' is true **or** '$x = y$' is true

2. '$x > y$' is true if and only if '$y < x$' is true

3. '$x \geq y$' is true if and only if **either** '$y < x$' is true **or** '$x = y$' is true.

The pointed ends of the symbols '$<$' and '$>$' point to the smaller of the two rational numbers in the inequality. In the other two symbols '$\leq$' or '$\geq$', the pointed end of the symbols also point to the smaller than, or equal to, rational number in the inequality, where the inequality is a true statement.

In relation to the other three order symbols, we can provide examples of finite valid strings of symbols using inequalities. For example: '$0 \leq 0$', '$1 > 2$', '$2 \times 4 > 2 + 4$', '$x + 2 \geq 3$', '$x \times y + 3 \leq 2 \times z$', '$x > 5$', '$x + y > 0$'. It is worth noting that the inequality '$1 > 2$' is a valid string of symbols, however, it is not a 'true' inequality as is demonstrated in the next section.

We will now extend our technique for solving the general linear equation '$ax + b = 0$' (recalling that this is the simplest of algebraic equations) to the case when the '$=$' is replaced by the less-than symbol '$<$'. Definition 8.3 provides the correct name for this generalisation.

Definition 8.3   General Linear Inequality

The **general linear inequality** in the variable 'x' (where 'x', 'a' and 'b' are rational number variables) is defined by:

$$ax + b < 0$$

Next, we further extend our technique for solving the general quadratic equation '$ax^2 + bx + c = 0$' to the case when the '$=$' is replaced by the '$<$'. Definition 8.4 provides the correct name for this generalisation.

Definition 8.4   General Quadratic Inequality

The **general quadratic inequality** in the variable 'x' (where 'x', 'a', 'b' and 'c' are rational number variables) is defined by:

$$ax^2 + bx + c < 0$$

In Definition 8.5 below, we will use the subtraction operator to define the '**directed distance**' from a point on the line with coordinate 'x' to a point on the same line with coordinate 'y'. That is, the number of unit steps to get from the point with coordinate 'x' (starting point) to the point with coordinate 'y' (finishing point).

Definition 8.5   Directed Distance between Two Points

Let 'x' be assigned the coordinate of the starting point (called 'P_{start}') and 'y' be assigned the coordinate of the finishing point (called 'P_{finish}') on the number line and let 'r' be a rational number variable. The **directed distance** 'r' from the point 'P_{start}' to the point 'P_{finish}' along the number line is the assignment to 'r' that makes the following equation true:

$$r = y - x$$

Note:

The directed distance has the following expected properties:

1. If 'x' and 'y' are the coordinates of the same point, then '$x = y$' is true and '$r = x - x = 0$' and therefore, the directed distance of a point from itself is zero.

2. If 'x' and 'y' are the coordinates of two points such that 'y' is to the right of 'x' on the number line, then we get '$y - x > 0$' and the directed distance from 'x' to 'y' is positive. For example, if 'y' is assigned '5' and 'x' is assigned '2' then '$r = 5 - 2 = 3 > 0$' is true and the directed distance from '2' to '5' is '3'. If 'y' is to the left of 'x' along the number line, then we get '$y - x < 0$' and the directed distance from 'x' to 'y' is negative. For example, if 'y' is assigned '5' and 'x' is assigned '7' then '$r = 5 - 7 = -2 < 0$' is true and the directed distance from '7' to '5' is '-2'.

An alternative way of measuring distance between two points along the number line is called the '**undirected distance**' between two points. This alternate measurement for the distance between two points will always be **positive or zero**, and corresponds to the physical process we use to measure the distance between two points using a ruler.

Definition 8.6 Undirected Distance between Two Points

The **undirected distance** 'd' between the two points 'P' and 'Q' along the number line (with coordinates 'x' and 'y' respectively) is given by the equations:

1. If '$x \geq y$' is true, then 'd' is assigned the value that makes the equation '$d = x - y$' true

   or

2. If '$x < y$' is true, then 'd' is assigned the value that makes the equation '$d = y - x$' true.

Note:

The undirected distance 'd' is defined in such a way that ensures it is always non-negative, that is, '$d \geq 0$'. For example, if you ask someone to tell you the distance between two points on a ruler, they will automatically take the larger number associated with one of the points and then subtract the smaller number associated with the other point in order to give you a positive answer! This positive answer is the undirected distance that Definition 8.6 encapsulates.

Definition 8.7 below provides us with a new operator called the 'absolute value operator' that denotes the undirected distance between two points.

Definition 8.7   Absolute Value Operator '| |'

The **absolute value operator** applied to the two rational numbers 'x' and 'y' is denoted by '$|x-y|$' and is always positive or zero. It is defined by the following statements:

1. '$|x-y| = x-y$' is true if and only if '$x-y > 0$' is true, **and**

2. '$|x-y| = -(x-y)$' is true if and only if '$x-y < 0$' is true.

_____

Note:

The absolute value is modelling the undirected distance (as stated in terms of points and a line) between two points with given coordinates on a line and is restated as follows:

1. Given two points with coordinates 'x' and 'y' on the number line, the absolute value operator gives the undirected distance 'd' between these two points by the following true equation:

$$d = |x-y|$$

2. To obtain the definition of the undirected distance of 'x' from the origin, '0', in the formula for 'd', we assign the the value '0' to the variable 'y' so that '$d = |x|$'.

The difference between the directed and undirected distance is that the directed distance uses positive and negative numbers to measure the direction and distance from one point to another, whereas the undirected distance uses positive numbers only to measure this distance.

Having now completed the definitions for this chapter, we can add the new symbols in these definitions and expand our alphabet of the Rational Number System to include them (see Table 8.4 below).

8.4 The Alphabet of the Rational Number System using Order	
Symbols	**Meaning from:**
'0', '1', '+', '×', '(', ')'	The Axioms of the Natural Number System
'=', '≠'	The Axioms of Logic
'2', '3', '4', '5', '6', '7', '8', '9'	Definition 2.1
'a', 'b', 'c', ..., 'x', 'y', 'z', ...	Rational number variables

8.4 The **Alphabet** of the Rational Number System using Order	
','	Definition 2.3
'−1'	The Axioms of the Integer Number System
'−'	Definitions 3.4 and 3.5
$\frac{1}{\square}$	The Axioms of the Rational Number System as Fractions
'$\frac{a}{b}$'	Definition 4.2
'÷'	Definition 4.4
'$\square^{-1}$'	The Axioms of Rational Number System using Index Form
'$(a \times b^{-1})^m$'	Definition 5.6
'.'	Definition 6.4
'$a_n \dots a_1 a_0 . d_1 \dots d_m$'	Definition 6.5
'±'	Definition 7.8
'<'	The Axioms of the Rational Number System using Order
'≤', '>', '≥'	Definition 8.2
'$\|x - y\|$'	Definition 8.7

Table 8.4

Our approach to developing the Rational Numbers using Order is a further extension of the set of rational numbers we covered in previous chapters. We will now continue our journey to developing the theory that follows on from this extension to our Rational Number System.

In the Theory Section below we provide formal proofs for results that we would expect the order symbols and associated definitions to model.

Theory of Rational Numbers using Order

The Axioms of Logic and the Axioms of Order allow us to manipulate arithmetic and algebraic inequalities in a way that preserves the truth of these inequalities and conforms to our intuitive notion of how these inequalities should behave.

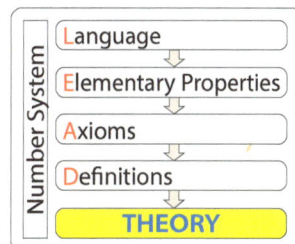

Number System
- Language
- Elementary Properties
- Axioms
- Definitions
- **THEORY**

Our current goal is to demonstrate that the Axioms of Order we discussed above align with our intuitive concept of order by proving that '−1', '0' and '1' satisfy the inequalities '−1 < 0', '0 < 1' and '−1 < 1'. You will recall in the note following Table 8.3 that the Axioms of Order can be applied without alteration to integers, as long as we use only the properties of integers.

Theorem 8.1   Basic Inequalities Theorem

The rational numbers '−1', '0' and '1' satisfy the following true inequalities:

$$-1 < 0$$

$$0 < 1$$

$$-1 < 1$$

Proof

Part 1:

We begin by assuming '0 < −1' is true and then show that a contradiction must follow.

	$0 < -1$	... by assumption
	$0 < -1 \times -1$	... by Multiplicative Axiom '<'
∴	$0 < 1$	... by simplifying terms
	$0 + -1 < 1 + -1$	... by Additive Axiom '<'
∴	$-1 < 0$	... by simplifying terms

The original assumption '0 < −1' and the result from the proof, namely '−1 < 0', **contradict** the Trichotomous Axiom (i.e. that only one of these inequalities can be true). Therefore, our original assumption must be false. However, we know that '0 = −1' is also a false equation (from the Not-Equal-to Axiom and Theorem 3.1), so the third option of the Trichotomous Axiom, namely '−1 < 0', must be a true inequality.

Part 2:

We now use this first result to derive the second result of the theorem.

	$-1 < 0$	... by Part 1 of this theorem
	$-1 + 1 < 0 + 1$	... by Additive Axiom '<'
∴	$0 < 1$	... by simplifying terms

Therefore, the result '0 < 1' must be a true inequality.

Part 3:

We now use the first two results to obtain the third result.

$$-1 < 0 \qquad \text{... by Part 1 of this theorem}$$

$$0 < 1 \qquad \text{... by Part 2 of this theorem}$$

$$\therefore \qquad -1 < 1 \qquad \text{... by Transitive Axiom '<'}$$

Therefore, the result '$-1 < 1$' must be a true inequality.

Here, we have proved that the rational numbers '-1', '0' and '1' satisfy the inequalities we would intuitively expect. In this way, our axioms are modelling the 'less-than' order of the rational numbers and so the theorem has been proved.

————————————————

We now wish to demonstrate that our axioms are correctly modelling our concept of 'positive' and 'negative' rational numbers. However, using Theorem 8.1, we first need to demonstrate that positive integers, as a subset of the Rational Number System, make the inequality '$0 < a$' true and negative integers make the inequality '$a < 0$' true when assigning positive and negative integers to the variable 'a', respectively.

Theorem 8.2   Positive and Negative Integers using Order Theorem

In the case of integers assigned to the variable 'a' (such that '$0 < a$' is true), all of these integers are positive integers. Similarly, for integers assigned to 'a' (such that '$a < 0$' is true), all of these integers are negative integers.

Proof

We begin this proof using mathematical deduction and Theorem 8.1 where we proved that '$0 < 1$' is true. From Definition 3.3, we know that '1' is a positive integer. First, we assume that we have assigned an integer to 'a' that makes the inequality '$0 < a$' true.

Next, we reason deductively as follows:

$$0 < 1 \qquad \text{... by Theorem 8.1}$$

$$0 + a < 1 + a \qquad \text{... by Additive Axiom '<'}$$

$$\therefore \qquad a < a + 1 \qquad \text{... by simplifying}$$

But: $\qquad 0 < a \qquad \text{... by assumption}$

And: $\qquad a < a + 1 \qquad \text{... by first part of proof}$

$$\therefore \qquad 0 < a + 1 \qquad \text{... by the Transitive Axiom}$$

Therefore, '$0 < a + 1$' is a true inequality.

Now we use proof by the Mathematical Induction Process. If we assume 'k' is an integer and the kth statement '$k > 0$' is true, then the first half of this proof shows that the $(k+1)^{th}$ statement '$k + 1 > 0$' is also a true inequality. Since '$0 < 1$' is a true inequality, it follows that '$0 < 1 + 1$' is also a true inequality. Therefore, using proof by the Mathematical Induction Process, it follows that all the positive integers 'k' must make the inequality '$0 < k$' true.

Since '$0 < k$' is a true inequality for positive integers, then:

$$0 < k \qquad \text{... by '}k\text{' is a positive integer}$$
$$0 - k < k - k \qquad \text{... by Additive Axiom '<'}$$
$$-k < 0 \qquad \text{... by simplifying}$$

Therefore, '$-k < 0$' is a true inequality and it follows that all the integers that make the inequality '$a < 0$' true are negative integers. This concludes the proof of the theorem.

From Theorem 8.2 we observe that we can order the positive integers using the true inequality '$a < a + 1$' as follows: '$1 < 2 < 3 < ...$'. By rearranging this inequality, we get '$-a - 1 < -a$' is also true and hence obtain the equivalent ordering of the negative integers as follows: '$... < -3 < -2 < -1$'.

Now that we know that the less-than symbol '<' can be used to identify and order positive and negative integers, we will use this result to identify and order rational numbers as well.

Theorem 8.3 Positive and Negative Rational Numbers using Order Theorem

For all integers that can be assigned to the variables 'a' and 'b' such that '$0 < a$' and '$0 < b$' are true inequalities, the rational number '$\frac{a}{b}$' makes the inequality '$0 < \frac{a}{b}$' true and so is a positive rational number. Likewise, the rational number '$-\frac{a}{b}$' makes the inequality '$-\frac{a}{b} < 0$' true and so is a negative rational number.

Proof

This proof will use the definition of the rational number '$\frac{a}{b}$', namely: '$\frac{a}{b} = a \times \frac{1}{b}$'. We use proof by mathematical deduction.

First, we must prove that the multiplicative inverse '$\frac{1}{b}$' is also a positive rational number for all positive integers assigned to 'b'.

$$0 < \frac{1}{b} \qquad \text{... by assumption}$$

$$0 < b \qquad \text{... given}$$

$$\therefore \quad 0 < \frac{1}{b} \times b \qquad \text{... by Multiplicative Axiom '<'}$$

$$\therefore \quad 0 < 1 \qquad \text{... by Inverse Axiom '×'}$$

The inequality '$0 < 1$' is true according to Theorem 8.1(Part 2). Therefore, by the Multiplicative Axiom, the assumption that '$\frac{1}{b}$' must be positive is true.

Using the Multiplicative Axiom for '<', if '$a > 0$' and '$\frac{1}{b} > 0$' are true, then their product '$a \times \frac{1}{b} = \frac{a}{b} > 0$' is also true. Therefore, for positive fractions '$\frac{a}{b}$' (Definition 4.5) we have '$\frac{a}{b} > 0$' is true. This result completes the first half of the proof.

We now prove the equivalent result for negative fractions (Definition 4.6).

$$0 < \frac{a}{b} \qquad \text{... by first half of proof}$$

$$0 - \frac{a}{b} < \frac{a}{b} - \frac{a}{b} \qquad \text{... by Additive Axiom '<'}$$

$$-\frac{a}{b} < 0 \qquad \text{... by simplifying terms}$$

Using the Additive Axiom for '<', the inequality '$-\frac{a}{b} < 0$' must be true. Hence, for negative fractions '$-\frac{a}{b}$' we have '$-\frac{a}{b} < 0$' is also true. This result completes the second half of the proof.

This concludes our proof that all rational numbers to the right of zero on the number line are positive and greater than zero while all rational numbers to the left of zero on the number line are negative and less than zero.

Theorem 8.3 outlined above demonstrates that our Axioms of Order using '<' model the concept of 'positive' and 'negative' rational numbers that we'd originally defined for integers in Chapter Three.

We will now use the Axioms of Order to help us model whether one rational number is greater than another. This process will permit us to obtain a strict order of the rational numbers along the number line.

Theorem 8.4   Ordering Rational Numbers Theorem

For all integers that make the inequalities '$0 < a$', '$0 < b$', '$0 < c$' and '$0 < d$' true, the rational numbers '$\frac{a}{b}$' and '$\frac{c}{d}$' make the inequality '$\frac{a}{b} < \frac{c}{d}$' true, if (and only if) they satisfy the inequality '$ad < bc$'.

Proof

This proof proceeds by mathematical deduction and involves a simple application of the axioms.

$$\frac{a}{b} < \frac{c}{d} \qquad \text{... given}$$

$$\frac{a}{b} - \frac{a}{b} < \frac{c}{d} - \frac{a}{b} \qquad \text{... Additive Axiom '<'}$$

$$\therefore \qquad 0 < \frac{c}{d} - \frac{a}{b} \qquad \text{... by simplifying a term}$$

Therefore, the inequality '$0 < \frac{c}{d} - \frac{a}{b}$' is true and our next step is to convert the rational numbers into integers since we have already observed (from Theorem 8.2) how to compare the value of two integers.

As '$0 < b$' and '$0 < d$' are true 'b' and 'd' are positive integers, we use the Multiplicative Axiom '<' to obtain the true inequality '$0 < bd$'. Using the Multiplicative Axiom '<', we now use the inequality '$0 < bd$' to simplify our previous inequality '$0 < \frac{c}{d} - \frac{a}{b}$' as follows:

$$0 < (\tfrac{c}{d} - \tfrac{a}{b})bd \qquad \text{... by Multiplicative Axiom '<'}$$

$$0 < \tfrac{c}{d} \times bd - \tfrac{a}{b} \times bd \qquad \text{... by Distributive Axiom}$$

$$0 < cb - ad \qquad \text{... by simplifying terms}$$

$$0 + ad < cb - ad + ad \qquad \text{... by Additive Axiom '<'}$$

$$ad < cb \qquad \text{... by simplifying terms}$$

All of the steps of this proof are reversible which implies that we could start with this last true inequality '$ad < cb$' and proceed to the first true inequality '$\frac{a}{b} < \frac{c}{d}$' by using deductive reasoning. In this way, we have completed the proof of Theorem 8.4 by proving the true inequalities '$ad < cb$' and '$\frac{a}{b} < \frac{c}{d}$' are equivalent.

————————————

Note:

We can easily obtain equivalent results for negative fractions as long as we use the Axioms of Order correctly.

Next we wish to derive some useful theorems that describe some general properties we would expect these inequalities to possess. In Theorem 8.5, we would expect that when we add the smaller numbers and the larger numbers of two true inequalities, '$a < b$' and '$c < d$', we will obtain a true inequality.

Theorem 8.5   Adding Inequalities Theorem

For all rational numbers assigned to the variables 'x', 'y', 'z' and 'w' such that '$x < y$' and '$z < w$' are true inequalities, and using the Axioms of Order, we can derive the following true inequality:

$$x + z < y + w$$

Proof

We begin with the first true inequality '$x < y$':

$$x < y \qquad \text{... given}$$

$$x - y < y - y \qquad \text{... by Additive Axiom '<'}$$

$$\therefore \qquad x - y < 0 \qquad \text{... by Inverse Axiom '+'}$$

Similarly, for the second true inequality '$z < w$' we get:

$$z < w \qquad \text{... given}$$

$$z - z < w - z \qquad \text{... by Additive Axiom '<'}$$

$$\therefore \qquad 0 < w - z \qquad \text{... by Inverse Axiom '+'}$$

You will note that we have made one term less than zero and the other term greater than zero in order to be able to apply the Transitive Axiom to these true inequalities. Using the Transitive Axiom, we now combine these two true inequalities '$x - y < 0$' and '$0 < w - z$' as follows:

$$x - y < w - z \qquad \text{... by Transitive Axiom '<'}$$

$$x - y + y + z < w - z + y + z \qquad \text{... by adding '$y + z$' to sides}$$

$$\therefore \qquad x + z < y + w \qquad \text{... by simplifying terms}$$

Consequently, '$x + z < y + w$' is a true inequality and this theorem is proved and demonstrates that we can 'add' inequalities of the same type.

——————————————

Next, we aim to derive a theorem that says when we multiply an inequality by a positive rational number, the resulting inequality will also be a true inequality. This concept is covered in Theorem 8.6 below.

Theorem 8.6   Multiplying an Inequality by a Positive Rational Number Theorem

For all rational numbers assigned to the variables 'x', 'y' and 'z' such that '$x < y$' is true and '$0 < z$' is true, then we can multiply the inequality '$x < y$' by 'z' to get the following true inequality:

$$x \times z < y \times z$$

Proof

We start by rearranging the true inequality '$x < y$' so that we can use the Multiplicative Axiom '$<$':

	$x < y$	... given
	$x - x < y - x$	... by Additive Axiom '$<$'
$\therefore$	$0 < y - x$	... by Theorem 4.11
Also:	$0 < z$	... given
Then:	$0 < (y - x) \times z$	... by Multiplicative Axiom '$<$'
	$0 < y \times z - x \times z$	... by Distributive Axiom '$\times$'
	$x \times z + 0 < y \times z - x \times z + x \times z$	... by Additive Axiom '$<$'
$\therefore$	$x \times z < y \times z$	... by simplifying terms

So this theorem proves that, provided 'z' is a positive rational number, we can 'multiply' both sides of a true inequality by a rational number variable 'z' and preserve the truth of the true inequality.

Note:

In Theorem 8.6 above if we assign 'z' a negative integer, then we have '$z < 0$'. We can rearrange '$z < 0$' to get '$0 < -z$', so that '$-z$' is positive and then apply this theorem to obtain:

$$-z \times x < -z \times y \qquad \text{... by Theorem 8.6}$$

Now, adding '$x \times z$' and '$y \times z$' to both sides of this inequality gives:

$\therefore$ $\qquad\qquad$ $y \times z < x \times z$ $\qquad\qquad\qquad$ ... by Additive Axiom '$<$'

Rearranging: $\quad$ $x \times z > y \times z$

Therefore, when multiplying an inequality by a negative integer, i.e. '$z < 0$', we have to reverse the inequality being multiplied.

Our next useful theorem (Theorem 8.7) ensures that the square of any rational number is always non-negative. For example, if 'x' is assigned the value '-3', then '$(-3)^2 = 9 \geq 0$'. We now prove this straightforward result.

Theorem 8.7 **Squares of Rational Numbers are Non-negative Theorem**

If 'x' is assigned any rational number, then '$x^2 \geq 0$' is always true.

Proof

Using the Trichotomous Axiom, we have '$0 < x$', '$x < 0$' or '$x = 0$'. We start this proof by assuming '$0 < x$' and proceed as follows:

Part 1:

$\qquad\qquad\qquad$ $0 < x$ $\qquad\qquad$ ... for the positive numbers

Then: $\qquad\qquad$ $0 < x \times x$ $\qquad\quad$ ... by Multiplicative Axiom '$<$'

$\therefore$ $\qquad\qquad\quad$ $0 < x^2$ $\qquad\qquad$ ... by simplifying terms

Therefore, '$0 < x^2$' is a true inequality.

Part 2:

$\qquad\qquad\qquad$ $x < 0$ $\qquad\qquad\qquad$ ... for the negative numbers

$\qquad\qquad$ $x - x < 0 - x$ $\qquad\qquad$ ... by Additive Axiom '$<$'

$\qquad\qquad\qquad$ $0 < -x$ $\qquad\qquad\quad$ ... by simplifying terms

$\qquad\qquad\quad$ $0 < -x \times -x$ $\qquad\quad$ ... by Multiplicative Axiom '$<$'

$\qquad\quad$ $0 < -1 \times x \times -1 \times x$ $\quad$ ... by Definition 3.4

$\qquad\quad$ $0 < -1 \times -1 \times x \times x$ $\quad$ ... by Commutative Axiom '$\times$'

$\qquad\qquad$ $0 < 1 \times x \times x$ $\qquad\quad$ ... by Theorem 3.3(3)

$\therefore$ $\qquad\qquad\quad$ $0 < x^2$ $\qquad\qquad\quad$ ... by simplifying a term

Therefore, '$0 < x^2$' is again a true inequality.

Part 3:

Where '$x = 0$', then '$x^2 = 0$' follows from '$0 \times 0 = 0$'.

In this way, we have proved that the square of any rational number is positive or zero (i.e. greater than or equal to zero).

We now continue to use the Axioms of Order to derive some basic theorems involving inequalities. Theorem 8.8 demonstrates how the negative sign '$-$' interacts with the inequality symbols in an inequality. This theorem formalises the note following Theorem 8.6.

Theorem 8.8   Negative of an Inequality Theorem

Given 'x' and 'y' are rational numbers, then '$y < x$' is true if (and only if) '$-x < -y$'.

Proof

This is a direct proof using the Axioms of Order as follows:

$$y < x \qquad \text{... given}$$

$$y - x - y < x - x - y \qquad \text{... by '} -x - y = -x - y\text{'}$$

$$\therefore \qquad -x < -y \qquad \text{... by simplifying both sides}$$

$$\therefore \qquad -y > -x$$

This end result leads to the unnecessary (but commonly taught) algorithm that states: "When multiplying both sides of an inequality by '-1', reverse the order symbol for that inequality". You will observe that once again this proof is completely reversible so that if '$-y > -x$' is true, then it follows '$y < x$' is true. This completes the proof.

Next, we prove another intuitive result; that is, a number that is greater than the sum of two positive rational numbers is always greater than either of the individual rational numbers. For example, for '$0 < 2$', '$0 < 3$' and '$2 + 3 < 6$', we have '$2 < 6$' and '$3 < 6$'. We will now prove this useful theorem.

Theorem 8.9 **Greater Than the Sum is Greater than Each Part of the Sum Theorem**

For all rational numbers 'x', 'y' and 'z' (where '$0 < x$', '$0 < y$' and '$x + y < z$' are true) we have: '$x < z$' and '$y < z$' are also true inequalities.

Proof

This proof follows by deductive reasoning as follows:

$$x + y < z \qquad \text{... given}$$

$$x < z - y \qquad \text{... by Additive Axiom '<'}$$

However, '$0 < x$' and '$x < z - y$' are ture inequalities, so that:

$$0 < z - y \qquad \text{... by Transitive Axiom '<'}$$

$$\therefore \qquad y < z \qquad \text{... by Additive Axiom '<'}$$

The same reasoning can be applied to making 'y' the subject of the true inequality '$x + y < z$' to show '$x < z$' is true as well. Hence, this proves the theorem that the rational number 'z' is greater than either of its parts 'x' or 'y'.

———————————

We will now prove the equivalent result to Theorem 8.5 for the multiplication operator. That is, we would expect that when we multiply the smaller numbers and the larger numbers of two inequalities we will obtain a true inequality if all the numbers are positive.

Theorem 8.10 **Multiplying Two Inequalities Theorem**

For all rational numbers 'x', 'y', 'z' and 'w' (where '$0 < x$', '$0 < y$', '$0 < z$', '$0 < w$' and '$x < y$' and '$z < w$' are true inequalities), we have the true inequality:

$$xz < yw$$

Proof

The proof by mathematical deduction proceeds as follows:

$$x < y \qquad \text{... given.}$$

$$0 < y - x \qquad \text{... by Additive Axiom '<'}$$

Similarly: $\qquad 0 < w - z \qquad \text{... by using '$z < w$'}$

$$\therefore \qquad 0 < (y - x)(w - z) \qquad \text{... by Multiplicative Axiom '<'}$$

$$0 < yw - yz - xw + xz \ldots \text{ by Distributive Axiom}$$

$$yz + xw - xz < yw \qquad \ldots \text{ by Additive Axiom '}<\text{'}$$

$$yz + xw - 2xz + xz < yw \qquad \ldots \text{ by adding '}-xz + xz\text{'}$$

$$(yz - xz) + (xw - xz) + xz < yw \qquad \ldots \text{ by collecting terms}$$

$$\therefore \quad z(y - x) + x(w - z) + xz < yw \qquad \ldots \text{ by Distributive Axiom}$$

Therefore, the two terms '$z(y - x)$' and '$x(w - z)$' on the left-hand side of the inequality shown above are both positive. Therefore, using Theorem 8.9, both of these positive terms can be dropped from the inequality as follows:

$$\therefore \qquad xz < yw$$

The other inequalities '$z(y - x) < yz$' and '$x(w - z) < yw$' are also true inequalities, however, they are not as intuitively obvious as '$xz < yw$'. Hence, this concludes the proof of the theorem.

———————

We now apply the Axioms of Order to illustrate how we can solve the general linear inequality.

Theorem 8.11 Solving the General Linear Inequality Theorem

For the integer variables 'a' and 'b' (where 'a' is not assigned '0') and the rational number variable 'x', we can write the two solutions of the linear inequality '$ax + b < 0$' as the true inequalities:

$$x < -\frac{b}{a} \text{ for '}0 < a\text{'}$$

or:

$$x > -\frac{b}{a} \text{ for '}a < 0\text{'}$$

Proof

We begin with the true general linear inequality '$ax + b < 0$' and apply the Axioms of Order to isolate 'x' as follows.

$$ax + b < 0 \qquad \ldots \text{ given}$$

$$ax + b - b < 0 - b \qquad \ldots \text{ by Additive Axiom '}<\text{'}$$

$$\therefore \qquad ax < -b \qquad \ldots \text{ by simplifying terms}$$

To isolate 'x' on the left-hand side of '$ax < -b$', we have to treat the two cases separately where 'a' could be positive or negative. First, we assume '$0 < a$' is true so that we can multiply both sides by '$\frac{1}{a} > 0$' as follows:

$$ax \times \frac{1}{a} < -b \times \frac{1}{a} \qquad \text{... by Multiplicative Axiom '<'}$$

$$\therefore \qquad x < -\frac{b}{a} \qquad \text{... by Identity Axiom '×'}$$

Second, by assuming '$a < 0$' is true, we then have to rearrange the general linear inequality to use '$-a$' so that '$0 < -a$' is true.

$$ax + b < 0 \qquad \text{... given}$$

$$ax + b - ax < 0 - ax \qquad \text{... by Additive Axiom '<'}$$

$$b < -ax \qquad \text{... by simplifying terms}$$

$$b \times \frac{1}{-a} < -ax \times \frac{1}{-a} \qquad \text{... by Multiplicative Axiom '<'}$$

$$\frac{b}{-a} < x \times 1 \qquad \text{... by simplifying terms}$$

$$\therefore \qquad -\frac{b}{a} < x \qquad \text{... by Theorem 4.18}$$

$$\therefore \qquad x > -\frac{b}{a} \qquad \text{... by Definition 8.2}$$

Therefore, the two possible true solutions of the general linear inequality are: '$x < -\frac{b}{a}$' if 'a' is positive and '$x > -\frac{b}{a}$' if 'a' is negative. This concludes the proof of the theorem.

When attempting to solve many practical mathematical problems, we often reduce that problem to finding the solutions that make the inequality '$0 < xy$' true for rational number variables 'x' and 'y'. The answer to this recurring question is provided in the following basic theorem (Theorem 8.12).

Theorem 8.12 Bilinear Inequality Theorem

If 'x' and 'y' are rational number variables and '$0 < xy$' is true, then either '$0 < x$' and '$0 < y$', or '$x < 0$' and '$y < 0$' must be true.

Proof

From Theorem 7.5, we know that if either '$x = 0$' or '$y = 0$' then '$xy = 0$' and, hence, the inequality cannot be true for these values.

Part 1:

We begin the proof by assuming that '$0 < y$' is true.

And:	$0 < xy$	... given
	$0 < \dfrac{1}{y}$	... by Theorem 8.3
	$0 < xy \times \dfrac{1}{y}$	... by Multiplicative Axiom '<'
$\therefore$	$0 < x$	... by simplifying terms

In this way, we have shown that if the inequality '$0 < xy$' is true, then '$0 < x$' and '$0 < y$' are also true.

Part 2:

To prove the second case, we begin the proof by assuming that '$y < 0$' so that '$0 < -y$' and, therefore, according to Theorem 8.3, '$0 < \dfrac{1}{-y}$'.

$0 < xy$	... given
$0 < \dfrac{1}{-y}$	... by Theorem 8.3
$0 < xy \times \dfrac{1}{-y}$	... by Multiplicative Axiom '<'
$0 < -x$	... by simplifying terms
$x < 0$	... by Additive Axiom '<'

Therefore, we have shown that if the inequality '$0 < xy$' is true, then '$x < 0$' and '$y < 0$' are true . In this way, we have shown both solutions to this inequality and so the theorem has been proved.

It is straightforward to apply the same reasoning shown above (but with the inequality reversed) to solve the true inequality for '$xy < 0$'.

We now use Theorem 8.12 to help us find the solution to one of the simplest quadratic inequalities.

Theorem 8.13   Solving a Quadratic Inequality Theorem

The inequality '$0 < x^2 - 1$' is true if and only if the following two inequalities are true:

'$x < -1$' or '$1 < x$'.

Proof

This proof involves a straightforward application of Theorem 7.3 and Theorem 8.12 as follows:

$$0 < x^2 - 1 \qquad\qquad \text{... given}$$

$$0 < (x-1)(x+1) \qquad \text{... by Theorem 7.3}$$

Therefore, either:

'$0 < (x-1)$' and '$0 < (x+1)$' are true by Theorem 8.12 where these two inequalities can be written as: '$1 < x$' and '$-1 < x$'. The values of 'x' that make both '$1 < x$ **and** '$-1 < x$' simultaneously true inequalities are expressed by the single inequality '$1 < x$'.

or:

'$(x-1) < 0$' and '$(x+1) < 0$' are true by Theorem 8.12 where these two inequalities can be written as: '$x < 1$' and '$x < -1$'. The values of 'x' that make both of the inequalities '$x < 1$' **and** '$x < -1$' simultaneously true inequalities are expressed by the single inequality '$x < -1$'.

Consequently, the solution to the quadratic inequality consists of the the values of 'x' that make either '$x < -1$' true **or** '$1 < x$' true. This completes the proof of the theorem.

We now demonstrate the usefulness of the absolute value operator by making use of this operator and then writing the solution in Theorem 8.13 more concisely. This representation of the solution forms Theorem 8.14.

Theorem 8.14   A Solution using the Absolute Value Operator Theorem

The values of 'x' which make either of the inequalities '$x < -1$' **or** '$1 < x$' true, can be written as the single true inequality:

$$|x| > 1$$

Proof

This proof is a straightforward application of the definition of the absolute value operator to the two inequalities as follows:

Part 1:

If '$1 < x$', then it follows that 'x' is positive and therefore we can substitute '$|x|$' for 'x' (from Definition 8.7) in the inequality '$1 < x$', and we get the inequality '$1 < |x|$'.

Part 2:

If '$x < -1$', then it follows that 'x' is negative and therefore we can substitute '$|x|$' for '$-x$' (from Definition 8.7) by rearranging '$x < -1$' into the inequality '$-x > 1$', and we get the inequality '$|x| > 1$'.

Therefore, using the absolute value operator we have a concise way of writing the two inequalities '$x < -1$' **or** '$x > 1$' as '$|x| > 1$'. In this way, the theorem is proved.

Note:

It is only possible to abbreviate two inequalities using the absolute value symbol as we have done in the previous theorem if both inequalities are of the same type, that is, they both come from the set of symbols $\{<, >\}$ or $\{\leq, \geq\}$. For example, the two inequalities '$x \leq -1$' and '$x > 1$' cannot be abbreviated using the absolute value symbols.

According to Theorem 8.14, the role of the absolute value operator can be generalised to express when a value assigned to a variable 'x' lies between the positive and negative values of a constant rational number. For example, we will show that we can generalise the result that '$-2 \leq x \leq 2$' can be written in the equivalent form '$|x| \leq 2$'.

Theorem 8.15   Express '$-a \leq x \leq a$' using the Absolute Value Operator Theorem

The values of 'x' which make both of the inequalities '$x \leq a$' **and** '$-a \leq x$' true simultaneously can be written in the compact notation '$-a \leq x \leq a$' or the equivalent compact notation:

$$|x| \leq a$$

Proof

We apply the definition of the absolute value operator to '$|x|$' as follows:

Part 1:

If '$0 \leq x$' **and** '$x \leq a$' then we can substitute '$|x|$' for 'x' in this last inequality and get: '$|x| \leq a$'.

Part 2:

If '$x < 0$' **and** '$-x \leq a$' then we can substitute '$|x|$' for '$-x$' in this last inequality and get: '$|x| \leq a$'.

Therefore, when using the absolute value operator, the inequality '$|x| \leq a$' expresses concisely the two inequalities given by '$x \leq a$' **and** '$-a \leq x$'. Consequently, the theorem is proved.

A very simple but useful theorem that follows on from Theorem 8.15 and that will be required in a later chapter of this book is the Linear Inequality Theorem for Rational Numbers (Theorem 8.16). This theorem says "that the sum of the absolute value of two numbers is always less than or equal to the sum of the absolute values of those two numbers individually". For example, if 'x' and 'y' are assigned the numbers '2' and '-3', respectively, we have: '$|2 + -3| \le |2| + |-3|$' which simplifies to '$|-1| \le |2| + |-3|$' or '$1 \le 2 + 3$'.

Theorem 8.16   Linear Inequality Theorem for Rational Numbers Theorem

For all rational numbers assigned to the variables 'x' and 'y', the following true inequality holds:

$$|x + y| \le |x| + |y|$$

Proof

From the definition of the absolute value, we observe that '$|x|$' is always greater than or equal to 'x' (i.e. '$x \le |x|$') and '$-|x|$' is always less than or equal to 'x' (i.e. '$-|x| \le x$'). Therefore, summarising this outcome in one inequality we have:

$$-|x| \le x \le |x| \qquad \text{... by summary notation}$$

Similarly:

$$-|y| \le y \le |y| \qquad \text{... by summary notation}$$

Adding these last two inequalities together gives the following:

$$-(|x| + |y|) \le x + y \le |x| + |y| \qquad \text{... by Theorem 8.5 for '}\le\text{'}$$

$$\therefore \qquad |x + y| \le |x| + |y| \qquad \text{... by Theorem 8.15}$$

Hence, the theorem is proved.

———————————

Another straightforward but useful theorem required in Chapter Nine, determines the product of the absolute value of two variables. An example of Theorem 8.17 would be: '$|2 \times -3| = |2| \times |-3|$' which is the same as '$|-6| = 2 \times 3$' or again '$6 = 2 \times 3$'.

Theorem 8.17   Product of Absolute Values Theorem

For all rational numbers assigned to the variables 'x' and 'y', the following true equality holds:

$$|x \times y| = |x| \times |y|$$

Proof

For the case where either '$x = 0$' or '$y = 0$' or both '$x = y = 0$', this equation clearly holds true since '$|0| = 0$'. We then consider the four cases where neither 'x' nor 'y' are equal to '0'. From the definition of the absolute value, we know that '$|x| = x$' if '$x > 0$' and '$|x| = -x$' if '$x < 0$'. So, at this point, our task is to consider the four possibilities for the positive and negative sign of 'x' and 'y' in their product, which are:

Case 1: '$x > 0$' and '$y > 0$'

Case 2: '$x > 0$' and '$y < 0$'

Case 3: '$x < 0$' and '$y > 0$'

Case 4: '$x < 0$' and '$y < 0$'

In the four cases below, we use bolding (e.g. '$\boldsymbol{x \times y}$') to emphasise the key result we will use to prove each case.

Case 1: According to the Multiplicative Axiom for '$<$', for '$x > 0$' and '$y > 0$' we have '$x \times y > 0$' so that, '$|x \times y| = \boldsymbol{x \times y}$'. Also, for '$x > 0$' and '$y > 0$', we have '$|x| = x$' and '$|y| = y$', so that, '$|x| \times |y| = \boldsymbol{x \times y}$'. Therefore, for this case: '$|x \times y| = |x| \times |y|$' by the Transitive Axiom for Rational Numbers.

Case 2: According to the Multiplicative Axiom for '$<$', for '$x > 0$' and '$-y > 0$' we have '$x \times -y > 0$' so that, '$|x \times y| = \boldsymbol{x \times -y}$'. Also, for '$|x| = x$' and '$|y| = -y$', we also have '$|x| \times |y| = \boldsymbol{x \times -y}$'. Therefore, for this case: '$|x \times y| = |x| \times |y|$' by the Transitive Axiom for Rational Numbers.

Case 3: This case is the same as Case 2 but with 'x' and 'y' interchanged. Consequently, for this case, we also have: '$|x \times y| = |x| \times |y|$'.

Case 4: For '$x < 0$' and '$y < 0$' we have '$0 < -x$' and '$0 < -y$' so that '$-x \times -y = x \times y > 0$' according to Theorem 3.3(3). Hence, '$|x \times y| = \boldsymbol{x \times y}$' and '$|x| = -x$' and '$|y| = -y$', so that we also have '$|x| \times |y| = -x \times -y = \boldsymbol{x \times y}$'. Therefore, for this case: '$|x \times y| = |x| \times |y|$' by the Transitive Axiom for Rational Numbers.

Hence, '$|x \times y| = |x| \times |y|$' in all cases and the theorem is proved for all possible cases.

———————————

Finally, it is often convenient to represent the solution to many inequalities graphically in order to create a simple picture of all those rational number values along a number line that make an inequality true. We will demonstrate this process by illustrating the solution for Theorem 8.14 in the following diagram:

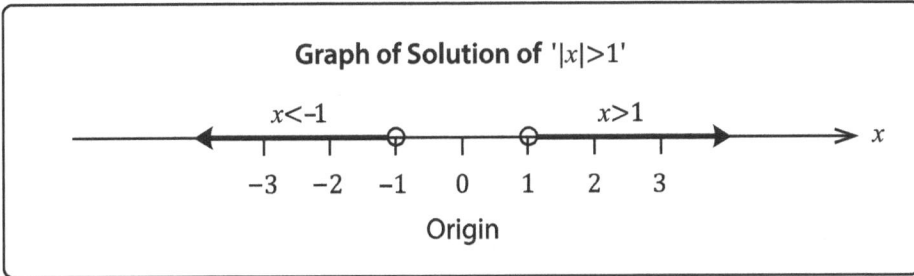

Graph of Solution of '|x|>1'

Figure 8.2

In Figure 8.2 we use an open circle around the points with coordinates '-1' and '1' to indicate that the coordinates of these points do not satisfy this inequality In contrast, we use two thick arrows to indicate all the other points with rational number coordinates that satisfy this inequality. In Chapter Eleven, we will be demonstrating that **not** every point along the number line can be represented by a rational number coordinate and, hence, this diagrammatic representation is somewhat misleading for rational numbers.

For the case where the point at the tail of the arrow is part of the solution of an inequality, we use a 'solid' circle as opposed to an 'open' circle. This result is demonstrated in Figure 8.3 below in which we illustrate the inequality consisting of the two inequalities '$-2 < x$' **and** '$x \leq 1$' which is often abbreviated to the single inequality '$-2 < x \leq 1$'.

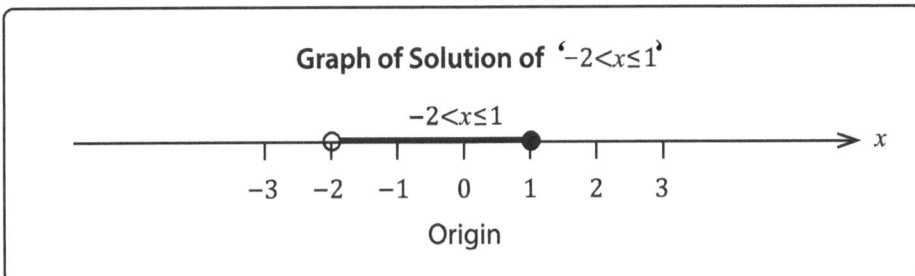

Graph of Solution of '$-2 < x \leq 1$'

Figure 8.3

In Figure 8.3, we used an open circle around the point with coordinate '−2' to indicate that this coordinate is not included in the inequality. In contrast, we use a solid circle around the point with coordinate '1' to indicate that this rational number is included in the inequality.

This pictorial representation of inequalities concludes our Theory Section of Rational Numbers using Order.

Applications of Rational Numbers using Order

In our everyday lives, there are many applications of the ordering operator with rational numbers we regularly use to make decisions. Ordering rational numbers along the number line is a key property of rational numbers that we will cover in the following applications.

Application 8.1

On average, your vehicle uses '8' litres of petrol per '100' kilometres and your petrol tank display indicates you have less than '$\frac{1}{2}$' of a '40'-litre tank of petrol remaining. How far can your vehicle travel with the available petrol in your tank?

First, we write the algebraic equation that gives the relationship of the litres 'L' of petrol in the tank to the distance 'd' in kilometers we can travel.

$$d = \frac{l \times 100}{8}$$

Note:

The distance given here is defined as undirected distance.

Since we have less than '$\frac{1}{2}$' a tank of petrol remaining, we know that '$l < 20$'. Therefore, the distance we can travel can be written as:

$$d < \frac{20 \times 100}{8}$$
$$\therefore \qquad d < 250$$

Therefore, our vehicle will be able to travel less than '250' kilometres.

_____

We now use the Axioms of Order to determine which number is the larger of two given rational numbers.

Application 8.2

How would you use an inequality to order the two rational numbers '$\frac{27}{31}$' and '$\frac{37}{41}$'?

First, we assume that one rational number is greater than another.

$$\frac{27}{31} < \frac{37}{41} \qquad \text{... by assumption}$$

$$\frac{27}{31} \times 41 \times 31 < \frac{37}{41} \times 41 \times 31 \qquad \text{... by Theorem 8.6}$$

$$27 \times 41 < 37 \times 31 \qquad \text{... by simplifying terms}$$

$$\therefore \qquad 1107 < 1147 \qquad \text{... by simplifying terms}$$

$$\therefore \qquad 1107 - 1107 < 1147 - 1107 \qquad \text{... by Additive Axiom '<'}$$

$$\therefore \qquad 0 < 40 \qquad \text{... by simplifying terms}$$

This is a true inequality according to Theorem 8.2. Therefore, our original assumption must have been true and the order of these two rational numbers can be given by the true inequality '$\frac{27}{31} < \frac{37}{41}$'.

Application 8.3

Write inequalities that model the following statements:

1. A rational number variable 'x' is assigned a number greater than two and a half or a number less than minus one and a half. What values of 'x' make this statement true?

2. A rational number variable 'y' is assigned a number less than or equal to two and a half and a number greater than minus one and a quarter. What values of 'y' make this statement true?

3. A rational number variable 'x' is assigned a number greater than two and a half and a number less than minus one and a quarter. What values of 'x' make this statement true?

4. A rational number variable 'y' is assigned a number less than or equal to two and a half or a number greater than minus one and a quarter. What values of 'y' make this statement true?

Question 1

The inequalities in 'x' that make Statement 1 shown above a true statement are: '$x > 2\frac{1}{2}$' or '$x < -1\frac{1}{2}$'.

Note:

With some work, we can write these two conditions in compact absolute value notation. We begin this process by rearranging these inequalities in 'x' where '$x > 0$' and '$-x$' where '$x < 0$'.

First, for '$x > 0$':

$$x > 2 + \frac{1}{2} \qquad \text{... by '} 2\frac{1}{2} = 2 + \frac{1}{2} \text{'}$$

$$\therefore \qquad x - \frac{1}{2} > 2 \qquad \text{... by adding '} -\frac{1}{2} \text{' to both sides}$$

Second, for '$x < 0$':

$$x < -1\frac{1}{2} \qquad \text{... given}$$

$$1\frac{1}{2} < -x \qquad \text{... by adding to both sides}$$

$$\therefore \qquad -x > 1\frac{1}{2} \qquad \text{... by Definition 8.2}$$

$$\therefore \qquad -x + \frac{1}{2} > 1\frac{1}{2} + \frac{1}{2} \qquad \text{... by adding '} \frac{1}{2} \text{' to both sides}$$

$$\therefore \qquad -(x - \frac{1}{2}) > 2 \qquad \text{... by simplifying both sides}$$

In this way, the two true inequalities '$x - \frac{1}{2} > 2$' **or** '$-(x - \frac{1}{2}) > 2$' satisfy the condition of the absolute value operator definition and so we can write (in compact notation) the conditions that must be satisfied by 'x' as: '$|x - \frac{1}{2}| > 2$'.

———————————

Question 2

The values of 'y' that make Statement 2 shown above a true statement are: '$y \leq 2\frac{1}{2}$' and '$y > -1\frac{1}{2}$'. These two inequalities can be combined in order notation as:

$$-1\frac{1}{2} < y \leq 2\frac{1}{2}$$

———————————

Question 3

The values of 'x' that make Statement 3 shown above a true statement are: '$x > 2\frac{1}{2}$' and '$x < -1\frac{1}{4}$'. Since these two true inequalities don't overlap and can't both be true simultaneously, there are no values of 'x' that make this statement true.

———————————

Question 4

The values of 'y' that make Statement 4 shown above a true statement are: '$y \leq 2\frac{1}{2}$' or '$y > -1\frac{1}{4}$'. For every value assigned to 'y', it must make one of these two inequalities true. Therefore, all values of 'y' make this statement a true statement.

———————

Application 8.4

Find the values of 'x' that make the inequality '$x^2 - 1 < 3$' a true inequality. Write the solutions to this inequality using the definition of the absolute value.

We write the left-hand side of this inequality in terms of its factors as follows:

$$x^2 - 1 < 3 \qquad \text{... given}$$

$$x^2 - 4 < 0 \qquad \text{... by the Additive Axiom '}<\text{'}$$

$$\therefore \quad (x-2)(x+2) < 0 \qquad \text{... by Theorem 7.3}$$

Since we know that the product of a positive and a negative number is the only way to get a negative number result from the product of two numbers, we have to consider the two cases:

Case 1: where '$x - 2$' is positive and '$x + 2$' is negative

Case 2: where '$x - 2$' is negative and '$x + 2$' is positive.

We examine the two cases:

1. When '$x - 2$' is positive **and** '$x + 2$' is negative.

   Rewriting these two conditions using the order symbol gives:

   $$\text{'}0 < x - 2\text{' and '}x + 2 < 0\text{'} \qquad \text{... given}$$

   $$\therefore \quad \text{'}2 < x\text{' and '}x < -2\text{'} \qquad \text{... by Additive Axiom '}<\text{'}$$

   There are no values we can assign to 'x' which will make both of these inequalities true at the same time (i.e. simultaneously).

2. When '$x - 2$' is negative **and** '$x + 2$' is positive.

   Rewriting these two conditions using the order symbol gives:

   $$\text{'}x - 2 < 0\text{' and '}0 < x + 2\text{'} \qquad \text{... given}$$

   $$\therefore \quad \text{'}x < 2\text{' and '}-2 < x\text{'} \qquad \text{... by Additive Axiom '}<\text{'}$$

Using Definition 8.7, we can rewrite these last two inequalities more compactly as: '$|x| < 2$'.

Therefore, the solutions to the inequality '$x^2 - 1 < 3$' are values of 'x' that satisfy the inequality '$|x| < 2$'.

———————

Application 8.5

Assume a person named Fred lives in a house beside a straight road which runs east-west with a numbering system (expressed in kilometres) which increases in the easterly direction. Fred has a friend Joe, who lives in a house '20km' due west of his house and another friend Mary, who lives in a house '50km' due east of his house. What is the directed and undirected distance from Mary's house to Joe's house?

We use Definitions 8.5 and 8.6 to calculate the directed and undirected distances, respectively. By using the local numbering system, we can summarise the information on the coordinates associated with Fred's, Joe's and Mary's houses which we will denote by 'x_F', 'x_J' and 'x_M', respectively. This coordinate information is:

$$x_J = x_F - 20 \qquad \text{... by Joe being west of Fred}$$

$$x_M = x_F + 50 \qquad \text{... by Mary being east of Fred}$$

The question asks for the directed distance from Mary's house (starting position 'x_M') to Joe's house (finishing position 'x_J'), which we denote by 'r'.

Applying Definition 8.5 gives:

$$r = x_J - x_M \qquad \text{... by Definition 8.5}$$
$$= (x_F - 20) - (x_F + 50) \qquad \text{... by substit. for 'x_J' and 'x_M'}$$
$$\therefore \quad r = -70 \qquad \text{... by simplifying terms}$$

Hence, the directed distance from Mary's house to Joe's house is '-70km'.

To calculate the undirected distance, we use the absolute value as follows:

$$|x_J - x_M| = -(x_J - x_M) \qquad \text{... by Definition 8.7}$$
$$= -(-70) \qquad \text{... by value of 'd' above}$$
$$\therefore \quad |x_J - x_M| = 70$$

Therefore, the undirected distance from Mary's house to Joe's house is '70km'.

———————

This completes the application of the ordering of rational numbers to some simple questions we may want to answer in our everyday lives.

Summary of Rational Numbers using Order

At this point, we have covered the objectives of this chapter and so can summarise our findings up to now. In terms of the basic concepts and theory of the ordering of the rational numbers, the most important definitions we have introduced in this chapter are:

1. Defining an **inequality** as a finite valid string of symbols that consists of a term on either side of an order symbol.

2. Implicitly defining the **less-than order symbol** symbol '$<$' through the Axioms of Order.

3. Defining **order symbols** that are used in inequalities to describe the order between two rational numbers. These order symbols are used to represent inequalities as follows:

   '$x \leq y$' is true if and only if '$x < y$' **or** '$x = y$'

   '$x > y$' is true if and only if '$y < x$'

   '$x \geq y$' is true if and only if '$y < x$' **or** '$x = y$'.

4. Defining the **general linear inequality** in the variable 'x' as the algebraic inequality '$ax + b < 0$'.

5. Defining the **general quadratic inequality** in the variable 'x' as the algebraic inequality '$ax^2 + bx + c < 0$'.

6. Defining the **directed distance** 'r' from a point with coordinate 'x' to a point with coordinate 'y' as '$r = y - x$'.

7. Defining the **undirected distance** 'd' from a point with coordinate 'x' to a point with coordinate 'y' as '$d = y - x$' if '$x < y$' and '$d = x - y$' if '$x \geq y$'.

8. Defining the **absolute value operator** for two rational number 'x' and 'y' as the same as the undirected distance between two points with those coordinates 'x' and 'y', and denoted by '$|y - x|$' or '$|x - y|$', which is the positive difference of these two coordinates.

These definitions require the Axioms of Order to assign meaning to their symbols. These Axioms of Order also provide the rules for manipulating arithmetic and algebraic inequalities. These ordering symbols, with their associated absolute value operator, will be crucial to developing the concepts in Chapter Nine.

We can now summarise the Axioms of Order as outlined in Table 8.5 below.

8.5 The Axioms of Order for the Rational Number System	
For the rational number variables 'x', 'y' and 'z', the order symbol '$<$' satisfies the following Axioms:	
1. Trichotomous Axiom for '$<$'	Either '$x < y$', '$y < x$' or '$x = y$' (and only one of these three statements can be **true** for each assignment of a rational number to each variable and the other two statements are **false**).
2. Transitive Axiom for '$<$'	Given '$x < y$' and '$y < z$' are true inequalities, then '$x < z$' is also a **true** inequality.
3. Additive Axiom for '$<$'	Given '$z = z$' and '$x < y$' then '$x + z < y + z$'. If one of these two inequalities is **true**, then the other is **true**. If one is **false**, then the other is **false**.
4. Multiplicative Axiom for '$<$'	Given '$0 < x$' and '$0 < y$' then '$0 < xy$'. If any two of these inequalities are **true**, then the other inequality is also **true**. If one is **true** and another is **false** then the other is **false**.

Table 8.5

The basic theorems in this chapter have been derived by applying the language, elementary properties, axioms and definitions (LEAD) approach to further develop our mastery over the Rational Number System. The axioms we have primarily focused on this chapter were the Axioms of Order. Once again, the Axioms of Logic have still not changed in form since Table 2.5.

This new order operation is fundamental to modelling the rational numbers and will be indispensable to developing the definition of the 'limit' in Chapter Nine. We will use this concept of the 'limit' to properly define 'repeating decimals' in Chapter Ten in order to provide a complete alternate representation of rational numbers as decimal numbers.

CHAPTER 9

SEQUENCES, SERIES AND LIMITS

Overview of Sequences, Series and Limits

The mathematical concepts we will be exploring in this chapter are 'Sequences', 'Series' and 'Limits'; these concepts are essential to understanding the Rational and Real Number Systems. We will continue to develop the tools we require to fully understand the rational numbers represented as either finite or repeating decimal numbers (which will be covered in Chapter Ten) and to allow us to investigate their natural extension to the 'Real Numbers' (which will be covered in Chapter Twelve).

Once again, we will be using the operations of '+' and '×' to manipulate the terms of 'Sequences', 'Series' and 'Limits' as a means of finding their sums or values. In this chapter, the terms of the 'Sequences', 'Series' and 'Limits' will be rational numbers and rational number variables. In this way, all our work in previous chapters on the Rational Number System, Algebra and Order will help us understand the nature of rational numbers when expressed as decimal numbers.

All sequences used in this book are represented as comma-separated lists of terms between parentheses – i.e. $(t_1, t_2, \ldots, t_n, \ldots)$ where 't_1', 't_2', up to 't_n' and so on are rational number terms. We use sequences of numbers on a daily basis without even realising it. For instance, a mobile telephone number is typically a sequence of ten digits and can be written as shown in the following example as: '0412 356 789'. In Mathematics, this sequence would be written in the specific format as the following sequence: $(0, 4, 1, 2, 3, 5, 6, 7, 8, 9)$. However, if the order of the terms in this sequence is altered, it becomes a different sequence and hence it would correspond to a different telephone number.

Likewise, we use series of numbers on a daily basis when we are adding up numbers. For example, if we purchase several items and need to know how much these will cost, we use a series in order to obtain the answer. For example: $1.50 plus $2.00 plus $1.75 would be written in mathematical notation as the series: '$1.50 + 2.00 + 1.75$'.

Background and Context of Sequences, Series and Limits

The mathematical background behind the use of 'sequences' and the sum of the terms of this sequence (which is called a 'series') harks back to the early development of Mathematics; to Euclid's Elements developed in the 3rd Century BC During this period, mathematicians sought to discover the surface area of solid figures such as cones and spheres.

The great mathematician Archimedes (287 BC – 212 BC) used sums of rational numbers to reveal certain areas of geometrical objects. As a result of Archimedes' work, we have the first recorded infinite sequence – $\left(1, 4^{-1}, 4^{-2}, 4^{-3}, \dots, 4^{-n}, \dots\right)$ with its corresponding sum – as the following series:

$$1 + 4^{-1} + 4^{-2} + 4^{-3} + \dots + 4^{-n} + \dots$$

The work done by Archimedes on adding infinitely smaller and smaller numbers predates (by nearly 2,000 years) the work of other great mathematicians such as Newton (1642 – 1627) who generalised this earlier mathematical work in order to develop his theory of mechanics and gravitation.

A very simple series is associated with a famous story about one of the greatest mathematicians of all time – Karl Friedrich Gauss (1777 – 1855). Apparently, while he was in his early years at school, his teacher asked his class to add up the sequence of natural numbers from '1' to '100' inclusive. After a few minutes Gauss handed his teacher the correct solution to this problem on a piece of paper; no doubt to their complete astonishment! After Definition 9.9, you will learn a simple way of adding up this sequence.

Approach to Sequences, Series and Limits

So far, our broad approach to the elementary number systems has been to use finite valid strings of symbols to represent mathematical terms, equations and inequalities. However, in Chapter Six we saw that this approach appears not to work when we tried to express the simple fraction, '$\frac{1}{3}$' as a decimal since it results in an 'infinite' string of numbers (i.e. '0.333...').

Following on from Archimedes' work, it took mathematicians a further 2,000 years to discover a way of adding up an unlimited (or informally an 'infinite') set of numbers by using finite operations.

The key elements of this discovery were:

- defining the concept of the 'limit'
- using the **finite** operations of the 'limit' concept and **finite** valid strings from the alphabet to add up an unlimited number of terms

- avoiding the formal use of the term 'infinite' (which is not a number) but, instead, using the infinity symbol '∞' informally to represent an unlimited number of terms.

Note:

The property of being unlimited in the Natural Numbers follows from the Closure Axiom for Addition which assures us that if 'k' is any natural number, then the next term '$k + 1$' is also a natural number.

Language of the Rational Number System

The symbols we require to extend the Rational Number System language to include sequences, series and limits will be outlined in the Definition Section later in this chapter. The current symbols of the alphabet are summarised in Table 9.1 below.

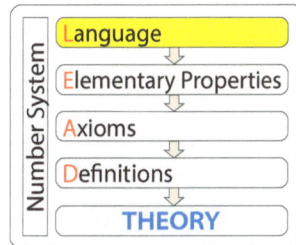

9.1 The Alphabet of the Rational Number System	
Symbols	**Meaning from:**
'0', '1', '+', '×', '(', ')'	The Axioms of the Natural Number System
'=', '≠'	The Axioms of Logic
'2', '3', '4', '5', '6', '7', '8', '9'	Definition 2.1
'a', 'b', 'c', ... , 'x', 'y', 'z', ...	The rational number variables
' '	Definition 2.3
'-1'	The Axioms of the Integer Number System
'$-$'	Definitions 3.4 and 3.5
'$\frac{1}{\Box}$'	The Axioms of the Rational Number System as Fractions
'$\frac{a}{b}$'	Definition 4.2
'÷'	Definition 4.4
'$\Box^{-1}$'	The Axioms of Rational Number System using Index Form
'$(a \times b^{-1})^m$'	Definition 5.6
'.'	Definition 6.4

9.1 The Alphabet of the Rational Number System	
'$a_n \ldots a_1 a_0 . d_1 \ldots d_m$'	Definition 6.5
'$\pm$'	Definition 7.8
'$<$'	The Axioms of the Rational Number System using Order
'$\leq$', '$>$', '$\geq$'	Definition 8.2
'$\lvert x - y \rvert$'	Definition 8.7

Table 9.1

Order of Operations

As we are using the Rational Number System to define sequences, series and limits, we use the same Order of Operations convention, '**PEMDAS**', that we applied to the Rational Number System in Chapters Four and Five.

The Elementary Properties of the Rational Number System

When considering the Rational Number System, there are no additional elementary properties required to represent sequences, series or limits. Consequently, we will continue to use the elementary properties we developed for Rational Numbers using Fractions and Index Form to manipulate the terms and equations of sequence, series and limits.

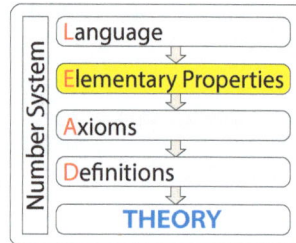

Number System
Language
Elementary Properties
Axioms
Definitions
THEORY

The Axioms of the Rational Number System

The basic operations of addition and multiplication of rational numbers using sequences, series and limits are given by the same axioms for the rational numbers used in Chapters 4 and 5. Therefore, we can use the Axioms of the Rational Number System using Fractions and Index Form to model the behaviour of the rational numbers in sequences, series and limits.

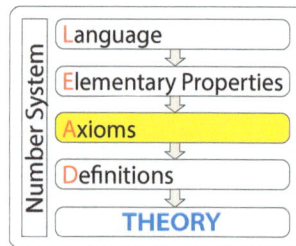

Number System
Language
Elementary Properties
Axioms
Definitions
THEORY

The Axioms of Logic

As in previous chapters, we require the Axioms of Logic to manipulate algebraic terms and equations. Since our sequences, series and limits will be expressed as arithmetic and algebraic terms or equations, we will be referring to the same

Axioms of Logic we used in earlier chapters of this textbook (i.e. Table 2.5 to manipulate sequences, series and limits).

Definitions of Sequences, Series and Limits

In this chapter, the most fundamental concept we will be exploring is a sequence. A sequence is constructed from basic building blocks known as 'sets'. It is an ordered set of terms which can be constructed with natural numbers, integers or rational numbers. In this chapter we will be focusing on sequences with terms constructed from the Rational Number System.

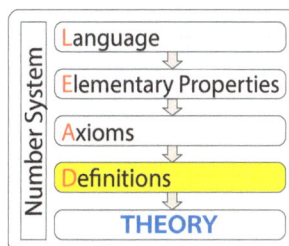

Definition 9.1   Set

A set is any collection of objects (which are also called elements) that can be identified as being different from one another. The elements in a set can consist of any objects such as numbers, terms, people, places, etc. However, in this textbook we will only focus on sets whose elements are mathematical '**terms**' such as: '7', '2×3', '25^2', '$\frac{n}{7}$', '3×10^{-n}'

A **set** of terms, e.g. $\{7, 2 \times 3, 25^2, \frac{n}{7}, 3 \times 10^{-n} \}$, has a left and right brace '{' and '}' respectively, at either end of it and the terms themselves are separated by commas. These terms are **not** required to be in any particular order.

For example, the numbers '1.5', '0.7', '6' and '3' can be written in set notation as: '$\{1.5, 0.7, 6, 3\}$'. However, since sets are not ordered, this set can also be written as:

$$\{1.5, 0.7, 6, 3\}, \{6, 0.7, 3, 1.5\}, \{3, 6, 0.7, 1.5\} \text{ and } \{3, 6, 1.5, 0.7\}$$

In other words, the set of terms can be written in any order without the identity of the set being altered. We will now define the concept of ordered sets where the terms of the set will be arithmetic or algebraic terms. These ordered sets are called 'sequences'.

Definition 9.2   Sequence

A sequence is a **set** of **terms** in a **specific order** (also called an **ordered set**). This order is achieved by assigning a unique label from the Natural Numbers to each of the terms in the sequence. The first **term** in the sequence is labelled as 'a_1', the second **term** in the sequence is labelled 'a_2', and so on through to the last **term** which is labelled as 'a_n' assuming there are 'n' **terms** in the sequence (we could have used 't_1', 't_2', etc.).

A sequence may be either finite or infinite as follows:

1. A sequence which has a **finite** number of terms is called a finite sequence and is written as the ordered set $(a_1, a_2, a_3, \dots, a_n)$. In this case, the last term in the sequence is the nth term in the sequence.

2. A sequence which has an **infinite** (i.e. unlimited) number of terms is called an infinite sequence and is written as the ordered set $(a_1, a_2, a_3, \dots)$. In this case, there is no last term in the sequence.

Note:

1. Sequences can also have the (subscripted) index of their terms start with '0' or any other natural number.

2. A sequence of terms has parentheses '$($' and '$)$' at either end of it and the terms themselves are separated by commas, for example $(a_1, a_2, a_3, \dots, a_n)$. This is the second way in which we use the parentheses symbols. You will recall the way in which we first used parentheses symbols was as grouping symbols to control the order of operations.

3. This definition ensures that the four **sets** of terms (according to Definition 9.1) are different when treated as sequences. That is, the following four sequences are all different even though they contain the same terms: $(1.5, 0.7, 6, 3)$, $(6, 0.7, 3, 1.5)$, $(3, 6, 1.5, 0.7)$ and $(3, 6, 0.7, 1.5)$.

As writing out a sequence is often a long and tedious process, mathematicians have introduced an abbreviated or shorthand notation for the sequence $(a_1, a_2, a_3, \dots, a_n)$ using the combination of symbols '$(a_k)_1^n$'. In the combination of symbols '$(a_k)_1^n$' the subscript 'k' indicates that all terms in the sequence are represented. The numbers to the lower and upper right of the parentheses indicate that the first term starts with a subscript of '1' and the last term finishes at the subscript 'n'. Some authors also use the notation '$(a_k)_{k=1}^{k=n}$' or '$(a_k)_{k=1}^n$' to represent '$(a_k)_1^n$'.

Some examples of finite sequences which use the shorthand notation described above for generating sequences include:

$(k^2)_{k=1}^{k=n}$ represents $(1^2, 2^2, 3^2, \dots, n^2)$ ... squares of natural numbers

$(2k-1)_{k=1}^6$ represents $(1, 3, 5, 7, 9, 11)$ ... six odd natural numbers

$(2k)\,_{k=1}^{n}$ represents $(2, 4, 6, \ldots, 2n)$ … 'n' even natural numbers

To allow us to write infinite sequences (i.e. no last term) in an abbreviated form, we will be introducing a new symbol in Definition 9.3 below.

Definition 9.3 Infinity Symbol '∞'

The **infinity** symbol '∞' is a symbol used to indicate that a sequence has an unlimited (i.e. infinite) number of terms, but it is **not** a number of any elementary number system.

Notes:

Now that we have the infinity symbol, we can abbreviate an infinite sequence as $(a_1, a_2, a_3, \ldots)$ as '$(a_k)_1^{\infty}$'.

Here, the infinity symbol '∞' has replaced the 'n' in the finite sequence notation so that the infinite sequence goes on to 'infinity' (i.e. it has no last term) rather than stopping at the nth term. Once again, some authors also use the notation '$(a_k)_{k=1}^{\infty}$' to represent '$(a_k)_1^{\infty}$'.

The simplest example of a sequence is the finite sequence of decimal digits used with the natural numbers – namely $(0, 1, 2, 3, \ldots, 9)$. The simplest example of an infinite sequence is the sequence of natural numbers: $(0, 1, 2, 3, \ldots, 9, \ldots)$.

One of the most useful examples of an infinite sequence that we will be exploring later in this chapter is the sequence of reciprocals of the natural numbers excluding zero – that is: $(\frac{1}{1}, \frac{1}{2}, \frac{1}{3}, \ldots, \frac{1}{9}, \frac{1}{10}, \ldots)$ which can also be abbreviated to '$(\frac{1}{k})_{k=1}^{\infty}$'.

Some other useful examples of infinite sequences are:

$(k^2)_{k=1}^{\infty}$ represents $(1^2, 2^2, 3^2, \ldots)$ … squares of natural numbers

$(2k+1)_{k=0}^{\infty}$ represents $(1, 3, 5, \ldots)$ … the odd natural numbers

$(2k)_{k=1}^{\infty}$ represents $(2, 4, 6, \ldots)$ … the even natural numbers

In summary, in the case of a sequence the order of its terms is critical, whereas in the case of a set the order of its terms is irrelevant. It is important to remember that 'sequences' are represented by the use of parentheses and that 'sets' are represented by the use of braces. Furthermore, the terms of the sequence are ordered by the natural numbers. However, we can start this ordering process with '0', '1' or any other natural number in the same way we did in the previous example above where 'k' started with the value '1'.

Now that we have defined sequences, we will define the operations of addition and multiplication of these sequences in Definition 9.4 below. We start this process by adding corresponding terms in the two sequences.

Definition 9.4   Sum of Two Sequences

Let $(a_1, a_2, a_3, \dots, a_n)$ and $(b_1, b_2, b_3, \dots, b_n)$ be two **finite** sequences of rational number terms. We define addition of these finite sequences by the sequence:

$$(a_1 + b_1, a_2 + b_2, a_3 + b_3, \dots, a_n + b_n)$$

Similarly, to add two **infinite** sequences, we simply add their corresponding terms.

––––––––––––––––––––

A further useful definition for manipulating sequences is to multiply a sequence by a constant term. This concept is outlined in Definition 9.5 below.

Definition 9.5   Product of a Sequence by a Constant

Let $(a_1, a_2, a_3, \dots, a_n)$ be a **finite** sequence of rational number terms and 'c' be a constant term. We define multiplication of this finite sequence by this constant term as follows:

$$(ca_1, ca_2, ca_3, \dots, ca_n)$$

Similarly, to multiply an **infinite** sequence by a constant, we simply multiply each term of that sequence by that constant.

––––––––––––––––––––

The third important operation for manipulating sequences is to multiply two sequences (i.e. multiply their corresponding terms). This concept is outlined in Definition 9.6 below.

Definition 9.6   Product of Two Sequences

Let $(a_1, a_2, a_3, \dots, a_n)$ and $(b_1, b_2, b_3, \dots, b_n)$ be two **finite** sequences of rational number terms. We define multiplication of these finite sequences by the sequence:

$$(a_1 b_1, a_2 b_2, a_3 b_3, \dots, a_n b_n)$$

Similarly, to multiply two **infinite** sequences, we simply multiply their corresponding terms.

––––––––––––––––––––

Now that we have covered general sequences and some operations defined on these sequences, we now divide these sequences into subsets called:

- Arithmetic sequences (both finite and infinite)
- Geometric sequences (both finite and infinite)
- Other sequences (both finite and infinite).

We will be defining the terms 'arithmetic' and 'geometric' sequence as they are the most common and useful sequences in the elementary number systems.

An **arithmetic sequence** is a sequence where the difference between each consecutive term in the sequence is a constant term. Definition 9.7 sets out the formal definition for an arithmetic sequence.

Definition 9.7   Arithmetic Sequence

Let a finite sequence of rational number terms be represented by $\left(a_1, a_2, a_3, \ldots, a_n\right)$. We call this sequence a **finite arithmetic sequence** if it can be written with a constant term '$a_1 = a$' the and subsequent terms can be derived by adding the non-zero constant term 'd' as follows:

$$\left(a_1, a_2, a_3, \ldots, a_n\right) = \left(a, a + d, a + 2d, \ldots, a + (n-1)d\right)$$

For example, the sequence $\left(1, 1+3, 1+2\times3, \ldots, 1+(50-1)\times3\right)$ where '$a = 1, d = 3, n = 50$' is a finite arithmetic sequence.

Similarly, an **infinite arithmetic sequence** is an infinite sequence represented by $\left(a_1, a_2, a_3, \ldots, a_n, \ldots\right)$ that can be written with a first term '$a_1 = a$' and subsequent terms can be derived by adding the constant term 'd' as follows:

$$\left(a, a + d, a + 2d, \ldots, a + (n-1)d, \ldots\right)$$

Notes:

1. Definition 9.7 is referred to as a recursive definition because we have specified how to generate each subsequent term in the sequence after the constant term 'a' by adding a constant term 'd'. This recursive relationship can be formalised by the equations '$a_1 = a$' and '$a_{k+1} = a_k + d$', where we assign 'k' the value '1' to find 'a_2' from the recursive equation as follows: '$a_{1+1} = a_2 = a_1 + d = a + d$'.

2. In the case of a **finite** arithmetic sequence, we can use an alternate notation for writing a sequence where 'a_k' is given by '$a + (k-1)d$', so that:

$$\left(a + (k-1)d\right)_{k=1}^{n} \text{ represents } \left(a, a + d, a + 2d, \ldots, a + (n-1)d\right)$$

3. In the case of an **infinite** arithmetic sequence, we can use the simplified notation introduced earlier so that:

$$\left(a + (k-1)d\right)_{k=1}^{\infty} \text{ represents } \left(a, a+d, a+2d, \dots, a+(n-1)d, \dots\right)$$

The other standard type of sequence is the '**geometric sequence**' which is outlined in Definition 9.8 below.

Definition 9.8   Geometric Sequence

Let a finite sequence of rational number terms be given by $\left(a_1, a_2, a_3, \dots, a_n\right)$. We call this sequence a **geometric sequence** if it can be written with a constant term 'a' (i.e. '$a_1 = a$') and subsequent terms of the sequence are generated by multiplying by a constant rational number 'r' as follows:

$$\left(a, ar^1, ar^2, \dots, ar^{n-1}\right)$$

For example, the sequence $\left(2, 2 \times 3^1, 2 \times 3^2, \dots, 2 \times 3^{50-1}\right)$ where '$a = 2, r = 3$, $n = 50$' is a finite geometric sequence.

———————————

Notes:

1. Once again, Definition 9.8 is a recursive definition as we have specified how to generate subsequent terms in the sequence from the preceding term by multiplying by a constant term 'r'. This recursive relationship can be formalised by the equations '$a_1 = a$' and '$a_{k+1} = a_k r$', up to the last term '$a_n = ar^{n-1}$'.

2. In the case of a **finite** geometric sequence, we can again use an alternate notation for writing this sequence as:

$$\left(ar^{k-1}\right)_{k=1}^{n} \text{ represents } \left(a, ar^1, ar^2, \dots, ar^{n-1}\right)$$

3. In the case of an **infinite** geometric sequence, we can use the simplified notation introduced earlier so that:

$$\left(ar^{k-1}\right)_{k=1}^{\infty} \text{ represents } \left(a, ar^1, ar^2, \dots, ar^{n-1}, \dots\right)$$

The simplest and most familiar example of an infinite geometric sequence is given by the decimal terms contained in the fraction '$\frac{1}{3}$', namely: $\left(3 \times 10^{-1}, 3 \times 10^{-2}, 3 \times 10^{-3}, \dots\right)$.

There are numerous other types of sequences that are not 'arithmetic' or 'geometric' sequences. The simplest example of this type of sequence (i.e. one that is not arithmetic or geometric) is the reciprocal of the positive integers given by the sequence $\left(\frac{1}{1}, \frac{1}{2}, \frac{1}{3}, \frac{1}{4}, \frac{1}{n}, ...\right)$.

The next key concept of this chapter, a 'series', is defined in Definition 9.9 below. A series is defined in terms of a sequence and simply refers to the sum of that sequence.

Definition 9.9 Series

A series is the sum of the terms of a sequence. A series can be either finite or infinite.

1. A finite series is the sum of a finite sequence $\left(a_1, a_2, a_3, ... , a_n\right)$ and gives the definition of the finite sum 'S_n' as a rational number by:

$$S_n = a_1 + a_2 + a_3 + ... + a_n$$

2. An infinite series represents the sum of an infinite sequence $\left(a_1, a_2, a_3, a_4, ... , a_n, ...\right)$ and defines the rational number 'S' if this sum exists (i.e. the sum may not be a rational number). In this case, we write:

$$S = a_1 + a_2 + a_3 + ... + a_n + ...$$

_____

Notes:

1. Our mathematical language only supports finite valid strings of symbols. Therefore, the notation in the second bullet point above (i.e. using 'S' to represent this infinite sum) is used to indicate our intention to assign meaning to 'S' by using a finite valid string of symbols in the 'limit process' (which we will be covering later in this chapter).

2. In the case where our infinite sequence is an infinite arithmetic sequence, the sum of the sequence does not exist because each term is increasing by the non-zero amount 'd' so the sequence increases indefinitely. This concept will be explained in more detail when we define a limit in Definition 9.10.

We now illustrate the definition of a finite series by demonstrating how we can sum the first 'n' terms of a sequence. In order to find the sum of the sequence of positive integers from '1' to '100' i.e. the sum of the sequence: $\left(1, 2, 3, ... , 99, 100\right)$, we begin by denoting the sum of this sequence in series notation as:

$$S_n = 1 + 2 + 3 + ... + n$$

A simple method of adding up this series (in the same way that Karl Friedrich Gauss discovered as a primary school student) is by writing the series in both forward and reverse order as follows:

$$S_{100} = 1 + 2 + 3 + ... + 99 + 100 \qquad \text{... first in forward order}$$

$$S_{100} = 100 + 99 + 98 + ... + 2 + 1 \qquad \text{... second in reverse order}$$

By adding the left-hand sides and right-hand sides of these equations together, we arrive at the following result:

$$S_{100} + S_{100} = (1 + 100) + (2 + 99) + (3 + 98) + ... + (99 + 2) +$$

$$(100 + 1)$$

$$2 \times S_{100} = (101 + 101 + 101 + ... + 101 + 101)$$

$$= 100 \times 101 \qquad \text{... by '100' terms of '101'}$$

$$S_{100} = \frac{100 \times 101}{2} \qquad \text{... by dividing both sides by '2'}$$

$$\therefore \qquad S_{100} = 5,050 \qquad \text{... by simplifying terms}$$

Thus, the sum of the original sequence (i.e. the series, '1 + 2 + 3 + ... + 99 + 100') is '5,050'. This clearly demonstrates Gauss' insight into how rearranging the numbers in this series would turn this tedious summation into a relatively straightforward multiplication.

The most common example of a series is the expression of an integer in decimal notation. You will recall from our work on finite decimals in Chapter Six that a simple decimal such as '123,456' is defined as the following series:

$$123,456 = 1 \times 10^5 + 2 \times 10^4 + 3 \times 10^3 + 4 \times 10^2 + 5 \times 10^1 + 6 \times 10^0$$

$$= 1 \times 100,000 + 2 \times 10,000 + 3 \times 1,000 + 4 \times 100 + 5 \times 10 + 6 \times 1$$

$$\therefore \quad 123,456 = 100,000 + 20,000 + 3,000 + 400 + 50 + 6$$

Therefore, we can say that the integer '123,456' is created by adding up the six terms: '100,00', '20,000', '3,000', '400', '50' and '6'. As stated above, this number (i.e. '123,456') is an abbreviated notation for a finite series.

An example of a series of fractions is:

$$1 + \frac{9}{10} + \frac{4}{100} + \frac{6}{1,000} + \frac{2}{10,000} + \frac{3}{100,000}$$

We now work out the sum of this series by setting up a common denominator as follows:

$$\frac{100{,}000}{100{,}000} + \frac{90{,}000}{100{,}000} + \frac{4{,}000}{100{,}000} + \frac{600}{100{,}000} + \frac{20}{100{,}000} + \frac{3}{100{,}000}$$

$$= \frac{100{,}000 + 94{,}623}{100{,}000} \qquad \text{... by a common denominator}$$

$$= 1.94623 \qquad \text{... by writing as a decimal}$$

$$\therefore \qquad 1.94623 = 1 + \frac{9}{10} + \frac{4}{100} + \frac{6}{1{,}000} + \frac{2}{10{,}000} + \frac{3}{100{,}000}$$

Therefore, a finite decimal fraction is a finite series of fractions.

The next key concept we will be covering in this chapter is the '**limit**' of a sequence. The prototype sequence that we will use to demonstrate the concept of a 'limit' is the sequence of reciprocals of the positive integers which is given by $\left(\frac{1}{1}, \frac{1}{2}, \frac{1}{3}, \frac{1}{4}, \ldots, \frac{1}{n} \ldots\right)$. Intuitively, we can see that we can make the terms of this sequence 'as small as we like' by going far enough along this sequence. This is the same as saying we can make the terms of this sequence 'as close to '0' as we like'.

Mathematicians say that the above sequence 'has a limit of '0''. We will formalise this concept of a 'limit' in Definition 9.10 below.

Definition 9.10   Limit of a Sequence

Let $\left(a_1, a_2, a_3, \ldots, a_N, \ldots, a_n, \ldots\right)$ be an infinite sequence of rational number terms where 'n' can be assigned any integer subscript of the terms in the sequence. This sequence is said to have a **limit** 'L' if and only if all of the following conditions are true:

1. 'L' is a given fixed rational number which is being tested as the limit of the sequence

2. 'ε' is any fixed small rational number that is given

3. Given 'ε', there **must** exist an 'N' (a fixed positive integer depending on 'ε') corresponding to a term in the sequence such that '$|a_N - L| < \varepsilon$'

4. Then all subsequent terms with an index 'n' assigned a number greater than 'N' **must** make the inequality '$|a_n - L| < \varepsilon$' true.

Definition 9.10 involves over 2, 000 years of careful thinking about the way to discuss the limit of an infinite sequence using finite operations and valid finite strings of symbols from our alphabet. This is quite an accomplishment and it takes some time to absorb the subtleties contained in the examples below.

Effectively, this definition says that if 'L' is the limit of the sequence given above, namely $\left(a_1, a_2, a_3, \ldots, a_N, \ldots, a_n, \ldots\right)$, then by going from left to right along this sequence far enough, to say the term 'a_N', all subsequent terms will be closer to 'L' (i.e. the terms '$|a_{N+1} - L|$', '$|a_{N+2} - L|$', etc.) than some small positive rational number represented by the symbol epsilon 'ε'. If we can show this is true no matter how small epsilon is made – that is we can find a relationship between 'ε' and 'n' that guarantees this is true – then we have demonstrated that 'L' truly is the limit of the sequence $\left(a_1, a_2, a_3, \ldots, a_N, \ldots, a_n, \ldots\right)$.

Note:

1. Informally, the number 'a_n' in the sequence $\left(a_1, a_2, a_3, \ldots, a_N, \ldots, a_n, \ldots\right)$ gets arbitrarily closer to 'L' as the subscript of the number 'a_n' becomes an arbitrarily large positive integer (i.e. approaches 'infinity'). The number 'L' is called the limit and this relationship is often abbreviated to: '$\lim\limits_{n \to \infty} a_n = L$'.

2. Our objective is to validate a specific limit of a specific sequence. To do this, we use the general formula '$|a_n - L| < \varepsilon$' with a specific sequence and a specific limit to find an inequality relation between 'N' and 'ε' using 'L'.

3. If the limit 'L' of a sequence $\left(a_1, a_2, a_3, \ldots, a_n, \ldots\right)$ exists, then we say "the sequence **converges** to that limit". If the sequence does not converge, then it is said to **diverge** and hence there is no limit.

It is important to note that not all sequences have a limit. For example, the simple sequence of the positive integers $\left(1, 2, 3, \ldots\right)$ progresses from smaller to larger numbers without stopping and hence does not have a limit. Likewise, the sequence of alternating numbers $\left(1, -1, 1, -1, 1, -1, \ldots\right)$ does not satisfy the definition of a limit because it does not converge to either '1' or '-1'.

By observation, if the distance between a term of our sequence and all its subsequent terms is not eventually getting smaller as we move (left to right) from term to term along the sequence, then a limit will not exist.

We now demonstrate the use of this limit formula for our prototype sequence $\left(\frac{1}{1}, \frac{1}{2}, \frac{1}{3}, \frac{1}{4}, \ldots, \frac{1}{n}, \ldots\right)$. First, we make the reasonable guess that the limit 'L' is '0'. We now apply the Definition 9.10 to find the relationship between 'N' and 'ε' using 'L' for the specific sequence $\left(\frac{1}{1}, \frac{1}{2}, \frac{1}{3}, \frac{1}{4}, \ldots, \frac{1}{N}, \ldots, \frac{1}{n}, \ldots\right)$.

$$\left|\tfrac{1}{N} - 0\right| < \varepsilon \qquad \text{... by '} a_n = \tfrac{1}{N} \text{' and '} L = 0 \text{'}$$

$$\therefore \qquad \tfrac{1}{N} < \varepsilon \qquad \text{... by '} \left|\tfrac{1}{N}\right| = \tfrac{1}{N} \text{' for '} N > 0 \text{'}$$

$$\therefore \qquad N > \tfrac{1}{\varepsilon} \qquad \text{... by rearranging this inequality}$$

For example, if epsilon was given by '$\varepsilon = 0.0001$' then we must have '$N > 10,000$' for the inequality '$N > \frac{1}{\varepsilon}$' to be true; so we let '$N = 10,001$'. Therefore, in the general case we have the equation '$N > \frac{1}{\varepsilon}$', and for all '$n > N$' we have '$|\frac{1}{n} - 0| < \varepsilon$' and we have confirmed '0' is the limit of the sequence $\left(\frac{1}{1}, \frac{1}{2}, \frac{1}{3}, \frac{1}{4}, \, ... \, , \frac{1}{N}, \, ... \, , \frac{1}{n}, \, ...\right)$. Now that we know that '0' is the limit of this sequence, we can abbreviate this result using summary notation as: '$\lim\limits_{n \to \infty} a_n = 0$'.

Our most important application of the limit will be to the elementary number systems. This application will allow us to find the sum of an infinite series using this limit process. However, before we can find the sum of an infinite series, we require the standard method of finding the sum of a finite series.

We can calculate the value of a finite series by progressively finding the partial sums of the series. We undertake this process by first identifying the sequence that is to be summed to create the series. Next, starting with the first term, we progressively add the terms of the sequence to a running total. These progressive totals (or summations) of the sequence are called 'partial' sums of the sequence. We can formalise this addition process for a sequence as shown in Definition 9.11 below.

Definition 9.11   Sequence of Partial Sums

Let a finite sequence of terms be given by $\left(a_1, a_2, a_3, \, ... \, , a_n\right)$ and let the sum of the sequence be denoted by the series 'S_n' so that: '$S_n = a_1 + a_2 + a_3 + ... + a_n$'.

We define the **partial sums** of this sequence as follows:

$S_1 = a_1$ ... the sum of first term

$S_2 = a_1 + a_2$ ... the sum of first two terms

$S_3 = a_1 + a_2 + a_3$ ... the sum of first three terms

$$\vdots$$

$S_n = a_1 + a_2 + a_3 + ... + a_n$ ... the sum of first 'n' terms

Having outlined these partial sums of the sequence $\left(a_1, a_2, a_3, \, ... \, , a_n\right)$, we can now form a new sequence called the sequence of partial sums. In this way, the sequence of partial sums of the sequence '$\left(a_1, a_2, a_3, \, ... \, , a_n\right)$' is given by:

$$\left(S_1, S_2, \, ... \, , S_n\right) \text{ represents } \left(a_1, a_1 + a_2, \, ... \, , a_1 + a_2 + a_3 + ... + a_n\right)$$

Note:

We have formally created the process of summing a sequence by writing out its partial sums 'S_n' and deriving the new sequence of partial sums $(S_1, S_2, S_3, \ldots, S_n)$. This sequence of partial sums is the approach we use for defining the sum of an infinite sequence.

We illustrate this aspect of Definition 9.11 by writing out the sequence of partial sums $(S_1, S_2, S_3, \ldots, S_n)$ of the sequence $(1, 2, 3, \ldots, 99, 100)$. Starting from the left-hand side of the sequence, we can write out these partial sums as follows:

$$S_1 = 1$$
$$S_2 = 1 + 2 = S_1 + 2 = 3$$
$$S_3 = 1 + 2 + 3 = S_2 + 3 = 6$$
$$\vdots$$
$$S_{100} = (1 + 2 + \ldots + 99) + 100 = S_{99} + 100$$

You will recognise this approach as the way most of the students in Karl Friedrich Gauss' class would have added up the numbers from '1' to '100'. Gauss circumvented this tedious approach by recognising the simple pattern formed by adding the first number to the last number to get '101', then by adding the second number to the second last number to also get '101', and so on.

We observe that the last term in the sequence of partial sums 'S_{100}' is the sum of the the the sequence $(1, 2, 3, \ldots, 99, 100)$. We will use this approach for finding the sum of a finite sequence by writing out its partial sums as the approach to define the sum of an infinite sequence (if it exists) in Definition 9.12 below.

Definition 9.12   Sum of an Infinite Sequence

Let the general infinite sequence of rational number terms be given by $(a_1, a_2, a_3, \ldots, a_n, \ldots)$. We define the **sum of this infinite sequence** as the limit 'S' (if it exists) of the sequence of partial sums $(S_1, S_2, S_3, \ldots, S_n, \ldots)$ where the nth term 'S_n' is defined by: '$S_n = a_1 + a_2 + a_3 + \ldots + a_n$'.

In this way, the **sum** 'S' of an infinite sequence $(a_1, a_2, a_3, \ldots, a_n, \ldots)$ is defined as the limit of its sequence of partial sums (if that sum exists) so that:

$$S = \lim_{n \to \infty} S_n$$

The definition of a 'limit' ensures that this limit of a sequence of partial sums can ONLY exist if the terms 'a_n' of the original sequence consist of smaller and smaller fractions which are becoming closer and closer to '0'.

Note:

Consequently, the sequence of partial sums $(S_1, S_2, S_3, ... , S_n, ...)$ is the key to defining the sum of an infinite sequence, $(a_1, a_2, a_3, ... , a_n, ...)$ where 'a_n' must be getting closer to '0' if this sequence is to have a finite limit. As we cannot directly add an infinite number of terms in a sequence (i.e. we cannot add '$a_1 + a_2 + a_3 + ... + a_n + ...$'), we can circumvent this infinite number of additions by finding the limit of the sequence of partial sums '$S_n = a_1 + a_2 + a_3 + ... + a_n$'.

The next step is to demonstrate the limit of the sequence of partial sums of an infinite sequence by providing a simple practical example of adding up an infinite decimal sequence associated with the fraction '$\frac{1}{3}$', that is, the sequence $(\frac{3}{10}, \frac{3}{100}, \frac{3}{1000}, ... , \frac{3}{10^n}, ...)$. This infinite decimal sequence could also be written using index notation as $(3 \times 10^{-1}, 3 \times 10^{-2}, 3 \times 10^{-3}, ... , 3 \times 10^{-n}, ...)$ or using decimal notation as $(0.3, 0.03, 0.003, ... , 0.000...3, ...)$ where '0.000...3' is the decimal representation of '3×10^{-n}'.

To add up the infinite sequence $(\frac{3}{10}, \frac{3}{100}, \frac{3}{1000}, ... , \frac{3}{10^n}, ...)$, we first find the limit of its sequence of partial sums. We choose to use fraction notation for this sequence because it is more compact to write and easier from a visual perspective to understand the operations being performed.

The formula for the nth partial sum of this sequence, 'S_n' is given by:

$$S_n = \frac{3}{10^1} + \frac{3}{10^2} + \frac{3}{10^3} + ... + \frac{3}{10^n} \qquad \text{... by definition of '}S_n\text{'}$$

$$\therefore \quad 10 \times S_n = \frac{3 \times 10}{10^1} + \frac{3 \times 10}{10^2} + \frac{3 \times 10}{10^3} + ... + \frac{3 \times 10}{10^n}$$

$$\text{... by multiplying terms by '10'}$$

$$\therefore \quad 10 \times S_n = 3 + \frac{3}{10^1} + \frac{3}{10^2} + ... + \frac{3}{10^{n-1}} \qquad \text{... by simplifying terms}$$

In order to find an expression for 'S_n' that only involves two terms, we use the difference between the two equations with left-hand sides of '$10 \times S_n$' and 'S_n'.

$$\therefore \quad 10 \times S_n - S_n$$

$$= \left(3 + \frac{3}{10^1} + \frac{3}{10^2} + \dots + \frac{3}{10^{n-1}}\right) - \left(\frac{3}{10^1} + \frac{3}{10^2} + \frac{3}{10^3} + \dots + \frac{3}{10^n}\right)$$

... by substitution for terms

$$\therefore \quad 9 \times S_n = 3 - \frac{3}{10^n} \qquad \text{... by simplifying terms}$$

$$\therefore \quad S_n = \frac{3}{9} - \frac{3}{9 \times 10^n} \qquad \text{... by division by '9'}$$

$$\therefore \quad S_n = \frac{1}{3} - \frac{1}{3 \times 10^n} \qquad \text{... by simplifying terms}$$

Having found the formula for the nth partial sum – namely '$S_n = \frac{1}{3} - \frac{1}{3 \times 10^n}$' – we can write out the sequence of partial sums as: $\left(\frac{1}{3} - \frac{1}{3 \times 10^1}, \frac{1}{3} - \frac{1}{3 \times 10^2}, \dots, \frac{1}{3} - \frac{1}{3 \times 10^n}, \dots\right)$. Our next task is to find the limit of this sequence of partial sums. It is logical to estimate that the limit of this sequence 'S' is given by '$S = \frac{1}{3}$', since the term (called the remainder) '$\frac{1}{3 \times 10^n}$' in 'S_n' is becoming smaller and smaller as we move from left to right along this sequence of partial sums.

If we apply the definition of the limit to this sequence of partial sums we must find an integer 'n' such that for all '$n > N$' and the inequality '$|S_N - \frac{1}{3}| < \varepsilon$' is true for some value of 'N'. We begin this process by simplifying the term '$|S_N - \frac{1}{3}|$' as follows:

$$|S_N - \frac{1}{3}| = |\left(\frac{1}{3} - \frac{1}{3 \times 10^N}\right) - \frac{1}{3}| \quad \text{... by subst. for 'S_N' and 'S'}$$

$$= |-\frac{1}{3 \times 10^N}| \qquad \text{... by simplifying terms}$$

$$\therefore \quad |S_N - \frac{1}{3}| = \frac{1}{3 \times 10^N} \qquad \text{... by '$|x| = -x$' for '$x < 0$'}$$

Our goal is now to find the value of 'N' that makes '$\frac{1}{3 \times 10^N} < \varepsilon$'. By comparing this inequality with the equivalent inequality for the prototype sequence $\left(\frac{1}{1}, \frac{1}{2}, \frac{1}{3}, \frac{1}{4}, \dots, \frac{1}{n}, \dots\right)$, we see that '$\frac{1}{3 \times 10^N} < \frac{1}{N}$' for all '$N > 0$'. Therefore, for '$N > \frac{1}{\varepsilon}$' we have found '$N$' so that for all '$n > N$', the inequality '$|S_n - \frac{1}{3}| < \varepsilon$' is a true statement.

Consequently, we can declare that the limit of the sequence of partial sums $\left(\left(\frac{1}{3} - \frac{1}{3 \times 10^1}\right), \left(\frac{1}{3} - \frac{1}{3 \times 10^2}\right), \dots, \left(\frac{1}{3} - \frac{1}{3 \times 10^n}\right), \dots\right)$ is '$\frac{1}{3}$'. As we progress along this sequence of partial sums from left to right, we can see the distance between each term and the limit '$\frac{1}{3}$' is becoming smaller by a factor of '$\frac{1}{10}$' for each step we advance.

Note:

Writing this sequence of partial sums in decimal notation, we obtain the alternative form of this sequence as '$(0.3, 0.33, 0.333, ...)$' which also converges to '$\frac{1}{3}$'. This example illustrates the application of the limit definition to a sequence of partial sums. An extension of this example, (which will be described later in this chapter) will demonstrate that the concept of a limit is essential to illustrate the fact that all finite and repeating decimals are equivalent to fractions and vice versa.

To complete this Definition Section of the chapter, we update the original version of Table 9.1 with the new symbols introduced in this section to produce Table 9.2 below.

9.2 The Alphabet of the Rational Number System with Limits	
Symbols	**Meaning from:**
'0', '1', '+', '×', '(', ')'	The Axioms of the Natural Number System
'=', '≠'	The Axioms of Logic
'2', '3', '4', '5', '6', '7', '8', '9'	Definition 2.1
'a', 'b', 'c', ... , 'x', 'y', 'z', ...	The rational number variables
','	Definition 2.3
'−1'	The Axioms of the Integer Number System
'−'	Definitions 3.4 and 3.5
'$\frac{1}{\Box}$'	The Axioms of the Rational Number System as Fractions
'$\frac{a}{b}$'	Definition 4.2
'÷'	Definition 4.4
'$\Box^{-1}$'	The Axioms of Rational Number System using Index Form
'$(a \times b^{-1})^m$'	Definition 5.6
'.'	Definition 6.4
'$a_n ... a_1 a_0 . d_1 ... d_m$'	Definition 6.5
'±'	Definition 7.8
'<'	The Axioms of the Rational Number System using Order

9.2 The **Alphabet** of the Rational Number System with Limits	
'$\leq$', '$>$', '$\geq$'	Definition 8.2
'$\lvert x - y \rvert$'	Definition 8.7
'{', '}'	Definition 9.1
'(', ')'	Definition 9.2 (overloading the parentheses symbols)
'∞'	Definition 9.3
'$\lim\limits_{n \to \infty} a_n = L$'	Definition 9.10 (refer to the Note 1 following this definition)

Table 9.2

In the Theory Section below we apply the definitions of sequences, series and limits to further investigate infinite sequences and their associated limits (if they exist).

Theory of Sequences, Series and Limits

Our overall objective for this chapter is to master operations involving sequences and series using limits. This objective is important because it will allow us to demonstrate that all finite and repeating decimals are equivalent to the fractions and vice versa. We begin this section by obtaining the sums of general infinite geometric sequences using limits. To make the proofs of some of the following theorems as simple and transparent as possible it will be necessary to prove some simple properties possessed by the definition of a limit of a sequence.

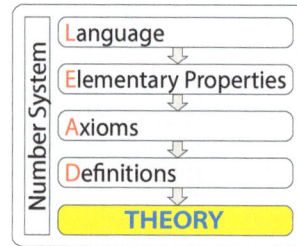

Number System:
Language → Elementary Properties → Axioms → Definitions → **THEORY**

The next step is to examine arithmetic sequences and generalise our method of adding the numbers from '1' through to '100' as we did earlier in this chapter. In this way, we are endeavouring to find the general formula for adding a finite arithmetic sequence. This process will be outlined in Theorem 9.1.

Theorem 9.1   Sum of a Finite Arithmetic Sequence

Let the sequence $(a, a + d, a + 2d, \ldots, a + (n-1)d)$, be the standard finite arithmetic sequence of rational number terms where there are 'n' terms with 'a' as the first term and 'd' is the common difference. If 'S_n' is the sum of this sequence, then we have:

$$S_n = \frac{n}{2}(2a + (n-1)d)$$

Proof

We use proof by mathematical deduction and begin by writing the sum of the arithmetic sequence in forward order first and then rewriting the sum of this sequence in reverse order to obtain:

$$S_n = a + (a + d) + \ldots + (a + (n - 2)d) + (a + (n - 1)d)$$

Also: $\qquad S_n = (a + (n - 1)d) + (a + (n - 2)d) + \ldots + (a + d) + a$

In this way, the second series is simply the first series written out in reverse order. Adding the left-hand sides and the right-hand sides of these series gives:

$$2S_n = (2a + (n - 1)d) + (2a + (n - 1)d) + \ldots + (2a + (n - 1)d)$$
$$+ (2a + (n - 1)d) \qquad \text{... by adding both series}$$

$\therefore \qquad 2S_n = n \times (2a + (n - 1)d) \qquad$ ... by adding 'n' terms

$\therefore \qquad S_n = \frac{n}{2}(2a + (n - 1)d) \qquad$ ... by dividing both sides by '2'

In the above equation, we can rewrite 'S_n' as:

$$S_n = \frac{n}{2}(a + a + (n - 1)d)$$

By recognising that 'a' is the first term of the sequence (called 'a_1' say) and '$a + (n - 1)d$' is the last term of the sequence (called 'a_n' say), and so we can then write our sum 'S_n' in the simple form:

$$S_n = \frac{n}{2}(a_1 + a_n)$$

Therefore, the result follows from basic arithmetic and the theorem is proved.

––––––––––

Our next step is to focus on the same process for adding the terms of a finite geometric sequence using a similar shortcut to the one we used on arithmetic sequences. We will derive the general format for this sum in Theorem 9.2 below.

Theorem 9.2   Sum of a Finite Geometric Sequence

Let the general finite geometric sequence of rational numbers be given by $(a, ar, ar^2, \ldots, ar^{n-1})$ where 'a' is the first term and '$r \neq 1$' is the common ratio. If 'S_n' is the sum of the first 'n' terms of this sequence, then we have:

$$S_n = \frac{a(1 - r^n)}{1 - r}$$

For the special case where '$r = 1$', this formula is not defined and the sequence becomes $(a, a, a, \dots, a)$. In this situation, the sequence has the simple sum of 'n' terms which results in '$S_n = n \times a$'.

Proof

To derive this proof we write out the explicit sum of the sequence $(a, ar, ar^2, \dots, ar^{n-1})$ to obtain:

$$S_n = ar^0 + ar^1 + ar^2 + \dots + ar^{n-1} \qquad \dots \text{given}$$

$$\therefore \quad rS_n = ar^1 + ar^2 + ar^3 + \dots + ar^{n-1} + ar^n \qquad \dots \text{by multiplying '}r\text{'}$$

$$S_n - rS_n = (a + ar + ar^2 + \dots + ar^{n-1}) - (ar + ar^2 + ar^3 + \dots + ar^{n-1} + ar^n)$$

$$(1 - r)S_n = a - ar^n \qquad \dots \text{by simplifying terms}$$

$$S_n = \frac{a - ar^n}{1 - r} \qquad \dots \text{by dividing by '}1 - r\text{'}$$

$$\therefore \quad S_n = \frac{a(1 - r^n)}{1 - r} \qquad \dots \text{by Distributive Axiom}$$

Consequently, this is a very useful result (as it will allow us to easily add any infinite repeating decimal (see Theorem 9.7)) and also follows from basic arithmetic and the theorem is proved.

———————————

Our next step is to generalise our method of summing a finite geometric sequence (as above) to include infinite geometric sequences. In this way, we want to find the general formula for adding an infinite geometric sequence. To enable us to perform this process we will need to use the standard operations of addition and multiplication as they apply to general sequences.

The first and simplest theorem involves finding the limit of a sequence when multiplied by a constant.

Theorem 9.3   Limit of a Product of a Sequence by a Constant

If '$c \neq 0$' is a constant rational number and $(a_1, a_2, a_3, \dots, a_n, \dots)$ is an infinite sequence of rational number terms with a limit 'L', then the limit of the sequence of the product of the constant and the sequence $(ca_1, ca_2, ca_3, \dots, ca_n, \dots)$ is 'cL'.

Symbolically, we write this result as: '$\lim\limits_{n \to \infty} ca_n = c \lim\limits_{n \to \infty} a_n$'.

Proof

To prove that the sequence $\left(ca_1, ca_2, ca_3, \dots, ca_n, \dots\right)$ has a limit 'cL', we have to demonstrate that for any given small positive rational number 'ε', we can find an index 'N' in this sequence such that for all '$n > N$' then '$|ca_n - cL| < \varepsilon$'.

We are given that the sequence $\left(a_1, a_2, a_3, \dots, a_n, \dots\right)$ has as limit 'L', so it follows that there exists a small positive rational number epsilon, denoted as 'ε_0', and a positive integer 'N_0' such that for all '$n > N_0$', so we have '$|a_n - L| < \varepsilon_0$'.

To prove the theorem, we start by simplifying the term '$|ca_n - cL|$':

$$|ca_n - cL| = |c \times (a_n - L)| \qquad \text{... by Distributive Axiom}$$

$$= |c| \times |a_n - L| \qquad \text{... according to Theorem 8.17}$$

$$\therefore \qquad |ca_n - cL| < |c| \times \varepsilon_0 \qquad \text{... substituting for '}|a_n - L| < \varepsilon_0\text{'}$$

We would like to make '$|c| \times \varepsilon_0$' equal to 'ε' which is the given small rational number that we have to make '$|ca_n - cL|$' less than in order to satisfy the definition of 'cL' as the limit of the sequence $\left(ca_1, ca_2, ca_3, \dots, ca_n, \dots\right)$. We can make '$\varepsilon_0$' as small as we like, because 'L' is the limit of the sequence $\left(a_1, a_2, a_3, \dots, a_n, \dots\right)$, so we can certainly make '$|c| \times \varepsilon_0 = \varepsilon$', so that '$\varepsilon_0 = \frac{\varepsilon}{|c|}$', (as the constant '$c \neq 0$'), in order to be guaranteed that we can find an 'N_0', so that for '$n > N_0$' then '$|a_n - L| < \varepsilon_0 = \frac{\varepsilon}{|c|}$)'.

Therefore, we can re-write the original limit as follows:

$$|a_n - L| < \varepsilon_0$$

$$|c| \times |a_n - L| < |c| \times \varepsilon_0 \qquad \text{... by multiplying by '}|c|\text{'}$$

$$|c| \times |a_n - L| < |c| \times \frac{\varepsilon}{|c|}) \qquad \text{... by substituting '}\varepsilon_0 = \frac{\varepsilon}{|c|}\text{'}$$

$$|ca_n - cL| < |c| \times \frac{\varepsilon}{|c|}) \qquad \text{... according to Theorem 8.17}$$

$$\therefore \qquad |(ca_n - cL)| < \varepsilon \qquad \text{... by simplifying terms}$$

In this way, by making 'N' from the index of the sequence $\left(ca_1, ca_2, ca_3, \dots, ca_n, \dots\right)$ the same as 'N_0' above, then we are assured that for all '$n > N$' then '$|(ca_n - cL)| < \varepsilon$' for any given arbitrarily small 'ε' and so we have proved the theorem: '$\lim_{n \to \infty} (ca_n) = c \lim_{n \to \infty} a_n$'.

In Theorem 9.4 we will observe that the limit of the sum of the terms of two sequences is the same as the sum of the limits of the sequences individually.

Theorem 9.4   Sum of Two Infinite Sequences

Let two infinite sequences of rational number terms $(a_1, a_2, a_3, ... , a_n, ...)$ and $(b_1, b_2, b_3, ... , b_n, ...)$ be sequences with limits 'L_1' and 'L_2', respectively. As a result, the limit of the sum of these two sequences '$(a_1 + b_1, a_2 + b_2, a_3 + b_3, ... , a_n + b_n, ...)$' is '$L_1 + L_2$'.

The sequence $(a_1, a_2, a_3, ... , a_n, ...)$ is abbreviated to '$(a_n)_{n=1}^{\infty}$' and its limit is written as '$\lim_{n\to\infty} a_n = L_1$', and the sequence $(b_1, b_2, b_3, ... , b_n, ...)$ is abbreviated to '$(b_n)_{n=1}^{\infty}$' and its limit is written as '$\lim_{n\to\infty} b_n = L_2$'.

Alternatively, we can now write the sum of these two sequences in abbreviated form as:

$$\lim_{n\to\infty} (a_n + b_n) = \lim_{n\to\infty} a_n + \lim_{n\to\infty} b_n$$

or:

$$\lim_{n\to\infty} (a_n + b_n) = L_1 + L_2$$

Proof

We prove this theorem by directly applying the definition of a limit. We assume that '$L_1 + L_2$' is the limit of the sequence $(a_1 + b_1, a_2 + b_2, a_3 + b_3, ... , a_n + b_n, ...)$. Next, we apply the definition of the limit of a sequence to test this assumption.

We can write that given a small number 'ε', a positive integer 'N' exists such that for all '$n > N$' then '$|(a_n + b_n) - (L_1 + L_2)| < \varepsilon$'.

To begin this proof we start with the left-hand side of this inequality.

$$|(a_n + b_n) - (L_1 + L_2)| = |(a_n - L_1) + (b_n - L_2)| \quad \text{... by rearranging terms}$$
$$\leq |a_n - L_1| + |b_n - L_2| \quad \text{... by Theorem 8.16}$$

Our aim now is to make the two terms, '$|a_n - L_1|$' and '$|b_n - L_2|$' add up to less than 'ε'. In this way, if we make each term less than '$\frac{\varepsilon}{2}$' since we can choose this number to be arbitrarily small, when added together they will be less than 'ε'.

Given that '$\lim_{n\to\infty} a_n = L_1$', we know if we select an arbitrarily small 'ε_1' (in particular '$\varepsilon_1 < \frac{\varepsilon}{2}$'), then a positive integer 'N_1' exists such that for all '$n > N_1$', '$|a_n - L_1| < \varepsilon_1$'. Similarly, given that '$\lim_{n\to\infty} b_n = L_2$', we know that if we select an arbitrarily small 'ε_2' (in particular '$\varepsilon_2 < \frac{\varepsilon}{2}$'), then a positive integer 'N_2' exists such that for all '$n > N_2$', '$|b_n - L_2| < \varepsilon_2$'. In this way, if we let 'n' be the bigger number of the two numbers 'N_1' and 'N_2', then we can be sure that both '$|a_n - L_1| < \varepsilon_1$' and '$|b_n - L_2| < \varepsilon_2$' are true statements.

Therefore, we can write:

$$|(a_n + b_n) - (L_1 + L_2)| \leq |a_n - L_1| + |b_n - L_2| \quad \text{... according to Theorem 8.16}$$

$$< \varepsilon_1 + \varepsilon_2 \qquad \text{... by substit. for '}|a_n - L_1|\text{' etc}$$

$$< \frac{\varepsilon}{2} + \frac{\varepsilon}{2} \qquad \text{... by substit. for '}\varepsilon_1\text{' and '}\varepsilon_2\text{'}$$

$$= \varepsilon \qquad \text{... by simplifying the term}$$

$$\therefore \quad |(a_n + b_n) - (L_1 + L_2)| < \varepsilon$$

Hence, we have found an 'N' (the maximum number of the two numbers 'N_1' and 'N_2') so that for all '$n > N$' then '$|(a_n + b_n) - (L_1 + L_2)| < \varepsilon$' for any given arbitrarily small 'ε'. Thus we have proved the theorem.

Note:

By using Theorem 9.3, as well as the result of this theorem and letting the variable '$b_n = -c_n$', the special case of '$\lim_{n \to \infty} a_n - c_n$' follows on easily.

In Theorem 9.5 below we will observe that the limit of the product of the terms of two sequences is the same as the product of the limits of the sequences individually.

Theorem 9.5 Product of Two Infinite Sequences

Let two infinite sequences of rational number terms $(a_1, a_2, a_3, ..., a_n, ...)$ and $(b_1, b_2, b_3, ..., b_n, ...)$ have limits 'L_1' and 'L_2', respectively. As a result, the product of these two sequences $(a_1 \times b_1, a_2 \times b_2, a_3 \times b_3, ..., a_n \times b_n, ...)$ has a limit '$L_1 \times L_2$' which can be written in limit notation as:

$$\lim_{n \to \infty} (a_n \times b_n) = \lim_{n \to \infty} a_n \times \lim_{n \to \infty} b_n$$

or:

$$\lim_{n \to \infty} (a_n \times b_n) = L_1 \times L_2$$

Proof

We can prove this theorem by directly applying the definition of a limit. We will assume '$L_1 \times L_2$' is the limit of the sequence '$(a_1b_1, a_2b_2, a_3b_3, \ldots, a_nb_n, \ldots)$' and then apply the definition of the limit to test this assumption.

We aim to prove that given a small number 'ε' (assume '$\varepsilon < 1$' without loss of generality), a positive integer 'n' exists such that for all '$n > N$' then '$|a_nb_n - L_1L_2| < \varepsilon$'.

We begin the proof with the left-hand side of this inequality and rewrite it in terms of '$|a_n - L_1|$' and '$|b_n - L_2|$' using a standard technique as follows:

$$
\begin{aligned}
|a_nb_n - L_1L_2| &= |a_nb_n - L_1b_n + L_1b_n - L_1L_2| &&\text{... by adding '} -L_1b_n + L_1b_n \text{'}\\
&\leq |a_nb_n - L_1b_n| + |L_1b_n - L_1L_2| &&\text{... according to Theorem 8.16}\\
&= |b_n||a_n - L_1| + |L_1||b_n - L_2| &&\text{... according to Theorem 8.17}
\end{aligned}
$$

$$\therefore \quad |a_nb_n - L_1L_2| \leq |b_n||a_n - L_1| + |L_1||b_n - L_2|$$

It will be advantageous later in this proof to demonstrate that we can even make '$|a_nb_n - L_1L_2| < \varepsilon$' by using a slightly larger term on the right-hand side of this last inequality. We simply replace '$|L_1|$' by '$(1 + |L_1|)$' in the second term as follows:

$$|a_nb_n - L_1L_2| \leq |b_n||a_n - L_1| + (1 + |L_1|)|b_n - L_2|$$

Note: we have to maintain the use of the '$\leq$' symbol in the above inequality just in case the sequence $(b_1, b_2, b_3, \ldots, b_n, \ldots)$ is a constant sequence and '$|b_n - L_2| = 0$'.

In our next step we will show that each of these two terms on the right-hand side of this last inequality above can be made less than '$\frac{\varepsilon}{2}$'. We want to be able to show:

Case 1: $\qquad |b_n||a_n - L_1| < \frac{\varepsilon}{2}$

Case 2: $\quad (1 + |L_1|)|b_n - L_2| < \frac{\varepsilon}{2}$

Given that '$\lim\limits_{n\to\infty} a_n = L_1$' and '$\lim\limits_{n\to\infty} b_n = L_2$', we know we can make '$\varepsilon_1$' and '$\varepsilon_2$' respectively as small as we choose (where we can also assume '$\varepsilon_1 < 1$' and '$\varepsilon_2 < 1$' without loss of generality) by making 'N_1' and 'N_2' sufficiently large so we have '$|a_n - L_1| < \varepsilon_1$' and '$|b_n - L_2| < \varepsilon_2$'.

Case 1:

Starting with the inequality '$|b_n||a_n - L_1| < \frac{\varepsilon}{2}$', we wish to re-write '$|b_n|$' in terms of '$L_2$' as follows:

$$|b_n| = |b_n - L_2 + L_2| \qquad \text{... by '} 0 = -L_2 + L_2 \text{'}$$

$$\therefore \qquad |b_n| \le |b_n - L_2| + |L_2| \qquad \text{... according to Theorem 8.16}$$

However, given that '$\varepsilon_2 < 1$', we can rewrite '$|b_n - L_2| < \varepsilon_2$' as:

$$|b_n - L_2| < 1 \qquad \text{... by Transitive Axiom '<'}$$

Substituting into the inequality '$|b_n| \le |b_n - L_2| + |L_2|$' above:

$$\therefore \qquad |b_n| < 1 + |L_2| \qquad \text{... by '}|b_n - L_2| < 1\text{'}$$

We now replace '$|b_n|$' and '$|a_n - L_1|$' in Case 1 and make 'ε_1' small enough so that:

$$(1 + |L_2|) \times \varepsilon_1 < \frac{\varepsilon}{2} \qquad \text{... by substituting into Case 1}$$

$$\therefore \qquad \varepsilon_1 < \frac{\varepsilon}{2(1 + |L_2|)} \qquad \text{... by making '}\varepsilon_1\text{' the subject}$$

After we have made 'ε_1' the subject of this inequality, all that remains for us to do is to ensure 'ε_1' is less than '$\frac{\varepsilon}{2(1 + |L_2|)}$'. This outcome is guaranteed since we can make 'ε_1' arbitrarily small and be assured an 'N_1' exists such that for all 'n' greater than 'N_1', we have '$|b_n||a_n - L_1|$' less than '$\frac{\varepsilon}{2}$'.

Case 2:

For this case, we now substitute 'ε_2' for '$|b_n - L_2|$' to obtain:

$$(1 + |L_1|) \times \varepsilon_2 < \frac{\varepsilon}{2} \qquad \text{... by subst. for '}|b_n - L_2|\text{'}$$

$$\therefore \qquad \varepsilon_2 < \frac{\varepsilon}{2(1 + |L_1|)}$$

After we have made 'ε_2' the subject of this inequality, all that remains for us to do is to ensure 'ε_2' is less than '$\frac{\varepsilon}{2(1 + |L_1|)}$'. This outcome is guaranteed since we can make 'ε_2' arbitrarily small and be assured an 'N_2' exists such that for all 'n' greater than 'N_2' we have '$(1 + |L_1|)|b_n - L_2|$' less than '$\frac{\varepsilon}{2}$'.

In order to be sure that both of these conditions in Cases 1 and 2 are satisfied simultaneously, the only requirement is to let 'n' be the maximum of 'N_1' and 'N_2'. We can then substitute for 'ε_1' and 'ε_2' in our original inequality to obtain:

$$|a_n b_n - L_1 L_2| \le |b_n| \times |a_n - L_1| + (1 + |L_1|) \times |b_n - L_2|$$

$$\text{... by original inequality}$$

$$< (1 + |L_2|) \times \varepsilon_1 + (1 + |L_1|) \times \varepsilon_2$$

$$\text{... by limits of '}a_n\text{' and '}b_n\text{'}$$

$$< (1 + |L_2|) \times \frac{\varepsilon}{2(1 + |L_2|)} + (1 + |L_1|) \times \frac{\varepsilon}{2(1 + |L_1|)}$$

$$\qquad\qquad = \frac{\varepsilon}{2} + \frac{\varepsilon}{2} \qquad\qquad\qquad \text{... by subst. for '}\varepsilon_1\text{' and '}\varepsilon_2\text{'}$$
$$\text{... by simplifying terms}$$
$$\therefore \qquad |a_n b_n - L_1 L_2| < \varepsilon$$

In this way, we have found an 'n' so that for all '$n > N$', then '$|a_n b_n - L_1 L_2| < \varepsilon$' for any given arbitrarily small 'ε'. Consequently, we have proved the theorem and can write:

$$\lim_{n \to \infty} (a_n \times b_n) = L_1 \times L_2$$

_____

Now that we have developed the basic theorems on adding and multiplying sequences, we will now explore the limit of a geometric sequence (which is essential for Theorem 9.7).

Theorem 9.6   Limit of a Geometric Sequence

For an infinite sequence of rational number terms given by the geometric sequence $(r, r^2, r^3, \dots, r^n, \dots)$ we can prove that the limit of this sequence is zero if '$|r| < 1$' (i.e. 'r' is a fraction less than '1' and greater than '-1'). Hence:

$$\lim_{n \to \infty} r^n = 0$$

Proof

We can prove this theorem by assuming '0' is the limit of the sequence and then applying the definition of a limit to test whether this assumption is true or false.

If '0' is the limit of this sequence then for any small rational number 'ε' we must have a large integer 'n' such that for all '$n > N$', the inequality '$|r^n - 0| < \varepsilon$' is a true statement.

By starting with the case where '$0 < r < 1$' (that is, 'r' is a positive fraction less than '1'), our formula simplifies to '$r^n < \varepsilon$'. Next, our approach will be to express 'r' in terms of integers so that we can derive a simpler expression for 'r^n'.

If we assume 'r' to be a positive rational number less than '1', we can express 'r' as the ratio of two positive integers 'p' and 'q' (i.e. '$r = \frac{p}{q}$'). We start the process of simplifying 'r^n' as follows:

$$r^n = (\tfrac{p}{q})^n \qquad\qquad \text{... by using '}r = \tfrac{p}{q}\text{'}$$
$$= \tfrac{p}{q} \times \dots \times \tfrac{p}{q} \qquad\qquad \text{... by Theorem 5.22}$$
$$= (p \times \tfrac{1}{q}) \times \dots \times (p \times \tfrac{1}{q}) \qquad \text{... by Definition 4.2}$$

$$= (p \times ... \times p) \times (\tfrac{1}{q} \times ... \times \tfrac{1}{q}) \text{ ... by Associative Axiom '}\times\text{'}$$

$$\therefore \qquad r^n = \frac{p^n}{q^n} \qquad\qquad\qquad \text{... by Theorem 5.22}$$

However, since '$\frac{p}{q} < 1$' and '$q > 0$' (given above), then it follows that '$p < q$'. Next, in order to simplify the fraction '$\frac{p^n}{q^n}$', we look for a fraction that is bigger than '$\frac{p^n}{q^n}$' and in a more manageable form. We observe that since 'p' and 'q' are integers and '$p < q$', it follows that '$p + 1 \le q$' and hence '$(p + 1)^n \le q^n$' by the Deduction Axiom (with all the terms being positive).

We can now substitute for 'q^n' with a less-than or equal-to integer in the equation for 'r^n'.

$$r^n = \frac{p^n}{q^n} \qquad\qquad \text{... by continuing with the proof}$$

$$\therefore \qquad r^n \le \frac{p^n}{(p + 1)^n} \qquad \text{... by substituting for '}q^n\text{'}$$

Simplifying this last term involves expanding the denominator '$(p + 1)^n$' into its first two terms and hence a smaller integer value. We first carry out this expansion for the first four values of the positive integer 'n'.

$$(p + 1)^1 = p + 1$$

$$(p + 1)^2 = p^2 + 2p + 1$$

$$(p + 1)^3 = p^3 + 3p^2 + 3p + 1$$

$$(p + 1)^4 = p^4 + 4p^3 + 6p^2 + 4p + 1$$

$$\vdots$$

$$(p + 1)^n = p^n + np^{n-1} + ... + np + 1$$

From the 'n^{th}' equation outlined above (i.e. the expansion of '$(p + 1)^n$'), we have the following inequality: '$(p + 1)^n > p^n + np^{n-1}$' for '$n > 1$' by dropping off all terms after the first two. This inequality provides us with a simpler form than '$(p + 1)^n$' which, in turn, allows us to simplify our original equation '$\frac{p^n}{(p + 1)^n}$'. We will now use this inequality to substitute into our original inequality as follows:

$$r^n \le \frac{p^n}{(p + 1)^n}$$

$$< \frac{p^n}{p^n + np^{(n-1)}} \qquad \text{... by '}(p + 1)^n > p^n + np^{n-1}\text{'}$$

$$= \frac{p}{p + n} \qquad\qquad \text{... by dividing by '}p^{n-1}\text{'}$$

$$\therefore \qquad r^n < \frac{p}{n} \qquad\qquad \text{... by the relation '}n < n + p\text{'}$$

The next step involves ensuring that '$r^n < \varepsilon$', to make '$\frac{p}{n} < \varepsilon$', that is, to make '$n > \frac{p}{\varepsilon}$'. In this way, if we let '$N = \frac{p}{\varepsilon}$', then for all '$n > N$' we are assured that '$|r^n - 0| < \varepsilon$' for any arbitrary small 'ε'. Consequently, we have shown that '$\lim\limits_{n \to \infty} r^n = 0$' for '$0 < r < 1$'.

By replacing 'r' with '$-r$' and then reworking the above proof the result for '$-1 < r < 0$' follows easily and we note the case where '$r = 0$' follows in a straight-forward manner also.

Hence, we have proved the theorem and '$\lim\limits_{n \to \infty} r^n = 0$' if '$|r| < 1$'.

––––––––––––

At this point, we have available to us all the necessary theorems to prove the key result of this chapter – namely that there is a simple formula that gives the sum of an infinite geometric sequence. We could refer to this as the 'Eureka statement' for this chapter!

Theorem 9.7   Sum of an Infinite Geometric Sequence

Let an infinite geometric sequence of rational number terms be given by $\left(a, ar, ar^2, ..., ar^{n-1}, ...\right)$ where 'a' and 'r' are rational numbers and '$0 < |r| < 1$'. Let $\left(S_1, S_2, S_3, ..., S_n, ...\right)$ be the corresponding sequence of partial sums of this infinite geometric sequence and the nth term of this sequence be given by: '$S_n = \frac{a(1 - r^n)}{1 - r}$'. The sum of the infinite sequence $(a, ar, ar^2, ..., ar^{n-1}, ...)$ is the limit of the sequence of partial sums and is given by:

$$\lim_{n \to \infty} S_n = \frac{a}{1 - r}$$

Proof

The proof of this general result follows in a straightforward way from the previous theorems in this chapter. We perform this proof using limit notation as follows:

$$\lim_{n \to \infty} S_n = \lim_{n \to \infty} \frac{a(1 - r^n)}{1 - r} \qquad \text{... according to Theorem 9.2}$$

$$= \lim_{n \to \infty} \left(\frac{a}{1 - r} - \frac{ar^n}{1 - r}\right) \qquad \text{... by Distributive Axiom}$$

$$= \lim_{n \to \infty} \frac{a}{1 - r} - \lim_{n \to \infty} \frac{ar^n}{1 - r} \qquad \text{... according to Theorem 9.4}$$

$$= \frac{a}{1 - r} - \lim_{n \to \infty} \frac{ar^n}{1 - r} \qquad \text{... by '} \lim_{n \to \infty} \frac{a}{1 - r} = \frac{a}{1 - r} \text{'}$$

$$= \frac{a}{1 - r} - \lim_{n \to \infty} \left(\frac{a}{1 - r} \times r^n\right) \text{... by properties of fractions}$$

$$= \frac{a}{1-r} - \frac{a}{1-r} \times \lim_{n \to \infty} r^n \quad \text{... according to Theorem 9.3}$$

$$= \frac{a}{1-r} - \frac{a}{1-r} \times 0 \quad \text{... according to Theorem 9.5}$$

$$\therefore \quad \lim_{n \to \infty} S_n = \frac{a}{1-r}$$

Hence, the theorem has been proved. We now have the very useful result that the sum of an infinite geometric sequence is given by:

$$\lim_{n \to \infty} S_n = \frac{a}{1-r}$$

In this equation 'a' is any rational number and 'r' is the rational number satisfying the inequality: '$0 < |r| < 1$'. This result completes the proof of the theorem

Theorem 9.7 provides us with a powerful tool for calculating the limit of infinite geometric sequences. We will be using this tool to find the limit of repeating decimals and hence to derive an important link between rational numbers as fractions and rational numbers as repeating decimals.

To illustrate the usefulness of Theorem 9.7, we again find the sum of the infinite geometric sequence $(\frac{3}{10}, \frac{3}{10^2}, \frac{3}{10^3}, ..., \frac{3}{10^n}, ...)$ which we identified by following Definition 9.12. The nth partial sum of this sequence is given by: '$S_n = \frac{3}{10} + \frac{3}{10^2} + ... + \frac{3}{10^n}$'. Applying Theorem 9.7, we assign 'a' the value '$\frac{3}{10}$' and 'r' the value '$\frac{1}{10}$'.

The limit of the sequence of partial sums is given by:

$$\lim_{n \to \infty} S_n = \frac{\frac{3}{10}}{1 - \frac{1}{10}} \quad \text{... according to Theorem 9.7}$$

$$= \frac{3}{10 - 1} \quad \text{... by multiplying by '10'}$$

$$\therefore \quad \lim_{n \to \infty} S_n = \frac{1}{3}$$

This outcome is the same result we derived earlier for the limit of the sequence of partial sums using the direct definition of the limit. Generally speaking, the above method is a simpler and quicker process for finding the sums of geometric sequences where '$0 < |r| < 1$'.

With this simple example of a repeating decimal as the sum of an infinite geometric sequence, the theory component of this chapter is now complete.

Applications of Sequences, Series and Limits

In this section we will be considering some applications of sequences, series and limits using rational numbers.

Application 9.1

Find the sum of the first twenty terms of the arithmetic sequence consisting of the odd numbers starting at the number '1'.

The arithmetic sequence is '$(1, 3, 5, \dots, a_n)$' where '$a_1 = 1$', '$d = 2$' and '$n = 20$' for the last term in the sequence. Substituting '$a_1 = 1$', '$d = 2$' into the formula for the nth term '$a_n = a + (n-1)d$' gives '$a_n = 2n - 1$' where 'n' takes on the integer values '1', '2' and so up to '$n = 20$'.

Using Theorem 9.1, which states that '$S_n = \frac{n}{2}(a_1 + a_n)$', we can now find the sum of this sequence.

The sum of the first twenty terms of this sequence is given by 'S_{20}' so that:

$$S_{20} = \frac{20}{2}(a_1 + a_{20}) \qquad \text{... according to Theorem 9.1}$$

$$= \frac{20}{2}(1 + 39) \qquad \text{... by '}a_{20} = 2 \times 20 - 1 = 39\text{'}$$

$$\therefore \qquad S_{20} = 400 \qquad \text{... by simplifying terms}$$

Therefore, the sum of the first twenty odd numbers starting at '1' is '400'.

In Application 9.2 below we will be applying Theorem 9.2 to find the sum of a finite Geometric sequence.

Application 9.2

Find the sum of the geometric sequence: $\left(10 \times (\frac{1}{2})^0, 10 \times (\frac{1}{2})^1, 10 \times (\frac{1}{2})^2, \dots, 10 \times (\frac{1}{2})^5\right)$ as a fraction.

In the case of this geometric sequence '$a = 10$', '$r = \frac{1}{2}$' and '$a^n = ar^{n-1}$', so '$n = 6$' for the last term.

Using Theorem 9.2, which states that '$S_n = \frac{a(1 - r^n)}{1 - r}$', we can now find the sum of this sequence.

The sum of this geometric sequence is given by 'S_6' so that:

$$S_6 = \frac{10\left(1 - (\frac{1}{2})^6\right)}{1 - \frac{1}{2}} \qquad \text{... according to Theorem 9.2}$$

$$= \frac{10\left(1-\frac{1}{64}\right)}{1-\frac{1}{2}} \qquad \text{... by } \text{‘}\left(\tfrac{1}{2}\right)^6 = \tfrac{1}{64}\text{’}$$

$$= \frac{10\left(\frac{63}{64}\right)}{\frac{1}{2}} \qquad \text{... by simplifying terms}$$

$$\therefore \qquad S_6 = \frac{315}{16} \qquad \text{... by simplifying terms}$$

Therefore, the sum of the sequence $\text{‘}\left(10 \times \left(\tfrac{1}{2}\right)^0,\ 10 \times \left(\tfrac{1}{2}\right)^1,\ 10 \times \left(\tfrac{1}{2}\right)^2,\ ...,\ 10 \times \left(\tfrac{1}{2}\right)^5\right)\text{’}$ is $\text{‘}\frac{315}{16}\text{’}$.

In Application 9.3 below we will be applying Theorem 9.3, which states $\text{‘}\lim\limits_{n\to\infty} ca_n = c \lim\limits_{n\to\infty} a_n\text{’}$, to find the limit of the product of an infinite geometric sequence and a constant.

Application 9.3

Find the limit of the infinite geometric sequence $\left(0.7,\ 0.77,\ 0.777,\ ...\right)$; i.e. the infinite sequence of partial sums of the infinite sequence $\left(0.7,\ 0.07,\ 0.007,\ ...\right)$, which is the infinite sequence $\left(0.1,\ 0.11,\ 0.111,\ ...\right)$ multiplied by the constant '7'. We write this 'infinite sequence and constant' as the product sequence: $\left(7 \times 0.1,\ 7 \times 0.11,\ 7 \times 0.111,\ ...\right)$.

We can apply Theorem 9.3 to find the limit of this sequence.

$$\lim_{n\to\infty}\left(0.7,\ 0.77,\ 0.777,\ ...\right)$$

$$= \lim_{n\to\infty}\left(7 \times 0.1,\ 7 \times 0.11,\ 7 \times 0.111,\ ...\right)$$

$$\text{... by taking out the product '7'}$$

$$= 7 \times \lim_{n\to\infty}\left(0.1,\ 0.11,\ 0.111,\ ...\right)$$

$$\text{... according to Theorem 9.3}$$

We use Theorem 9.7 to find the limit of the sequence $\left(0.1,\ 0.11,\ 0.111,\ ...\right)$. For this sequence $\text{‘}a = \tfrac{1}{10}\text{’}$ and $\text{‘}r = \tfrac{1}{10}\text{’}$, so that $\lim\limits_{n\to\infty}\left(0.1,\ 0.11,\ 0.111,\ ...\right)$ is

$$\text{‘}\frac{a}{1-r} = \frac{\frac{1}{10}}{1-\frac{1}{10}} = \frac{1}{9}\text{’}.$$

$$\therefore \qquad \lim_{n\to\infty}\left(0.7,\ 0.77,\ 0.777,\ ...\right) = 7 \times \tfrac{1}{9} \quad \text{... by first half of application}$$

Or,

$$\lim_{n\to\infty}\left(0.7,\ 0.77,\ 0.777,\ ...\right) = \frac{7}{9} \qquad \text{... by simplifying terms}$$

Therefore, when multiplied term by term by the constant '7', the limit of the infinite geometric sequence $(0.1, 0.11, 0.111, ...)$ is the rational number '$\frac{7}{9}$'.

In Application 9.4 below we will be applying Theorem 9.4, which states that '$\lim\limits_{n\to\infty} (a_n + b_n) = L_1 + L_2$', to find the sum of two infinite geometric sequences.

Application 9.4

Find the limit of the sum of the infinite geometric sequence $(0.7, 0.77, 0.777, ...)$ and the constant infinite sequence $(\frac{2}{9}, \frac{2}{9}, \frac{2}{9}, ...)$ using Theorem 9.4.

We write this limit of the sum of these two infinite sequences as:

$$\lim_{n\to\infty} \left(\tfrac{2}{9} + 0.7, \tfrac{2}{9} + 0.77, \tfrac{2}{9} + 0.777, ...\right)$$

$$= \lim_{n\to\infty} \left(\tfrac{2}{9}, \tfrac{2}{9}, \tfrac{2}{9}, ...\right) + \lim_{n\to\infty} \left(0.7, 0.77, 0.777, ...\right)$$

... by Theorem 9.4

$$= \tfrac{2}{9} + \tfrac{7}{9}$$

... by Application 9.3

$$\therefore \lim_{n\to\infty} \left(\tfrac{2}{9} + 0.7, \tfrac{2}{9} + 0.77, \tfrac{2}{9} + 0.777, ...\right) = 1$$

... by simplifying terms

Therefore, when added to the constant sequence $(\frac{2}{9}, \frac{2}{9}, \frac{2}{9}, ...)$, the limit of the infinite geometric sequence $(0.7, 0.77, 0.777, ...)$ is the rational number '1'.

In Application 9.5 below we will be finding the sum of an infinite geometric sequence (which was the famous example quoted at the beginning of the chapter). This result is the first recorded sum of an infinite sequence and Archimedes used a form of this reasoning to find an area inside a two-dimensional figure.

Application 9.5

Find the sum of the infinite geometric sequence $(1, 4^{-1}, 4^{-2}, ... , 4^{-n}, ...)$ using Theorem 9.7.

From the definition of a geometric sequence: '$a = 1$' and '$r = \frac{1}{4} < 1$', we can write the limit of the sequence of partial sums $(S_1, S_2, S_3, ... , S_n, ...)$ as:

$$\lim_{n\to\infty} S_n = \frac{a}{1 - r}$$

... by Theorem 9.7

$$= \frac{1}{1 - \frac{1}{4}}$$

... by subst. for 'a' and 'r'

$\therefore$ $\qquad\qquad \lim\limits_{n \to \infty} S_n = \frac{4}{3}$ $\qquad\qquad$ ... by simplifying terms

Therefore, the sum of the infinite geometric sequence $\left(1,\ 4^{-1},\ 4^{-2},\ ...\ ,\ 4^{-n},\ ...\right)$ is the rational number '$\frac{4}{3}$'.

_____

This formal sum of an infinite sequence as the limit of the sequence of partial sums of the sequence gives us a simple way of proving Archimedes result and is an appropriate conclusion to the Application Section of this chapter.

Summary of Sequences, Series and Limits

In this chapter we have introduced the basic concepts and theory of sequences, series and limits of rational numbers. At this point, we have covered the key objectives of this chapter and so can summarise our findings up to now. The most important definitions we have covered in this chapter are:

1. Defining a 'sequence' as a set of **terms** in a **specific order**.

2. Defining a finite arithmetic sequence as a sequence that is written with a first term '$a_1 = a$' and subsequent terms derived by adding the constant term 'd' as follows:

    $$\left(a,\ a + d,\ a + 2d,\ ...\ ,\ a + (n-1)d\right)$$

3. Defining a finite geometric sequence as a sequence that is written with a first constant term 'a' (i.e. '$a_1 = a$') and subsequent terms generated by using the common ration 'r' as follows:

    $$\left(a,\ ar^1,\ ar^2,\ ...\ ,\ ar^{n-1}\right)$$

4. Defining a series as the sum of the terms of a sequence.

5. Defining the limit of the sequence $\left(a_1,\ a_2,\ a_3,\ ...\ ,\ a_N,\ ...\ ,\ a_n,\ ...\right)$ as 'L' if (and only if) in the case of an arbitrary small positive rational number 'ε', a positive integer 'n' exists such that for every integer 'n' where '$n > N$', the relation '$|a_n - L| < \varepsilon$' is a true inequality.

6. Defining the partial sums of a sequence $\left(a_1,\ a_2,\ a_3,\ ...\ ,\ a_n\right)$ as the sequence $\left(S_1,\ S_2,\ ...\ ,\ S_n\right)$ where the nth term, 'S_n' is defined by:

    $$S_n = a_1 + a_2 + ... + a_n$$

7. Defining the sum of an infinite sequence $\left(a_1,\ a_2,\ a_3,\ ...\ ,\ a_n,\ ...\right)$ as the limit 'S' (if it exists) of its sequence of partial sums $\left(S_1,\ S_2,\ S_3,\ ...\ ,\ S_n,\ ...\right)$ where this sum 'S' is given by the limit, as follows:

    $$S = \lim\limits_{n \to \infty} S_n$$

The three key results associated with the definitions of a series, sequences and limits are summarised in Table 9.3 below.

9.3 The Key Results for Sequences, Series and Limits	
1. **Sum of a Finite Arithmetic Sequence**	If $(a_1, a_2, \dots, a_n)$ is a finite arithmetic sequence of rational number terms and $(S_1, S_2, \dots, S_n)$ its sequence of partial sums, then we have: $$S_n = \frac{n}{2}(a_1 + a_n)$$ where 'a_1' is the first term, '$a_n = a_1 + (n-1)d$' is the nth term and 'd' is the common difference between adjacent terms.
2. **Sum of a Finite Geometric Sequence**	If $(a, ar^1, \dots, ar^{n-1})$ is a finite geometric sequence of rational number terms and $(S_1, S_2, \dots, S_n)$ its sequence of partial sums, then we have: $$S_n = a + ar + ar^2 + \dots + ar^{n-1} = \frac{a(1-r^n)}{1-r}$$ where 'a' is the first term, 'ar^{n-1}' is the nth term and 'r' is the common ratio between adjacent terms.
3. **Sum of an Infinite Geometric Sequence**	If $(S_1, S_2, \dots, S_n, \dots)$ is the infinite sequence of partial sums of the infinite geometric sequence $(a, ar, ar^2, \dots, ar^{n-1}, \dots)$ where 'a' and 'r' are rational number terms and '$|r| < 1$' and '$S_n = \frac{a(1-r^n)}{1-r}$' then it follows that: $$\lim_{n \to \infty} S_n = \frac{a}{1-r}$$

Table 9.3

There are many simple practical examples of the uses of sequences, series and limit. Some of these key examples are:

1. Calculating the interest on savings in a bank account uses a geometric sequence formula to calculate interest earned on savings each day, month or year.

2. For paying a mortgage on a house, the bank uses a geometric sequence to calculate the payments required every month at a certain interest rate to pay off the loan within a certain number of years.

3. To make a serious business decision a manager may use a geometric sequence to calculate the value of that decision today (using Net Present Value, i.e. NPV) from estimates of future payments, future earnings, future costs, depreciation, long-term bond rates, etc. to compare the value of pursuing various business strategies.

4. Although 'limits' are indispensable for understanding later chapters of this book, they are also required in secondary school mathematics for the following processes:

   • Calculus to define rates of change of simple physical quantities such as slopes of graphs, displacements, velocities, acceleration, force, energy, momentum, temperature, currents, costs, earnings, productivity, etc.

   • Calculus to define summations of quantities, such as areas, volumes and a vast array of other physical quantities.

These are simple examples which will affect the future of most students at some time during their life.

In this chapter we have explored all of the definitions and theorems necessary to use the concept of a limit in order to overcome the conflict associated with arithmetic or algebraic terms with an infinite number of operations. Developing arithmetic proficiency in order to manipulating these 'infinite' arithmetic and algebraic terms will be essential to mastering the remaining chapters in this textbook.

In Chapter Ten we will be extrapolating the theory we learnt about in this chapter to close the gap between finite decimals (which we introduced in Chapter Six) and rational numbers by demonstrating how we can represent all rational numbers as repeating decimal numbers.

CHAPTER 10

THE RATIONAL NUMBER SYSTEM USING REPEATING DECIMALS

Overview of Rational Number System using Repeating Decimals

In Chapter Nine we continued our study of the Rational Number System using sequences, series and limits and applied them to decimal numbers. In particular, we saw how an infinite geometric sequence can be added progressively to obtain the limit of an infinite sum.

You will now be aware that we can form infinite geometric sequences from 'repeating' decimals. From this starting point, we can go on to prove some simple results such as the sum of the infinite series '$0.333... = 0.3 + 0.03 + 0.003 + ...$' is '$\frac{1}{3}$' and sum of the infinite series '$0.999... = 0.9 + 0.09 + 0.009 + ...$' is '1'. You will recall from Chapter Nine that the use of the terms '$0.333...$' and '$0.999...$' in the previous two equations is short-hand notation for using the limit process as we can only deal with finite valid strings of symbols when writing terms and equations. So, in this (and later) chapters when the words 'infinite decimal' are used, you will be aware that this is a cue to say to yourself, "At this point I am dealing with an unlimited number of digits using the limit process"

In this chapter we will use the result from Chapter Nine showing that '$0.999...$' can be identified with the number '1' to change the format of a finite decimal to a 'repeating' decimal format. In this way, we will be changing the finite decimal '3.0' to the 'repeating' decimal '$2.999...$' (This concept will be demonstrated later in Theorem 10.2).

If we now combine the set of finite decimals from Chapter Six with the 'repeating' decimals, that is, those decimals originating from fractions such as '$\frac{1}{3}$', '$\frac{12}{99}$', we refer to the combined set of decimals as the set of **repeating decimals**. Consequently, if a number can be written as a repeating decimal (even though it may be given in finite format), then it can be considered to be a repeating decimal number.

Different Representations of the
same Rational Number 'one-third'

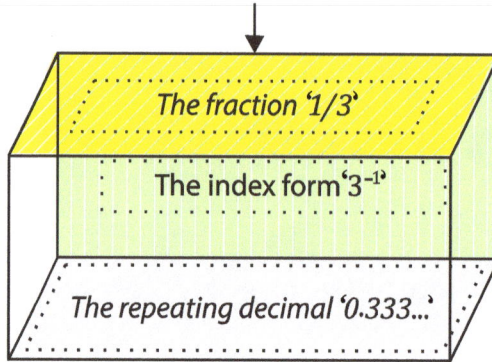

By extending the theory we developed in Chapter Six, we can demonstrate that every finite decimal (e.g. '3.0') also has a multiplicative inverse (e.g. '$\frac{1}{3}$'). We can then naturally build on this idea to develop the 'Repeating Decimal System'.

In this chapter we will be demonstrating that every rational number using fractions can be expressed as a repeating decimal and that every repeating decimal can be expressed as a fraction. This premise clearly illustrates that the 'Repeating Decimal System' is simply the Rational Number System using Repeating Decimals.

In Chapter Nine we demonstrated the way in which repeating decimals can be manipulated with the operations of addition and multiplication using the limit process. In this chapter we will explore how the operations of addition, multiplication, subtraction and division are defined for these repeating decimals by their equivalent operations acting on their equivalent rational numbers as fractions.

Developing an awareness that every number in the set of repeating decimals has an equivalent fraction (and vice versa) enhances our comprehension of the rational numbers, as well as providing insight into the next extension of the Rational Number System – the 'Real Number System' which we will be studying in Chapter Twelve.

Another way of expressing a rational number is through the use of Percentage Notation. Percentages are simply a way of representing a number relative to a base of '100'. The concept of percentages is quite straightforward and, hence, does not require a chapter of its own. The notation and operations for percentages will be provided at the end of each section in this chapter.

In Chapter Eleven we will be examining those decimals with an unlimited number of digits in their decimal fraction that do not have a repeating pattern; these decimals will be defined as 'irrational numbers'. Also in Chapter Eleven we will be considering the theory behind the famous proof that since these irrational numbers cannot be written as fractions, they therefore cannot be rational numbers.

Pictorial Representation of Rational Numbers using Repeating Decimals

Like in previous chapters where we discussed rational numbers as fractions and using index form, we will provide a pictorial representation of repeating decimals. In Figure 10.2 below you will see that we have labelled several points along the number line with their repeating decimal notation (including several integer values).

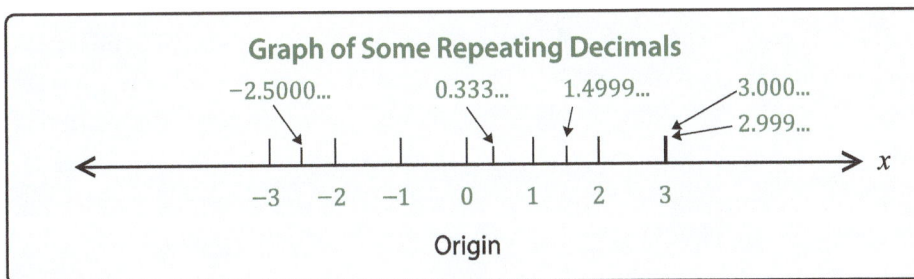

Graph of Some Repeating Decimals

Figure 10.2

In Figure 10.2 above, we have labelled the point that is furthest right with the coordinate '3', '3.000...' and '2.999...'. As mentioned above, Theorem 9.6 gives these coordinates as equivalent alternative representations of the same rational number '$\frac{3}{1}$' using repeating decimals.

Background and Context of Rational Numbers using Repeating Decimals

In Chapter Six we established the background and the standard notation for **finite decimals** which can be used to carry out many mathematical calculations. However, as the numbers we studied in that chapter were finite, they were of limited value in understanding and working with rational numbers. In this chapter we will create the 'standard' decimal system by extending the theory covered in Chapter Six to include repeating decimal numbers.

The introduction of repeating decimal notation in this chapter provides us with a third **major** way of representing rational numbers. Although the finite decimals allow us to perform the three operations of addition, subtraction and multiplication of rational numbers, they do not permit the operation of division ($\div$). The full operation of division of rational numbers using decimals is only possible as a result of introducing repeating decimals.

To accommodate the new way of expressing rational numbers as repeating decimals, we use the ellipsis notation introduced in Chapter Two and continued in

Chapter Nine. For example, the notation that we use to represent the fractions '$\frac{1}{3}$' and '$\frac{12}{99}$' as repeating decimals is '0.333...' and '0.121212...'.

The fourth and final **major** way of representing rational numbers is with Percentages. Even before the decimal system was invented, the Ancient Romans performed some of their mathematical calculations with fractions using '100' as the standard number of units. Using a base of '100' was known as the Percentage System, whereby expressing a fraction of a number with a base of '100' results in a percentage. The notation that identifies a percentage string is the 'percent' symbol '%'. The string of symbols '25%' is simply referred to as 'twenty five per cent'.

Approach to the Rational Number System using Repeating Decimals

In this section our aim is to assign meaning to the alphabetic symbols of the language of repeating decimals. This language is an extension of the language we used to represent finite decimals, rational numbers as fractions and rational numbers in index form.

Our approach in this chapter is to establish the language and definitions required to prove that repeating decimals offer an alternative third equivalent way of expressing rational numbers.

To maintain consistency with our definition of terms and equations as finite valid strings of symbols, we will use the results of Chapter Nine to find a valid way of representing '$\frac{1}{3}$' in decimal notation.

Within this chapter you will learn that for every fraction, index form or repeating decimal, there is a corresponding percentage. Similarly, for every percentage there is a corresponding fraction, index form and repeating decimal. Finally, you will discover how repeating decimal numbers and percentages can be combined using the basic operations of addition and multiplication.

Language of the Rational Number System using Repeating Decimals

The symbols necessary to create the language of the repeating decimals are outlined in the Definition Section below. This language is an extension of the language we first introduced in Chapter Six. We will also be extending the alphabet to allow us to express any rational number as a percentage using the '%' symbol.

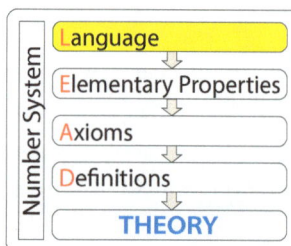

Number System

Language
⇩
Elementary Properties
⇩
Axioms
⇩
Definitions
⇩
THEORY

The current symbols of the alphabet are summarised in Table 10.1 below.

10.1 The Alphabet of the Rational Numbers as Repeating Decimals	
Symbols	**Meaning from:**
'0', '1', '+', '×', '(', ')'	The Axioms of the Natural Number System
'=', '≠'	The Axioms of Logic
'2', '3', '4', '5', '6', '7', '8', '9'	Definition 2.1
'a', 'b', 'c', ... , 'x', 'y', 'z', ...	The rational number variables
','	Definition 2.3
'-1'	The Axioms of the Integer Number System
'$-$'	Definitions 3.4 and 3.5
'$\frac{1}{\square}$'	The Axioms of the Rational Number System as Fractions
'$\frac{a}{b}$'	Definition 4.2
'÷'	Definition 4.4
'$\square^{-1}$'	The Axioms of Rational Number System using Index Form
'$(a \times b^{-1})^m$'	Definition 5.6
'.'	Definition 6.4
'$a_n ... a_1 a_0 . d_1 ... d_m$'	Definition 6.5
'$\pm$'	Definition 7.8
'$<$'	The Axioms of the Rational Number System using Order
'$\leq$', '$>$', '$\geq$'	Definition 8.2
'$|x - y|$'	Definition 8.7
'{', '}'	Definition 9.1
'$(\cdot, \cdot)$'	Definition 9.2 (overloading the parentheses symbols)
'∞'	Definition 9.3
'$\lim\limits_{n \to \infty} a_n$'	Definition 9.10

Table 10.1

Once again, we use variables to represent these decimal numbers in this language of the Rational Numbers using Repeating Decimals. In this way, a variable (such as 'x', 'y' or 'z') is a symbol to which we can assign any repeating decimal number.

The use of repeating decimals and variables is summarised in Table 10.2 below.

10.2 The Language of Rational Numbers as Repeating Decimals	
In this Table let 'x', 'y' and 'z' represent repeating decimal variables and let 'a', 'b', and 'c' be restricted to representing integer variables.	
1. **Arithmetic Terms**	An individual repeating decimal (or a sum and/or product of repeating decimals constructed from the repeating decimals alphabet) is referred to as an **arithmetic term** or simply a **term**. Hence: '$4.777... \times 21.12333... + 10^{-1}$', '$54.5 \times 10^{-1} \times 1.21\,2121... \times 10^{-1}$', '$7.1 + 5 - 5.987987987... \times 10^{-1}$', '$-5.23 + 4.999... \times 10^{-1}$', '$123.46$', '$-3.142... \times 10^{-1}$', '$(2 + 10^{-1}) \times 10^{-1}$', '$5.2346$', '$-10^{-1} + -0.999... \times 4$' are all valid strings of symbols called **arithmetic terms**.
2. **Arithmetic Equations**	An equation with arithmetic terms on either side of an '$=$' or '$\neq$' sign, is called an **arithmetic equation**. For example: '$4 \times (-1.999... + 5 \times 10^{-1} + 1.25) = -1$' is a valid string of symbols called an **arithmetic equation**.
3. **Algebraic Terms and Equations**	If any of the above terms or equations contains one or more variables, then we refer to them as **algebraic** terms or equations. For example, '$3.999... \times 10^{-1} \times x$' is a valid string of symbols called an **algebraic term**. Likewise, '$y \times 1.5 \times 10^{-1} = z \times 2.999...$', '$x \times y \times z \times 10^{-1} = 5.123999... \times x$' and '$x = a \times 0.333... + b$' are valid strings of symbols called **algebraic equations**.

Table 10.2

Note:

This table includes decimals in 'repeating decimal notation' which is introduced in Definition 10.1 below.

The operations of addition and multiplication will be defined in Definition 10.3 below.

Order of Operations

We can extend the same Order of Operations convention we used for the Rational Number System using Fractions and using Index Form and apply it to repeating decimal numbers.

The same Order of Operations convention, '**PEMDAS**', that we applied to the Rational Number System in Chapters 4 and 5 is used for Rational Numbers using Repeating Decimals.

The Elementary Property of Multiplication for Rational Numbers using Repeating Decimals

In the same way that an Inverse Property of Multiplication exists for the Rational Number System using Fractions and using Index Form, there is an Inverse Property of Multiplication in the Rational Number System using Repeating Decimals. This multiplicative inverse is outlined in Definition 10.4 below and relies on the one-to-one correspondence between repeating decimals and rational numbers as fractions. You will recall from previous chapters that we have always defined the division operation in terms of the multiplicative inverses of numbers. In this way, it is apparent that the concept of multiplicative inverses is more fundamental than the concept of the division operation between numbers.

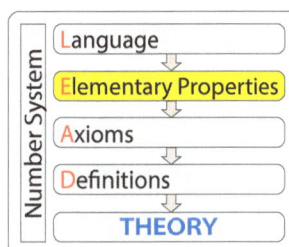

The Axioms of the Rational Number System using Repeating Decimals

The basic operations of addition and multiplication of repeating decimals are defined in terms of their equivalent operations on fractions. Therefore, we can use the Axioms of the Rational Number System using Fractions to model the behaviour of repeating decimals.

In other words, the Repeating Decimal System is not a 'standalone' system the way we introduce it here as it relies on the Rational Number System using Fractions to assign meaning to its limits. In particular, the ellipsis symbol in the repeating decimal '0.121212...' is an abbreviation for the limit process which was rigorously defined in Chapter Nine. This process will be enhanced in Chapter Twelve after you have become more familiar with infinite decimals.

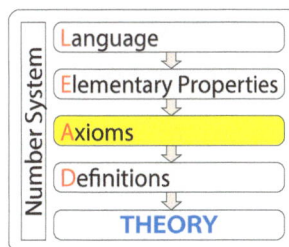

The Axioms of Logic

To reason correctly within repeating decimal equations, we again need to use the Axioms of Logic which offer us the properties of reasoning that we learnt about in Chapter Two, Table 2.5. This can also be expressed by saying that The Axioms of Logic allow us to manipulate arithmetic and algebraic terms and equations also written in the language of Repeating Decimals.

Definitions of the Rational Number System using Repeating Decimals

The fundamental concept of this chapter centres on the repeating decimal. In the Theory Section below, we will explore the relationship between this concept and the previous expression of rational numbers as fractions and using index form.

The definition of 'repeating decimal numbers' will allow us to formalise the direct relationship between fractions and repeating decimal numbers.

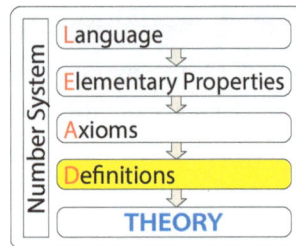

Definition 10.1   Positive Repeating Decimal Numbers

The series of digits '$a_n \dots a_1 a_0.d_1 \dots d_m d_1 \dots d_m \dots$' is a **positive repeating decimal** if (and only if) the following statements are true:

1. The sequence of symbols $(a_n, \dots, a_0)$ represent the '$n + 1$' digits to the left of the decimal point and the sequence of symbols $(d_1, \dots, d_m)$ represent the 'm' digits directly to the right of the decimal point.

2. The ellipsis after the second group of digits '$d_1 \dots d_m$' in the repeating pattern means there is an unlimited number of repetitions of this group of digits.

3. Repeating decimals can be written in two basic formats:

    i. either where the repeating pattern commences directly after the decimal point '$a_n \dots a_1 a_0.d_1 \dots d_m d_1 \dots d_m \dots$' or

    ii. where the repeating pattern does not commence directly after the decimal point, i.e. '$a_n \dots a_q.a_{q-1} \dots a_1 a_0 d_1 \dots d_m d_1 \dots d_m \dots$'. Initially, we will define a positive repeating decimal where the repeating pattern commences directly after the decimal point.

4. The use of positive repeating decimal numbers also requires meaning to be assigned to the unlimited number of repetitions of the digits '$d_1 \dots d_m$' in the repeating pattern. Once again, we will be using the limit process we learnt about in Chapter Nine to undertake this activity.

First, we form an infinite sequence from the repeating pattern of digits after the decimal point in the series '$a_n \dots a_1 a_0.d_1 \dots d_m d_1 \dots d_m \dots$' as:

$$\left(d_1 \dots d_m \times (10^{-m})^1, \, d_1 \dots d_m \times (10^{-m})^2, \, \dots, \, d_1 \dots d_m \times (10^{-m})^p, \, \dots \right).$$

Next, we write out the pth sum, i.e. 'S_p', of the sequence we have just formed as follows:

$$S_p = d_1 \dots d_m \times (10^{-m})^1 + d_1 \dots d_m \times (10^{-m})^2 + \dots + d_1 \dots d_m \times (10^{-m})^p$$

5. The term '$\lim\limits_{p \to \infty} S_p$' represents the decimal fraction sum of the unlimited string of digits to the right of the decimal point. Therefore, we can write the positive repeating decimal as an integer part and a fraction part as follows:

$$a_n \dots a_1 a_0.d_1 \dots d_m d_1 \dots d_m \dots = a_n \dots a_1 a_0 + \lim\limits_{p \to \infty} S_p$$

Note:

To define a positive repeating decimal, we use the limit process to find the infinite sum of the decimal fraction part of the decimal. This method allows us to express a repeating decimal as the sum of an integer and a fraction.

The simplest application of Definition 10.1 is to use the positive repeating decimal '0.999...'. By applying this definition, we get:

$$0.999\dots = 0 + \lim\limits_{p \to \infty} S_p$$

$$0.999\dots = \lim\limits_{p \to \infty} S_p$$

Here: $\qquad S_p = 9 \times 10^{-1} + 9 \times 10^{-2} + \dots + 9 \times 10^{-p}$

$\therefore \qquad \lim\limits_{p \to \infty} S_p = \dfrac{a}{1-r} = \dfrac{0.9}{1 - \frac{1}{10}} = \dfrac{9}{10-1} = 1 \qquad$ ... by Theorem 9.6

$\therefore \qquad 0.999\dots = 1 \qquad$ (where '0.999...' is defined by the limit process)

In this way, our definition of a positive repeating decimal forces the digits after the decimal point into fractional form by using the formula '$\lim\limits_{p \to \infty} S_p = \dfrac{a}{1-r}$' (Theorem 9.6). Using the same application of Definition 10.1, we are also aware that the expected answer for the repeating decimal '0.333...' is '$\dfrac{1}{3}$'.

It is **crucial** to emphasise that the '0.999...' is simply an abbreviated notation for the meaningful term '$\lim\limits_{p \to \infty} S_p$'. So we can say that '0.999...' is only equal to '1' or, '1/1' as a fraction, in the context of the concept of limits.

The next step is to define the second format for a repeating decimal where the repeating decimals have a repeating pattern that does not commence directly after the decimal point. We will define these numbers as a 'general decimal with a repeating pattern'.

Definition 10.2   Positive Repeating Decimal (where the pattern does not start directly after the decimal point)

The series of digits '$a_n \dots a_q.a_{q-1} \dots a_1 a_0 d_1 \dots d_m d_1 \dots d_m \dots$' is also a positive repeating decimal where the repeating pattern doesn't start directly after the decimal point. That is, a positive repeating decimal is a decimal number where the digits start to repeat a finite number of decimal places to the right of the left-most digit.

The above definition of a repeating decimal can be converted into the simpler form of a repeating decimal (Definition 10.1) by multiplying it by a power of '10'. Doing so will allow you to move the decimal point to the start of the repeating pattern.

An example of this conversion is the repeating decimal '123.4567898989...' that can be written as '$1234567.898989\dots \times 10^{-4}$' and, hence, the repeating pattern '898989...' starts immediately to the right of the decimal point in this notation.

Now that we have a definition for these new repeating decimal numbers, it is straightforward for us to define addition and multiplication of these numbers.

Definition 10.3   Addition and Multiplication of Positive Repeating Decimals

If we have two positive repeating decimals '$a_n \dots a_1 a_0.d_1 \dots d_m d_1 \dots d_m \dots$' and '$b_l \dots b_1 b_0.e_1 \dots e_n e_1 \dots e_n \dots$', created from the four sets of symbols $\{a_n, \dots, a_1, a_0\}$, $\{d_1, \dots, d_m\}$, $\{b_l, \dots, b_1, b_0\}$ and $\{e_1, \dots, e_n\}$, we can define addition and multiplication of these positive repeating decimals by:

$$(a_n \dots a_1 a_0.d_1 \dots d_m d_1 \dots d_m \dots) + (b_l \dots b_1 b_0.e_1 \dots e_n e_1 \dots e_n \dots)$$
$$= (a_n \dots a_1 a_0 + \lim_{p \to \infty} S_p) + (b_l \dots b_1 b_0 + \lim_{p \to \infty} T_p)$$

and,

$$(a_n \dots a_1 a_0.d_1 \dots d_m d_1 \dots d_m \dots) \times (b_l \dots b_1 b_0.e_1 \dots e_n e_1 \dots e_n \dots)$$
$$= (a_n \dots a_1 a_0 + \lim_{p \to \infty} S_p) \times (b_l \dots b_1 b_0 + \lim_{p \to \infty} T_p)$$

where the pth partial sums associated with the decimal fraction parts are given by:

$$S_p = d_1 \dots d_m \times (10^{-m})^1 + \dots + d_1 \dots d_m \times (10^{-m})^p$$

and:

$$T_p = e_1 \dots e_n \times (10^{-n})^1 + \dots + e_1 \dots e_n \times (10^{-n})^p$$

To give a consistent representation of the sum and product as repeating decimals, the terms '$\lim_{p \to \infty} S_p$' and '$\lim_{p \to \infty} T_p$' can always be converted back to repeating decimal format.

A simple example involves the addition of the repeating decimals '0.333...' and '0.121212...'. Using Definition 10.3 we get:

$$0.333\dots + 0.1212\dots \quad = \frac{1}{3} + \frac{12}{99} \qquad \text{... by Definition 10.1}$$

$$= \frac{33}{99} + \frac{12}{99} \qquad \text{... by common denominator}$$

$$= \frac{45}{99} \qquad \text{... by adding numerators}$$

$$= 0.454545\dots \qquad \text{... by Definition 10.1}$$

Therefore:

$$0.333\dots + 0.1212\dots \quad = 0.454545\dots \qquad \text{... written as a repeating decimal}$$

In order to define negative repeating decimals and inverses of repeating decimals, we will now define the additive and multiplicative inverses for these repeating decimals using the multiplication operation. This process will be undertaken in Definition 10.4 below.

Definition 10.4   Additive and Multiplicative Inverses of Repeating Decimals

Let 'x' represent the positive repeating decimal '$a_k \dots a_1 a_0.d_1 \dots d_m d_1 \dots d_m \dots$'. We can define the additive inverse of 'x' as '$-x$' where:

$$-x = -1 \times x$$

We can define the multiplicative inverse of 'x' as 'x^{-1}' where:

$$x \times x^{-1} = 1$$

Once again, in the same way that we've demonstrated the other representations of the Rational Number System, it is easy to define the subtraction and division operations in terms of these additive and multiplicative inverses, respectively.

Definition 10.5   Subtraction and Division of Repeating Decimals

Let 'x' and 'y' represent repeating decimals.

We can define 'x' **subtract** 'y' as '$x - y$' where:

$$x - y = x + -y$$

We can define 'x' **divided** by 'y' as '$x \div y$' where:

$$x \div y = x \times y^{-1}$$

————————————

To complete this Definition Section, we will examine how percentages are related to rational numbers. Percentages can be thought of as a way of representing a rational number as a fraction of '100'. Definition 10.6 below formalises our intuitive use of percentage numbers.

Definition 10.6   Percentage Numbers

The percentage symbol is defined by the following equation:

$$1 = 100\%$$

————————————

The way that we normally use percentages is in answering such questions as: What is 'a' percent of the number 'b'? For example, what is '25%' of the number '16'? The key word in this sentence is 'of' which can be translated into the equivalent mathematical symbol '$\times$'. In this way, to evaluate this term we follow the process outlined below:

$$25\% \times 16 = \frac{25}{100} \times 16 \qquad \text{... by '}1\% = 1/100\text{'}$$

$$= \frac{1}{4} \times 16 \qquad \text{... simplifying the fraction}$$

$$\therefore \quad 25\% \times 16 = 4 \qquad \text{... simplifying the product}$$

In the Theory Section of this chapter we will demonstrate how the main operations of '+' and '$\times$' can be applied to this new representation of numbers.

We can now append this new symbol relating to percentage notation to our Alphabet of the Rational Numbers using Repeating Decimals (see Table 10.3 on the following page).

10.3 The Alphabet of the Rational Numbers as Repeating Decimals	
Symbols	**Meaning from:**
'0', '1', '+', '×', '(', ')'	The Axioms of the Natural Number System
'=', '≠'	The Axioms of Logic
'2', '3', '4', '5', '6', '7', '8', '9'	Definition 2.1
'a', 'b', 'c', ..., 'x', 'y', 'z', ...	The rational number variables
','	Definition 2.3
'−1'	The Axioms of the Integer Number System
'−'	Definitions 3.4 and 3.5
'$\frac{1}{\Box}$'	The Axioms of the Rational Number System as Fractions
'$\frac{a}{b}$'	Definition 4.2
'÷'	Definition 4.4
'$\Box^{-1}$'	The Axioms of Rational Number System using Index Form
'$(a \times b^{-1})^m$'	Definition 5.6
'.'	Definition 6.4
'$a_n ... a_1 a_0.d_1 ... d_m$'	Definition 6.5
'±'	Definition 7.8
'<'	The Axioms of the Rational Number System using Order
'≤', '>', '≥'	Definition 8.2
'$|x-y|$'	Definition 8.7
'{', '}'	Definition 9.1
'(', ')'	Definition 9.2 (overloading the parentheses symbols)
'∞'	Definition 9.3
'$\lim_{n\to\infty} a_n$'	Definition 9.10 (refer to the Note following this definition)
'$a_n ... a_1 a_0.d_1 ... d_m ...$'	Definition 10.1
'%'	Definition 10.6

Table 10.3

In this chapter we introduced the notation of a repeating decimal number. This now allows us to generalise the Finite Decimal System presented in Chapter Six so that we can represent all rational numbers in decimal notation.

Theory of Rational Numbers as Repeating Decimals

In this section we will derive a simple general equation for expressing any repeating decimal as a fraction. We begin this process by first expressing the simplest form of a repeating decimal as a fraction. This result is expressed in Theorem 10.1 below.

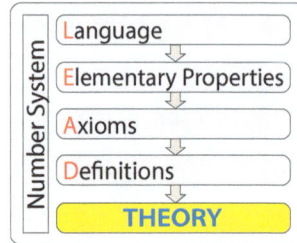

Number System
Language
⇩
Elementary Properties
⇩
Axioms
⇩
Definitions
⇩
THEORY

Theorem 10.1   Repeating Decimals as Fractions

Let a repeating decimal less than '1' (excluding the case of '0.999 ...') be given by the decimal expansion: '$0.d_1d_2 \ldots d_md_1d_2 \ldots d_md_1d_2 \ldots d_m \ldots$' where there are '$m$' digits in the repeating pattern. This repeating decimal can be written in fraction form as follows:

$$0.d_1d_2 \ldots d_md_1d_2 \ldots d_md_1d_2 \ldots d_m \ldots = \frac{d_1d_2 \ldots d_m}{99 \ldots 9}$$

In the fraction above there are 'm' nines in the denominator.

Proof

The repeating decimal expansion above can be written as the sum of the corresponding infinite geometric sequence:

'$(d_1d_2 \ldots d_m \times (10^{-m})^1, \ d_1d_2 \ldots d_m \times (10^{-m})^2, \ d_1d_2 \ldots d_m \times (10^{-m})^3, \ldots)$'

The pth partial sum of this geometric sequence is given by:

$$S_p = d_1d_2 \ldots d_m \times (10^{-m})^1 + d_1d_2 \ldots d_m \times (10^{-m})^2 + d_1d_2 \ldots d_m \times (10^{-m})^3 + \ldots + d_1d_2 \ldots d_m \times (10^{-m})^p$$

We can now identify the terms 'a' and 'r' of the infinite geometric sequence:

$$a = d_1d_2 \ldots d_m \times (10^{-m})^1 \quad \text{... by Definition 9.8}$$

$$r = 10^{-m} \quad\quad\quad\quad\quad\quad \text{... by Definition 9.8}$$

According to Definition 10.1, we have the following equation:

$$0.d_1d_2 \ldots d_md_1d_2 \ldots d_m \ldots = \lim_{p \to \infty} S_p \quad\quad \text{... by Definition 10.1}$$

$$= \frac{a}{1-r} \qquad \text{... by Theorem 9.6}$$

$$= \frac{d_1 d_2 \dots d_m \times 10^{-m}}{1 - 10^{-m}} \qquad \text{... by substit. for '}a\text{' and '}r\text{'}$$

$$= \frac{d_1 d_2 \dots d_m}{10^m - 1} \qquad \text{... by multiplying by '}\frac{10^m}{10^m}\text{'}$$

$$\therefore \quad 0.d_1 d_2 \dots d_m d_1 d_2 \dots d_m \dots = \frac{d_1 d_2 \dots d_m}{99 \dots 9} \qquad \text{... by simplifying denominator}$$

In this last equation, you can see that there are 'm' nines in the denominator. The general case of a repeating decimal can be written using this result as follows:

$$a_n \dots a_1 a_0 . d_1 \dots d_m d_1 d_2 \dots d_m \dots$$

$$= a_n \dots a_1 a_0 + 0.d_1 d_2 \dots d_m d_1 d_2 \dots d_m \dots \quad \text{... by Definition 10.1}$$

$$= a_n \dots a_1 a_0 + \frac{d_1 d_2 \dots d_m}{99 \dots 9} \qquad \text{... by first part of the Theorem}$$

$$\therefore a_n \dots a_1 a_0 . d_1 \dots d_m d_1 d_2 \dots d_m \dots = a_n \dots a_1 a_0 + \frac{d_1 d_2 \dots d_m}{99 \dots 9}$$

This completes the proof of the theorem and its generalisation.

––––––––––––––––

Note:

In cases where the repeating pattern does not commence directly after the decimal point, we can use a power of ten to convert it to the above form and then multiply by the inverse operation. Subsequently, we can convert the fraction part of the decimal in the same way that we did in the equation above.

For example, the repeating decimal '123.4567898989...' can be written as '1234567.898989... $\times 10^{-4}$' in such a way that the repeating pattern '898989...' starts immediately to the right of the decimal point. Hence, the result can be written as '$\frac{89}{99}$'.

By using Theorem 10.1 we can be certain that every repeating decimal can always be written with an integer part and a fraction part. Consequently, every repeating decimal number can be written as a rational number.

We now list several repeating decimals and give their corresponding representations as rational numbers to illustrate the limit of the sequence of partial sums for these general decimals:

$$0.333\dots = \lim_{n \to \infty} \left(\frac{3}{10} + \frac{3}{10^2} + \dots + \frac{3}{10^n} \right) = \frac{1}{3}$$

$$0.1666\ldots = \lim_{n\to\infty}\left(\frac{1}{10} + \frac{6}{10^2} + \ldots + \frac{6}{10^n}\right) = \frac{1}{6}$$

$$0.142857142857\ldots = \lim_{n\to\infty}\left(\frac{142857}{10^6} + \frac{142857}{10^{12}} + \ldots + \frac{142857}{10^{6n}}\right) = \frac{1}{7}$$

$$1.444\ldots = 1 + \lim_{n\to\infty}\left(\frac{4}{10} + \frac{4}{10^2} + \ldots + \frac{4}{10^n}\right) = 1 + \frac{4}{9} = \frac{13}{9}$$

$$2.090909\ldots = 2 + \lim_{n\to\infty}\left(\frac{9}{10^2} + \frac{9}{10^4} + \ldots + \frac{9}{10^{2n}}\right) = 2 + \frac{9}{99} = \frac{23}{11}$$

In the Introduction Section of this chapter, we established the fact that every finite decimal could be written as a repeating decimal. Theorem 10.2 below justifies this claim.

Theorem 10.2   Finite Decimals as Repeating Decimals

Every finite decimal can be expressed as a repeating decimal.

Proof

In Chapter Six we saw how every positive finite decimal could be written as '$a_n \ldots a_1 a_0.d_1 \ldots d_m$' where the set of digits $\{a_n, \ldots, a_1, a_0\}$ gives the value of the integer before the decimal point and the set of digits $\{d_1, \ldots, d_m\}$ gives the value of the decimal fraction after the decimal point.

Case 1:

There is a last non-zero digit to the right of the decimal point given by 'd_m'.

We will use the equation '$1 = 0.999\ldots$' to convert the finite decimal '$a_n \ldots a_1 a_0.d_1 \ldots d_m$' into a general decimal as follows:

1.   Reduce 'd_m' by '1' (since non-zero) and define 'd'_m' so that: '$d'_m = d_m - 1$'.

2.   Include a string of nines after 'd'_m' to create a general decimal expansion.

According to Definition 6.4, this result is justified since the digit 'd_m' contributes to the finite decimal '$a_n \ldots a_1 a_0.d_1 \ldots d_m$' the value '$d_m \times 10^{-m}$'. Hence, we can write:

$$d_m = d'_m + 1 \qquad\qquad \text{... by definition of '}d'_m\text{'}$$

$$= d'_m + 0.999\ldots \qquad\qquad \text{... by equation '}1 = 0.999\ldots\text{'}$$

$$\therefore \quad d_m \times 10^{-m} = d'_m \times 10^{-m} + 0.999\ldots \times 10^{-m} \text{ ... by multiplying by '}10^{-m}\text{'}$$

Consequently, the general decimal expansion of '$a_n \ldots a_1 a_0.d_1 \ldots d_m$' is given by:

$$a_n \ldots a_1 a_0.d_1 \ldots d_m = a_n \ldots a_1 a_0.d_1 \ldots d'_m\, 999 \ldots$$

Note:

Any trailing zeroes to the right of the last non-zero digit can be ignored as they don't contribute to the value of the decimal number.

Case 2:

There is no digit to the right of the decimal point.

Given that this is a finite decimal where there is no digit to the right of the decimal point, it must only be an integer decimal i.e. of the form '$a_n...a_1 a_0$'. We can then convert this result to a repeating decimal as follows:

$$a_n ... a_1 a_0 = a_n ... a_1 a_0 + 0 \qquad \text{... by Identity Axiom '+'}$$

$$= a_n ... a_1 a_0 + -1 + 1 \qquad \text{... by Inverse Axiom '+'}$$

$$= (a_n ... a_1 a_0 - 1) + 0.999... \qquad \text{... by eqn. '}1 = 0.999...\text{'}$$

$$\therefore \qquad a_n ... a_1 a_0 = (a_n ... a_1 a_0 - 1).999...$$

This last equation proves that any finite decimal (such as '30') with no decimal fraction component can be written as a repeating decimal (i.e. as '29.999...'). This completes the proof of Theorem 10.2.

––––––––––––––

We are now going to prove the inverse of the result we obtained in Theorem 10.1, that is, that every fraction can be expressed as a repeating decimal. In Chapter Eleven you will learn about 'infinite' decimals that aren't repeating decimals and how they represent numbers like '$\sqrt{2}$' which are 'irrational' numbers.

Theorem 10.3   Fractions as Repeating Decimals

Every rational number '$\frac{p}{q}$' (where 'p' and '$q \neq 0$' are integers) can be expressed as a repeating decimal.

Proof

Case 1:

The fraction '$\frac{p}{q}$' gives a finite decimal when 'q' is divided into 'p'.

Applying the long-division algorithm gives '$\frac{p}{q}$' as a finite decimal. Using Theorem 10.2 this decimal can be converted to a repeating decimal.

Case 2:

The fraction '$\frac{p}{q}$' gives a repeating decimal when 'q' is divided into 'p'.

By using the long-division algorithm, the fraction '$\frac{p}{q}$' must create a repeating pattern no further than '$q-1$' places after the decimal point since when dividing 'q' into 'p' the remainders can only be one of the numbers '1, 2, 3, ... , $q-1$' before the pattern starts repeating itself. Note that a remainder of '0' would indicate that the decimal expansion has terminated which can't be the case for a repeating decimal.

This completes the proof of Theorem 10.3.

———————————

An example which illustrates Case 2 above involves dividing the number '1' by the number '17'. We know that the pattern must start repeating no later than '$17 - 1 = 16$' digits. This outcome can be seen by the actual expansion:

$$\text{'}\frac{1}{17} = 0.05882352941176470588235294117647...\text{'}$$

We can now write specific equations for the operations of addition and multiplication using percentages.

Theorem 10.3   Addition of Two Percentages

To express the percentage notation '$a\%$' and '$b\%$' for the rational number variables 'a' and 'b', the percentage sum can be written as follows:

$$a\% + b\% = (a+b)\%$$

Proof

Using Definition 10.6, we can add two percentages as follows:

$$a \times 1\% = a \times \frac{1}{100}$$

$$a\% + b\% = \frac{a}{100} + \frac{b}{100} \qquad \text{... by Definition 10.6}$$

$$= \frac{a+b}{100} \qquad \text{... by common denominator}$$

$$= (a+b) \times \frac{1}{100} \qquad \text{... by Definition 4.2}$$

$$= (a+b) \times 1\% \qquad \text{... by Definition 10.6}$$

$$= (a+b)\% \qquad \text{... by Distributive Axiom}$$

$$\therefore \qquad a\% + b\% = (a+b)\%$$

In this way, adding two percentages simply involves adding together their number parts and then add the percentage sign as a suffix.

For example, we can add '70%' to '40%' using the arithmetic equation '70% + 40% = (70 + 40)% = 110%'. Next, we will prove the equivalent theorem for multiplying two percentages.

Theorem 10.4 Multiplication of Two Percentages

To express the percentage notation 'a%' and 'b%' for the rational number variables 'a' and 'b', the percentage product can be written as follows:

$$a\% \times b\% = \frac{a \times b}{100}\%$$

Proof

Using Definition 10.6, we can multiply two percentages as follows:

$$a\% \times b\% = \frac{a}{100} \times \frac{b}{100} \qquad \text{... by Definition 10.6}$$

$$= \frac{a \times b}{100 \times 100} \qquad \text{... by Theorem 4.14}$$

$$= \frac{a \times b}{100 \times 100} \times 1 \qquad \text{... by Identity Axiom '}\times\text{'}$$

$$= \frac{a \times b}{100 \times 100} \times \frac{100}{1}\% \qquad \text{... by Definition 10.6}$$

$$= \frac{a \times b}{100}\% \qquad \text{... by Theorem 4.14}$$

$$\therefore \qquad a\% \times b\% = \frac{a \times b}{100}\%$$

In this way, multiplying two percentages simply involves multiplying their number parts, then dividing this result by '100' and including the percentage sign as a suffix. For example, we can multiply '70%' by '40%' using the arithmetic equation '70% × 40% = (70 × 40)% = $\frac{2800}{100}$% = 28%'.

Finally, we can show the most common ways of representing a rational number in Theorem 10.5 below.

Theorem 10.5 Rational Numbers in Different Notations

Rational numbers have equivalent representations as fractions, as an index form, as finite or repeating decimals and as percentages.

Proof

For a simple integer such as '2' we are able to illustrate the various representations of a rational number. These different forms of '2' include:

$$2 = \frac{2}{1} \qquad\qquad \text{... by fraction form}$$

$$= 2 \times 1^{-1} \qquad\qquad \text{... by index form}$$

$$= 2.0 \qquad\qquad \text{... by finite decimal}$$

$$= 1.999... \qquad\qquad \text{... by repeating decimal}$$

$$= 200\% \qquad\qquad \text{... by percentage}$$

In Chapter Four we defined rational numbers as those numbers which can be written in the form '$\frac{a}{b}$' where 'a' and 'b' are integers and '$b \neq 0$'. In Chapter Five we showed the total equivalence of using the index form '$a \times b^{-1}$' to express rational numbers. We were then able to demonstrate the usefulness of this form by extending the values we could assign to the index to include all integer indices. Finally, in this chapter we have shown the equivalence of fractions to finite or repeating decimals and percentages.

This explanation completes the illustration of Theorem 10.5.

———————————

Applications of Rational Numbers as Repeating Decimals

In this section we again consider some simple applications of the theory covered in this chapter that have relevance to arithmetic and algebraic questions. These simple applications will highlight the way in which rational numbers expressed as repeating decimals can be easily manipulated.

Application 10.1

From first principles, find the fraction that represents the repeating decimal '0.121212...'.

This repeating decimal corresponds to the sum of the infinite geometric sequence given by: $(0.12, 0.0012, 0.000012, ...)$. We begin the process by expressing this sequence in index notation to find its partial sums, first term and common ratio.

This geometric sequence can be written in the form:

$$(0.12, 0.0012, 0.000012, ...) = (12 \times 10^{-2}, 12 \times 10^{-4}, 12 \times 10^{-6}, ...)$$

Note:

We use index form because it allows us to express the nth partial sum using an index of 'n'.

From this form of the sequence we can see that the partial sums of this sequence are given by '$S_n = 12 \times 10^{-2} + 12 \times 10^{-4} + ... + 12 \times 10^{-2n}$' and that '$a = 12 \times 10^{-2}$' and '$r = 10^{-2}$'.

According to Definition 10.1:

$$0.121212... = \lim_{n \to \infty} S_n$$

$$= \frac{a}{1-r} \qquad \text{... by Theorem 9.6}$$

$$= \frac{12 \times 10^{-2}}{1 - 10^{-2}} \qquad \text{... by replacing '}a\text{' and '}r\text{'}$$

$$= \frac{12}{100-1} \qquad \text{... by multiplying by '}\frac{100}{100}\text{'}$$

$$\therefore \qquad S = \frac{12}{99}$$

Note:

By using the long-division algorithm on the fraction '$\frac{12}{99}$', we now recover the decimal fraction '0.121212...'.

This application completes our illustration of the application of the limit of the sequence of partial sums '$\lim_{n \to \infty} S_n$' for finding the fraction equivalent '$S = \frac{12}{99}$' to a repeating decimal '0.121212...' from first principles.

In Application 10.2, we will use Definition 10.3 to add two repeating decimals and demonstrate the validity of Theorem 9.4 using these decimals.

Application 10.2

Find the sum of the two repeating decimals '0.121212...' and '0.345345345...'.

We use Theorem 9.4 to find the sum of two repeating decimals and then go on to find the limits of the sequences of partial sums separately for both of these sequences. The infinite geometric sequences that correspond to these two repeating decimals are:

$$\left(0.12, 0.0012, 0.000012, ...\right) = \left(12 \times 10^{-2}, 12 \times 10^{-4}, 12 \times 10^{-6} + ...\right)$$

and,

$$\left(0.345, 0.00345, 0.0000345, ...\right) = \left(345 \times 10^{-3}, 345 \times 10^{-6} + ...\right)$$

The sums of these geometric sequences are given by '$S = \lim\limits_{n\to\infty} S_n$' and '$T = \lim\limits_{n\to\infty} T_n$' respectively. These results are the limits of the sequences of partial sums as follows:

$$S = \frac{12}{99} \qquad \text{... by Theorem 9.6}$$

$$T = \frac{345}{999} \qquad \text{... by Theorem 9.6}$$

$$\therefore \quad S + T = \frac{12}{99} + \frac{345}{999} \qquad \text{... by Definition 10.3}$$

$$= \frac{121212}{999999} + \frac{345345}{999999} \qquad \text{... by common denominator}$$

$$\therefore \quad S + T = \frac{466557}{999999}$$

We observe that '$S + T$' represents the repeating decimal '0.466557466557...'. This outcome demonstrates the usefulness of Theorem 9.4 since we can add the two sequences directly to get the same result. Consequently, we have identified the sum of the two repeating decimals.

_____

In order to now multiply two repeating decimals we only have to multiply their limits using Definition 10.3.

In Application 10.3 below we will demonstrate how we can use Definition 10.3 to multiply repeating decimals.

Application 10.3

Find the product of the two repeating decimals '0.121212...' and '0.345345345...'.

The product of these geometric sequences (refer to Application 6.2) is given by '$S \times T$', where '$S = \lim\limits_{n\to\infty} S_n$' and '$T = \lim\limits_{n\to\infty} T_n$' respectively (and '$S$' and '$T$' are the limits of the sequences of partial sums) so that:

$$S = \frac{12}{99} \qquad \text{... by Theorem 10.1}$$

$$T = \frac{345}{999} \qquad \text{... by Theorem 10.1}$$

$$\therefore \quad S \times T = \frac{12}{99} \times \frac{345}{999} \qquad \text{... by Definition 10.3}$$

$$= \frac{4140}{98901} \qquad \text{... by Theorem 4.14}$$

$$\therefore \quad S \times T = 0.041860041860... \qquad \text{... by division algorithm}$$

$$\therefore \quad S \times T = \frac{41860}{999999} \qquad \text{... by Theorem 10.1}$$

Therefore, the product of the repeating decimals '0.121212...' and '0.345345345...' is the repeating decimal '0.041860041860...'.

In order to multiply repeating decimals less than '1', we simply need to multiply their representations as fractions. In the case of decimals greater than '1' we have to multiply their integer parts and decimal fraction parts using the Distributive Axiom, for example. '$(1.2323...) \times (4.567567...) = (1 + \frac{23}{99}) \times (4 + \frac{567}{999})$'.

In Application 10.4 below, we will illustrate the method of adding two decimal numbers – one as a finite decimal and the other with a repeating decimal.

Application 10.4

Find the sum of '−1' and the repeating decimal '0.999...'.

The fractional equivalent of the repeating decimal '0.999...' is the limit of the sequence of partial sums as follows:

$$0.999... = \frac{9}{9} = 1 \qquad \text{... by Theorem 10.1}$$

∴ $\quad -1 + 0.999... = -1 + 1 \qquad$ ... by adding '−1' to both sides

∴ $\quad -1 + 0.999... = 0 \qquad$ ... by Identity Axiom '+'

Therefore, we can represent '0' as an integer and a repeating decimal.

In this way, the repeating decimal '0.999...' has allowed us to express the Additive Identity Axiom in a different form. Application 10.5 below will demonstrate how we can represent any finite decimal as a repeating decimal.

Application 10.5

How can we represent the finite decimal '0.1928' as a repeating decimal?

The simple way to do this is to recognise that we can use '1 = 0.999...' as follows:

$$1 = 0.999... \qquad \text{... by Definition 10.1}$$
$$0.0001 = 0.0000999... \qquad \text{... by dividing by '10,000'}$$

But: $\quad 0.1928 = 0.1927 + 0.0001 \qquad$ ... by writing '0.1928' as a sum

$$= 0.1927 + 0.0000999... \qquad \text{... by substituting for '0.0001'}$$

∴ $\quad 0.1928 = 0.1927999...$

Consequently, we have represented the finite decimal '0.1928' as the repeating decimal '0.1927999...'.

By now you will be aware that it is easy to show that every integer can be written as a repeating decimal. In fact, it is apparent that '1928 = 1927.999...' by multiplying the above equation by '10^4'. Likewise, for negative integers it is straightforward to illustrate the equivalent result that:

$$-1928 = -1927 - 1 \qquad \text{... by defn. '}1928 = 1927 + 1\text{'}$$
$$= -1927 - 0.999... \qquad \text{... by substituting '}1 = 0.999...\text{'}$$
$$\therefore \qquad -1928 = -1927.999...$$

All of the applications illustrated above have used the equivalence of repeating decimals to fractions. We used this equivalence to define the arithmetic operations of addition and multiplication applied to repeating decimals. In addition, we demonstrated several key features of these repeating decimals. .

Summary of Rational Numbers as Repeating Decimals

In this chapter we have shown the total equivalence between representing a rational number as a fraction, or repeating decimal, and a percentage. This important observation will allow us to extend the Rational Number System using 'infinite decimals' in Chapter Twelve.

At this point, we have covered the objectives of this chapter and so can summarise our findings up to now. The most important definitions that we have introduced in this chapter are:

1. Defining a **repeating decimal number** from the two sequences of symbols $(a_n, \dots, a_0)$ and $(d_1, \dots, d_m)$ as the number '$a_n \dots a_0.d_1 \dots d_m d_1 \dots d_m \dots$' given by:

$$a_n \dots a_0.d_1 \dots d_m d_1 \dots d_m \dots = (a_n \dots a_0 + \lim_{n \to \infty} S_p)$$

where 'S_p' is the general term in the sequence of partial sums given by:

$$S_p = d_1 \dots d_m \times (10^{-m})^1 + \dots + d_1 \dots d_m \times (10^{-m})^p$$

2. Defining **addition** and **multiplication** of two repeating decimals 'x' and 'y', where 'x' and 'y' represent the repeating decimals:

'$a_n \dots a_0.d_1 \dots d_m d_1 \dots d_m \dots$' and '$b_l \dots b_0.e_1 \dots e_n e_1 \dots e_n \dots$', respectively, by:

$$x + y = a_n \dots a_1 a_0 + \lim_{p \to \infty} S_p + b_l \dots b_1 b_0 + \lim_{p \to \infty} T_p$$

And:

$$x \times y = (a_n \dots a_1 a_0 + \lim_{p \to \infty} S_p) \times (b_l \dots b_1 b_0 + \lim_{p \to \infty} T_p)$$

where 'S_p' and 'T_p' are the partial sums given by:

$$S_p = d_1 \dots d_m \times (10^{-m})^1 + \dots + d_1 \dots d_m \times (10^{-m})^p$$
$$T_p = e_1 \dots e_n \times (10^{-n})^1 + \dots + e_1 \dots e_n \times (10^{-n})^p$$

3. Defining a **percentage** number using the following conversion equation: '$1 = 100\%$'.

As every fraction can be expressed as a repeating decimal number and every repeating decimal number can be expressed as a fraction, there is a corresponding expression for the cases of addition and multiplication of repeating decimals (given in Theorems 9.4 and 9.5). Hence the Axioms of the Rational Number System can be applied equivalently to both fractions and repeating decimals.

This chapter has given us the opportunity to develop mastery of the Rational Number System by demonstrating the several equivalent ways of expressing rational numbers and the correspondence between their operations of addition and multiplication.

In Chapter Eleven we will commence our extension of the Rational Number System by identifying irrational numbers. These numbers (such as '$\sqrt{2}$' and 'π') occur naturally but cannot be expressed in number form in the Rational Number System.

CHAPTER 11

IRRATIONAL NUMBERS USING INFINITE DECIMALS

Overview of Irrational Numbers using Infinite Decimals

In this chapter we commence our investigation of the natural extension of the rational numbers to include 'irrational numbers'. Put simply, these 'irrational numbers' are the missing coordinates of points on the number line that aren't able to be represented by rational number coordinates. Up until this stage, we have been able to assign coordinates to points that can be represented by rational numbers (that is, numbers which have finite or repeating decimal values). Irrational numbers are distinguished from other numbers as having infinite (non-repeating) decimal values.

Our goal in this chapter is to demonstrate how these 'irrational numbers' arise conceptually in our everyday experiences and provide some insight into their properties. Irrational numbers were **formally** recognised as having different properties from rational numbers around the 6th century BC in the school of Pythagoras, although, around 400 years earlier, the Babylonians had discovered approximate values for some of these irrational numbers and may have suspected that these numbers were different.

Pictorial Representation of Irrational Numbers using Infinite Decimals

Like in previous chapters where we discussed rational numbers using fractions, index form and repeating decimals, we will provide a pictorial representation of irrational numbers using infinite decimals. In Figure 11.1 depicted on the following page you will see that we have labelled several points along the number line with their infinite decimals.

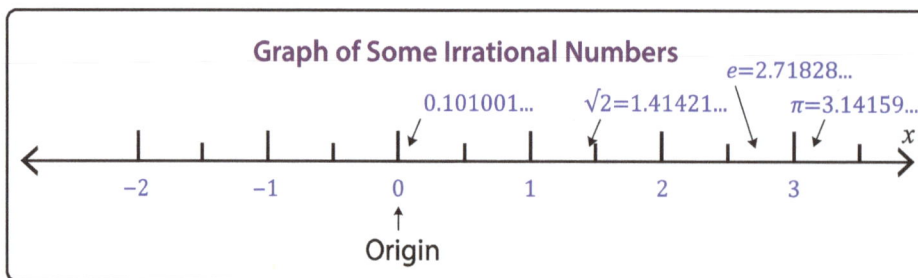

Graph of Some Irrational Numbers

$e=2.71828...$

$0.101001...$ $\sqrt{2}=1.41421...$ $\pi=3.14159...$

-2   -1   0   1   2   3 x

↑
Origin

Figure 11.1

Background and Context of Irrational Numbers using Infinite Decimals

Although the first approximate values for some irrational numbers were documented by the Babylonians on clay tablets over 3,000 years ago, these numbers were also known to mathematicians in China (and probably other locations in Asia) during this same period.

One ancient Babylonian clay tablet (dated around 1800–1600 BC) shows an approximate calculation for the diagonal of a square called the hypotenuse of a right-angle triangle. This calculation is equivalent to finding an approximate solution of the algebraic equation '$c^2 = 2$'. Other clay tablets from this era illustrate that the Babylonians were aware of the concept that is now known as Pythagoras' Theorem.

The simplest construction that demonstrates the existence of an irrational number is derived from a simple right-angle triangle with two equal sides. We can assume that the sides of this triangle are '1' unit in length (as shown in Figure 11.2). Given that we know the length of two sides of this triangle, we now want to determine the length of the longest side (known as the hypotenuse).

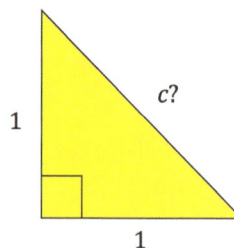

$c?$

1

1

Figure 11.2

First, we must establish the area of the triangle. To do this, we replicate the triangle once and place both of their hypotenuses adjacent to each other. In this way, we create a square with all sides being '1' unit in length (as shown in Figure 11.3).

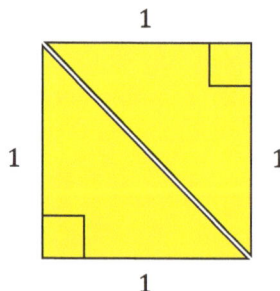

1

1   1

1

Figure 11.3

Note:

The first step above assumes that a square has four right angles and four equal sides and that cutting the square from one corner to the opposite corner results in two triangles which, when placed one on top of the other (with their right angles coinciding), are the same. Therefore, each of the two angles that aren't marked as right angles must be able to be added together to produce a right angle.

Now that we have created a square, we can calculate its area as follows:

$$\text{Area of square} = base \times height = 1 \times 1 = 1 \text{ square unit}$$

The area of each triangle must be equal to '$\frac{1}{2}$' a square unit since there are two identical triangles in this square.

Next, we will use this result (area of a right-angle triangle) to find the length of the hypotenuse of a right-angle triangle. To do this, we produce three replicas of this right-angle triangle and place all four of the triangles with their right angles in the centre. In this way, as indicated in Figure 11.4, the length of the hypotenuse 'c' of each of the triangles becomes the side of a square.

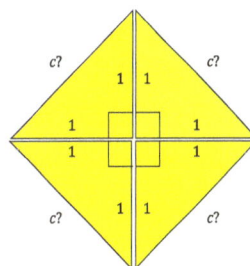

Figure 11.4

We know the area of the right-angle triangle in Figure 11.2 is equal to '$\frac{1}{2}$' a square unit. Consequently, in Figure 11.4, the area of the square is given by '$4 \times \frac{1}{2} = 2$' square units. From the explanation above, we know that the area of a square is calculated by multiplying the side of a square by itself. Therefore, we have:

$$c^2 = 2$$

Therefore, we have found that the length of the hypotenuse of this right-angle triangle with sides of '1' unit is that positive number which when squared gives the number '2'. This positive number – which is called 'the square root of '2' – may or may not be a rational number. We will require a proof to demonstrate whether or not it is rational.

What is a Square Root?

The 'square root' of any positive rational number 'n' is defined to be the positive number denoted by '$\sqrt{n}$', so that when this positive number is multiplied by itself it results in the number 'n'. Therefore, the square root of '2' is denoted by '$\sqrt{2}$'. In index notation, calculating the square root of a number is the same as raising that number to the index '$\frac{1}{2}$'. Hence: '$\sqrt{c} = c^{\frac{1}{2}}$'.

There are two infinite decimal numbers that make the equation '$c^2 = 2$' true; namely, '1.414...' and '−1.414...' (where the ellipsis symbol indicates that there are an unlimited number of digits following the last '4' but doesn't indicate that the pattern is repeating). When squared, these two numbers result in the number '2'. It is the positive number '1.414...' that is defined as '$\sqrt{2}$'. In the Approach Section below, we will be calculating a more accurate decimal approximation for the value of '$\sqrt{2}$'.

In practical terms, we only require the positive square root from the equation '$c^2 = 2$' to determine the length of the hypotenuse of the right-angle triangle (shown in Figure 11.2) which is '$c = \sqrt{2}$'.

It is possible the Babylonians might have taken this method for calculating the hypotenuse of a right-angle triangle a step further in order to determine how to find the hypotenuse of a right-angle triangle with sides '1' and '2' units long (see Figure 11.5).

Similar to forming the square in Figure 11.3, we assume that we can place two identical copies of this right-angle triangle (without equal sides) next to each other to form a rectangle which has four right angles and two opposite equal sides. Therefore, the other two angles of the each of these triangles that aren't the right angles must add together to produce a right angle.

Once again, we use the method described above to calculate the area of the triangle in Figure 11.5.

$$\tfrac{1}{2} \times base \times height = \tfrac{1}{2} \times 1 \times 2 = 1 \text{ square units}$$

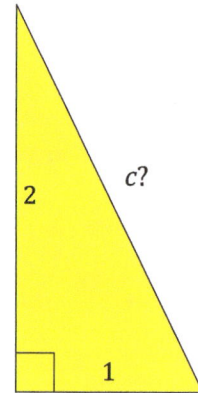

Figure 11.5

Following the reasoning we used for Figure 11.4, we decide how we might rearrange four of these right-angle triangles (without equal sides) to again produce a square with each of their hypotenuses 'c' on the side of a square (see Figure 11.6). However, this square now has a small square of sides '1' unit long missing in the middle of the square.

To calculate the total area in Figure 11.6, we add the area of the four triangles (four times '1' square unit) to the area of the smaller square that has been formed in the middle of the bigger square. This calculation of the total area 'c^2' of the square is achieved by adding these areas together to give:

$$c^2 = 4 \times 1 + 1 \times 1$$

$$\therefore \qquad c^2 = 5 \text{ square units}$$

We have now discovered that the length of the hypotenuse 'c' of Figure 11.5 is a positive number whose square is '5'. Once again, we can represent this number by the historical notation:

$$c = \sqrt{5}$$

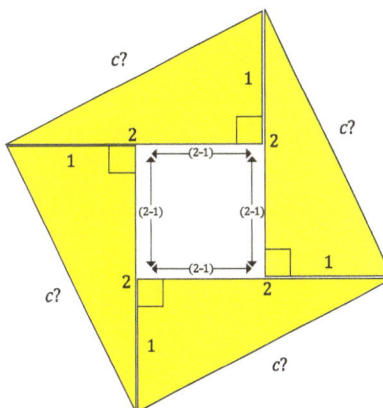

Figure 11.6

This length of the hypotenuse of '$\sqrt{5}$' is also irrational. At this point we have still not demonstrated that '$\sqrt{2}$' is an irrational number – only that it physically exists as the hypotenuse of the simplest right-angle triangle. The first theorem in our Theory Section will prove that '$\sqrt{2}$' is an irrational number and hence an infinite non-repeating decimal.

Since this second example of the calculation of a right-angle triangle gives an irrational number, we will now give the simplest example of a triangle which has an hypotenuse with a rational number as its length. The Babylonians also gave this example of a right-angle triangle which has sides of length '3' and '4' and results in an hypotenuse of '5'. We will be using the same calculation we used in Figure 11.6 to determine these new dimensions. That is:

Total area:
$$c^2 = 4 \times (\tfrac{1}{2} \times base \times height) + area\ of\ inner\ square$$

$$= 4 \times (\tfrac{1}{2} \times 3 \times 4) + 1$$

$$= 24 + 1$$

∴ $$c^2 = 25$$

∴ $$c = 5 \qquad \text{... taking the positive value}$$

This '3 – 4 – 5' triangle has special significance because it allows us to easily construct a right-angle triangle and hence a right angle. The Babylonians may have shown that Figure 11.6 could be generalised by assigning arbitrary lengths of 'a' and 'b', and a hypotenuse of 'c' to the sides of the triangle.

In order to calculate the general relationship between the sides of a right-angle triangle and the hypotenuse, we use the following equation:

Total area:
$$c^2 = 4 \times (\tfrac{1}{2} \times base \times height) + area\ of\ inner\ square$$

$$= 4 \times (\tfrac{1}{2} \times a \times b) + (a - b)^2$$

$$= 2ab + a^2 - 2ab + b^2$$

$$= a^2 + b^2$$

$$\therefore \qquad c^2 = a^2 + b^2$$

This relationship that 'the square of the hypotenuse is equal to the sum of the squares of the other two sides' for a right-angle triangle has been present as an inherent characteristic of the triangles described above, however, it only emerged when we defined the general case.

The equation, '$c^2 = a^2 + b^2$', for which we have just provided a proof by construction, is the most famous equation in Mathematics. It is attributed to Pythagoras around 600 BC and is known as Pythagoras' Theorem. However, the first theoretical proof appears in the most famous Mathematics text of all time – Euclid's Elements which was published around 300 BC.

Using Pythagoras' Theorem, we can also scale-up a right-angle triangle by multiplying all sides by the value 's' so that we obtain the following equation: '$(s \times a)^2 + (s \times b)^2 = (s \times c)^2$'. A property of this scaled-up triangle that remains constant is the ratio of its sides to each other. We say these triangles have the same shape or are 'similar'. This property will be demonstrated to be useful later in this chapter.

The Ancient Greek school of Pythagoras is also attributed with providing the first proof of the existence of 'irrational numbers' by showing that '$\sqrt{2}$' could not be a rational number. Some historical accounts actually claim the proof was put forward by one of Pythagoras' students and the existence of these mysterious numbers was concealed for many years. We will prove 'irrational numbers' exist and discuss their properties in the sections below.

Approach to Irrational Numbers using Infinite Decimals

We will begin our investigation of irrational numbers by recalling the relationship between fractions and repeating decimals (Theorem 10.1 and Theorem 10.2). This relationship illustrates the one-to-one affiliation between **every** rational number as a fraction, and as a repeating decimal. We will now demonstrate that there are simple infinite (non-repeating) decimals that cannot be written as rational numbers. These are the irrational numbers.

A simple example of a non-repeating infinite decimal is: '0.101001000100001...'. As this decimal number is not finite and doesn't have a repeating pattern, it cannot be written in the form '$\frac{a}{b}$' and hence it is not a rational number. According to Definition 11.1, this infinite decimal is an irrational number.

We recall Definition 6.4 from Chapter Six where we established that writing a rational number in decimal notation simply involves adding up a sequence of fractions with denominators of some power of '10' (for example, '$1.2345 = 1 + \frac{2}{10} +$

$\frac{3}{100} + \frac{4}{1000} + \frac{5}{10000}$'). Alternatively, we could write a sequence of partial sums that accomplishes the same task; namely, the sequence $(1, 1.2, 1.23, 1.234, 1.2345)$. In the case where the sequence of partial sums continues on indefinitely, we have to use the limit process to accomplish the equivalent of an infinite number of additions!

For the infinite decimal '0.101001000100001...', that is, the limit of the partial sums of the non-geometric sequence '$(10^{-1}, 10^{-3}, 10^{-6}, ..., 10^{\frac{-n(n+1)}{2}}, ...)$' where '$n$' is assigned the positive integer values '1, 2, 3, ...' to generate the terms of this sequence, the nth term in the sequence of partial sums is:

$$S_n = 10^{-1} + 10^{-3} + 10^{-6} + ... + 10^{\frac{-n(n+1)}{2}}$$

This approach is the same as the one we used to find the sum of an infinite repeating decimal, so that we could prove its limit was a fraction and hence it was a rational number.

Our preliminary claim is that '$\lim_{n\to\infty} S_n$' is not a rational number. As we are attempting to add a non-geometric sequence, the ratio of each term to its subsequent term is not constant from each term to the next in the sequence. This non-repeating pattern of the number '0.101001000100001...' suggests this number is not rational. However, it turns out to be very difficult to formally prove that it is an irrational number. Fortunately, our next irrational number '$\sqrt{2}$' has an ingenious proof that demonstrates that it can't be a rational number. This proof is attributed to a student of Pythagoras.

Our first irrational number, '$\sqrt{2}$', can be calculated from its definition by successively more accurate decimal approximations. From the information about areas of triangles outlined above, we know that '$\sqrt{2}$' possesses the property that '$(\sqrt{2})^2 = 2$'. We will now approximate '$\sqrt{2}$' by positive finite decimal numbers which when multiplied by themselves get closer and closer to '2' from below and above. By doing this we obtain the following lower rational (on the left-hand side of the page) and upper rational (on the right-hand side of the page) approximations for '$\sqrt{2}$':

$1^2 = 1$	$2^2 = 4$
$1.4^2 = 1.96$	$1.5^2 = 2.25$
$1.41^2 = 1.9881$	$1.42^2 = 2.0164$
$1.414^2 = 1.999396$	$1.415^2 = 2.002225$
$1.4142^2 = 1.99996164$	$1.4143^2 = 2.00024449$
$1.41421^2 = 1.9999899241$	$1.41422^2 = 2.0000182084$
$1.414213^2 = 1.999998409369$	$1.414214^2 = 2.000001237796$

After studying the pattern of the squares of numbers above, we see that the square of numbers down the left-hand side are becoming progressively closer to the number '2' from below (i.e. less than) '2' and the square of the numbers down the right-hand side of the page are becoming progressively closer to the number '2' from above (i.e. greater than) '2'. Consequently, the numbers down the left-hand side of the page that are being squared are providing a better and better approximation of the decimal expansion of '$\sqrt{2}$'.

This concept is something we haven't seen in our mathematical studies so far! Here is a number with a longer and longer decimal expansion which appears to have no repeating pattern emerging. Regardless of how many decimal places we continue to find, and how close we get to the number represented by '$\sqrt{2}$', there appears to be no repeating pattern in the decimal expansion. For example, if we write out the first fifty decimal places of '$\sqrt{2}$' we get:

1.41421356237309504880168872420969807856967187537694

However, this result does not constitute a proof that the decimal expansion does not repeat eventually – after a thousand places, say. The decimal expansion of '$\sqrt{2}$' above **does** cause us to question the fact that numbers like '$\sqrt{2}$' may not have repeating decimal expansions and hence may not be rational numbers. As a result, we may not be able to write numbers like '$\sqrt{2}$' as a fraction in the form of '$\frac{a}{b}$' where 'a' and 'b' are integers and '$b \neq 0$'. In Theorem 11.1 in the Theory Section below, we will prove that '$\sqrt{2}$' is not a rational number.

Another famous irrational number is the number 'pi', which is written as the letter 'π' from the Greek alphabet. The length of the circumference of a circle (the distance around the circle) divided by the length of its diameter (the straight line from one side of the circle to the other passing through its centre) is defined as 'π'. If we assume that normal arithmetic of rational numbers applies where the circumference has a length 'c' and the diameter has a length 'd', then 'π' would be given by the equation:

$$\pi = \frac{c}{d}$$

Note:

The number 'π' can also be written in the equivalent form as the number of radii that can be laid around half the circumference of this circle (as depicted in Figure 11.7). The diameter 'd' of a circle is given by '$d = 2 \times r$', so that:

$$\pi = \frac{c}{2r}$$

Rearranging this definition of 'π' gives us the well-known formula for the length of the circumference of a circle as: '$c = 2\pi r$'.

By observing Figure 11.7 you will note that there are slightly more than three radius lengths that can be laid down around half a circle. The exact number is obtained by the infinite decimal 'π' where:

$$\pi = 3.14159265358...$$

An important property of 'π' to be aware of is that it is not dependent on the size of the circle that we use to measure the radius and half the circumference.

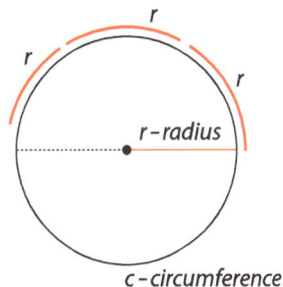

Figure 11.7

As with Pythagoras' Theorem, the Ancient Greeks reasoned that if you expand a circle by a certain factor 'S', then it remains the same shape and hence its radius and circumference always expand by the same factor. In this way, the ratio of the circumference 'c' of a circle to the diameter 'd' of that circle will always remain constant irrespective of the size of the circle. Consequently, the value of 'π' does not change when the circle changes in size.

In general, it is apparent that it is quite difficult to determine whether numbers are actually 'irrational'.

Language of Irrational Numbers using Infinite Decimals

The symbols required for creating the language of the irrational numbers in this chapter are outlined in the Definition Section later in this chapter. The current symbols of the alphabet are summarised in Table 11.1 below.

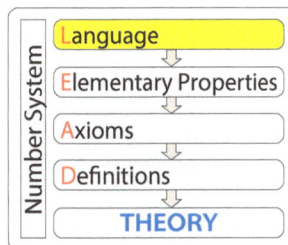

11.1 The Symbols of the Alphabet of the Rational Number System and Irrational Numbers	
Symbols	**Meaning from:**
'0', '1', '+', '×', '(', ')'	The Axioms of the Natural Number System
'=', '≠'	The Axioms of Logic
'2', '3', '4', '5', '6', '7', '8', '9'	Definition 2.1
'a', 'b', 'c', ..., 'x', 'y', 'z', ...	The rational number variables
','	Definition 2.3
'−1'	The Axioms of the Integer Number System
'−'	Definitions 3.4 and 3.5

11.1 The Symbols of the **Alphabet** of the Rational Number System and Irrational Numbers	
$`\frac{1}{\square}`$	The Axioms of the Rational Number System as Fractions
$`\frac{a}{b}`$	Definition 4.2
$`\div`$	Definition 4.4
$`\square^{-1}`$	The Axioms of Rational Number System using Index Form
$`(a \times b^{-1})^m`$	Definition 5.6
$`.`$	Definition 6.4
$`a_n \dots a_1 a_0 . d_1 \dots d_m`$	Definition 6.5
$`\pm`$	Definition 7.8
$`<`$	The Axioms of the Rational Number System using Order
$`\leq, >, \geq`$	Definition 8.2
$`|x - y|`$	Definition 8.7
$`\{, \}`$	Definition 9.1
$`(, )`$	Definition 9.2 (overloading the parentheses symbols)
$`\infty`$	Definition 9.3
$`\lim_{n \to \infty} a_n`$	Definition 9.10 (refer to the note following this definition)
$`a_n \dots a_1 a_0 . d_1 \dots d_m \dots`$	Definition 10.1
$`\%`$	Definition 10.3
$`\sqrt{}`$	$`\sqrt{}`$ is called the 'square root' symbol

Table 11.1

The Elementary Properties of the Irrational Numbers

When considering irrational numbers, it is apparent that there are no additional elementary properties required in order to represent operations on these numbers. We will continue to use the elementary properties we developed for the Rational Number System and the

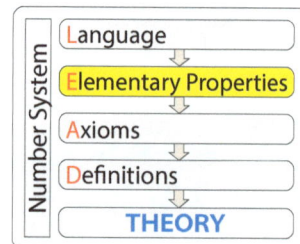

Number System
Language → Elementary Properties → Axioms → Definitions → THEORY

property that, using rational numbers, we can approach every irrational number as close as we like. That is, a decimal approximation to an irrational number can be made as accurate as we choose to make it.

The Axioms of Irrational Numbers using Infinite Decimals

In the next chapter we will be showing that the Axioms of the Rational Number System can be derived as theorems for the irrational numbers. However, we will first have to define what we mean by 'addition' and 'multiplication' of these irrational numbers and then proceed to fully extend the Rational Number System. In this chapter we will be using the Axioms of the Rational Number System to understand properties of the irrational numbers.

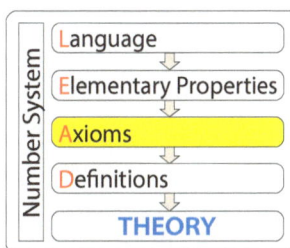

The Axioms of Logic

In this chapter we will be using the Axioms of Logic to manipulate rational number terms and equations in order to understand the properties of the irrational numbers.

Definitions of Irrational Numbers using Infinite Decimals

The fundamental concept of this chapter is that an irrational number is a **non-repeating** infinite decimal number. In the Theory Section below we will explore the relationship between this concept and the previous expression of rational numbers as repeating decimals.

First, we will provide the definition of an irrational number.

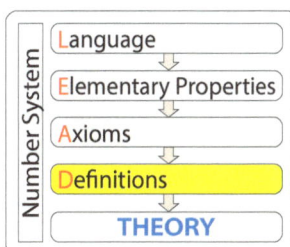

Definition 11.1   Irrational Number

An irrational number is any number that is not a rational number. In this way, an irrational number is a non-repeating infinite decimal number and hence cannot be expressed in the form of a fraction '$\frac{a}{b}$' where 'a' and 'b' are integers and '$b \neq 0$'.

From Chapter Ten we know that every rational number can be written as a repeating infinite decimal number (and vice versa). In Definition 11.1 we now discover that every irrational number can be written as a non-repeating infinite decimal number.

In order to expand our Rational Number System to a new number system that includes irrational numbers we must define what we mean by an 'infinite decimal number' that can be either repeating or non-repeating. In Definition 11.2 we commence this process by looking at positive 'infinite decimal numbers'.

Definition 11.2   Positive Infinite Decimal Numbers

Let the two sets of symbols $\{a_n, \dots, a_0\}$ and $\{d_1, \dots, d_p, \dots\}$ represent the '$n + 1$' digits to the left of the decimal point and an unlimited set of digits to the right of the decimal point, respectively. We will now define the **positive infinite decimal number** 'x' to be a number formed by these two sets as: '$a_n \dots a_1 a_0.d_1 \dots d_p \dots$' where there must be a rule for calculating the pth digit of 'x' after the decimal point for all positive integers 'p'.

Note:

1. This definition does not include negative numbers. We will have to extend it in Chapter Twelve to obtain the full definition of infinite decimal numbers.

2. The ellipsis after 'd_p' in the infinite decimal '$a_n \dots a_1 a_0.d_1 \dots d_p \dots$' indicates that there is a rule for calculating all digits after the decimal point. With this definition we can calculate every digit after the decimal point without talking about 'infinity' as though it were a number.

Now that we know what an infinite decimal number looks like, we can start to introduce operations on these numbers. However, we will have to wait until Chapter Twelve before defining the operations of addition and multiplication for these new numbers. Nevertheless, there is a very simple operation we can perform on these infinite decimal numbers which will be described in Definition 11.3 below.

Definition 11.3   Truncation Operator on Positive Infinite Decimal Numbers

Let 'x' be assigned the infinite decimal number '$a_n \dots a_1 a_0.d_1 \dots d_p \dots$'. We then define the **truncation operator** '$T(x, p)$' as the rational number generated by truncating 'x' after the 'p^{th}' place, so that:

$$T(x, p) = a_n \dots a_1 a_0.d_1 \dots d_p$$

All digits after the pth digit in the decimal expansion are removed.

Notes:

1. This is the same truncation process present in most spreadsheet software packages (which is usually called the '$TRUNC(x, p)$' function or an equivalent name). We will be using '$T(x, p)$' here to make our work more readable.

2. The truncation function '$T(x, p)$' allows us to determine the pth term in the sequence of partial sums of the infinite sequence $\left(a_n \ldots a_1 a_0, d_1 \times 10^{-1}, d_2 \times 10^{-2}, \ldots, d_p \times 10^{-p}, \ldots\right)$.

One conspicuous property of the sequence of partial sums '$T(x, p)$' of an infinite decimal is that '$T(x, p)$' does not decrease in value as 'p' increases. For example, if we let '$T(x, 3) = 13.123$', '$T(x, 4) = 13.1230$' and '$T(x, 5) = 13.12301$', then we have '$T(x, 3) \leq T(x, 4) \leq T(x, 5)$, which is a non-decreasing sequence'. This premise leads to our final definition of this chapter – Definition 11.4 below.

Definition 11.4   Non-decreasing and Non-increasing Sequences

If $\left(a_1, \ldots, a_m, \ldots\right)$ is an infinite sequence of rational numbers such that '$a_m \leq a_{m+1}$' for all positive integers 'm', then the sequence is said to be **non-decreasing**. Similarly, if this infinite sequence of rational numbers is such that '$a_{m+1} \leq a_m$' for all positive integers 'm', then the sequence is said to be **non-increasing**.

The complete set of symbols required for this chapter are summarised in Table 11.2 below.

11.2 The Symbols of the Alphabet of the Rational Number System and Irrational Numbers	
Symbols	**Meaning from:**
'0', '1', '+', '×', '(', ')'	The Axioms of the Natural Number System
'=', '≠'	The Axioms of Logic
'2', '3', '4', '5', '6', '7', '8', '9'	Definition 2.1
'a', 'b', 'c', ... , 'x', 'y', 'z', ...	The rational number variables
','	Definition 2.3
'−1'	The Axioms of the Integer Number System
'−'	Definitions 3.4 and 3.5

11.2 The Symbols of the Alphabet of the Rational Number System and Irrational Numbers	
'$\frac{1}{\square}$'	The Axioms of the Rational Number System as Fractions
'$\frac{a}{b}$'	Definition 4.2
'$\div$'	Definition 4.4
'$\square^{-1}$'	The Axioms of Rational Number System using Index Form
'$(a \times b^{-1})^m$'	Definition 5.6
'.'	Definition 6.4
'$a_n \dots a_1 a_0.d_1 \dots d_m$'	Definition 6.5
'$\pm$'	Definition 7.8
'$<$'	The Axioms of the Rational Number System using Order
'$\leq$', '$>$', '$\geq$'	Definition 8.2
'$|x - y|$'	Definition 8.7
'{', '}'	Definition 9.1
'(,)'	Definition 9.2 (overloading the parentheses symbols)
'∞'	Definition 9.3
'$\lim_{n \to \infty} a_n$'	Definition 9.10 (refer to the note following this definition)
'$a_n \dots a_1 a_0.d_1 \dots d_m \dots$'	Definition 10.1
'%'	Definition 10.3
'$\sqrt{}$'	'$\sqrt{}$' is called the 'square root' symbol
'$T(x, p)$'	Definition 11.3

Table 11.2

In this chapter we have introduced irrational numbers which motivated the definition of positive infinite decimal numbers which don't terminate and may or may not have a repeating pattern of digits after the decimal point.

Theory of Irrational Numbers as Infinite Decimals

The main focus of this section will be to prove that the number '$\sqrt{2}$' is an irrational number. In addition, we will provide some observations about '$\sqrt{2}$' that will be useful in Chapter Twelve.

Our main purpose of this chapter is to use the proof by contradiction to reach our intended outcome. We will assume that '$\sqrt{2}$' is a rational number and so can be added and multiplied in the same way that rational numbers usually are. We will then go on to prove that this assumption results in a contradiction and hence '$\sqrt{2}$' is not a rational number and so must be irrational (according to Definition 11.1).

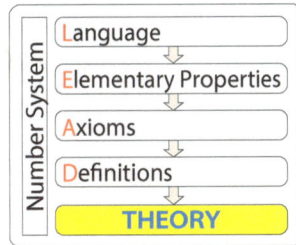

Theorem 11.1 '$\sqrt{2}$' is an Irrational Number

We want to prove that the number '$\sqrt{2}$' is an irrational number.

Proof

We assume '$\sqrt{2}$' is a positive rational number and we use proof by contradiction. If '$\sqrt{2}$' is a positive rational number, then we can write it in the form '$\sqrt{2} = \frac{a}{b}$' where 'a' and 'b' are assumed to be positive integers and '$b \neq 0$'. Furthermore, we can assume that 'a' and 'b' do not share a common factor and therefore '$\frac{a}{b}$' is a fraction in its simplest form.

The assumption above is justified as we know every fraction (after a finite number of cancellations) can be reduced to its simplest form. That is, if a fraction '$\frac{c}{d}$' is equivalent to the fraction '$\frac{a}{b}$', then there must be a finite number of factors in the numerator and denominator of '$\frac{c}{d}$' that can be cancelled to yield '$\frac{a}{b}$' in its simplest form. In symbols, we have '$\frac{c}{d} = (a \times f)/(b \times f) = \frac{a}{b}$' where '$f$' is the product of the finite number of factors.

By combining the two assumptions outlined above into one assumption, we get:

$$\text{'}\sqrt{2} = \frac{a}{b}\text{' and '}\frac{a}{b}\text{' is a fraction in its simplest form.}$$

We can then substitute this fraction form for '$\sqrt{2}$' as follows:

$$(\tfrac{a}{b})^2 = 2 \qquad \text{... by property '}(\sqrt{2})^2 = 2\text{'}$$
$$\frac{a^2}{b^2} = 2 \qquad \text{... by product '}\frac{a}{b} \times \frac{a}{b} = \frac{a^2}{b^2}\text{'}$$
$$\therefore \qquad a^2 = 2b^2 \qquad \text{... by multiplying sides by '}b^2\text{'}$$

If 'a' is an even number, then we could write it in the form '$a = 2k$' where 'k' can take on the values from the set, $\{1, 2, 3, ...\}$. Now, if we multiply 'a' by itself then:

$$a^2 = (2k)^2 \qquad \text{... by ‘} a = 2k \text{’}$$
$$= 2k \times 2k \qquad \text{... by the square of a number}$$
$$= 4k^2 \qquad \text{... by product of terms}$$
$$\therefore \quad a^2 = 2(2k^2) \qquad \text{... by taking out a factor of ‘2’}$$

Hence, ‘a^2’ must be even.

If ‘a’ is an odd number, then we could write it in the form ‘$2k+1$’ where ‘k’ can take on the values from the set: $\{0, 1, 2, 3, ...\}$. We now square ‘a’ and observe what happens to its value:

$$a^2 = (2k+1)^2 \qquad \text{... by ‘} a = 2k+1 \text{’}$$
$$= (4k^2 + 4k) + 1 \qquad \text{... by Distributive Axiom}$$
$$\therefore \quad a^2 = 2(2k^2 + 2k) + 1 \qquad \text{... by rewriting its form}$$

In this way, if ‘a’ is any odd integer, then its square is also an odd integer. However, from the equation ‘$a^2 = 2b^2$’ we know that ‘a^2’ is even. Consequently, ‘a’ cannot be an odd integer; it can only be an even integer.

Since ‘a’ is an even integer, we write it in the form ‘$2k$’ where ‘k’ can take on the values from the set: $\{1, 2, 3, ...\}$, that is, ‘$a = 2k$’. As a result, we can substitute for ‘a’ in the equation ‘$a^2 = 2b^2$’ and obtain:

$$(2k)^2 = 2b^2 \qquad \text{... by ‘} a = 2k \text{’}$$
$$4k^2 = 2b^2 \qquad \text{... by substitution}$$
$$\therefore \quad 2k^2 = b^2 \qquad \text{... by dividing both sides by ‘2’}$$

According to this last equation, ‘b^2’ must be an even integer. Likewise, our previous reasoning for ‘a’ indicates that ‘a’ must also be an even integer. This implies both ‘a’ and ‘b’ share a common factor of ‘2’; a result which contradicts our initial assumption that ‘a’ and ‘b’ did not share a common factor. Therefore, the assumption that ‘$\sqrt{2}$’ is a rational number – that is, it can be written in the form of ‘$\frac{a}{b}$’ and satisfy the equation ‘$(\frac{a}{b})^2 = 2$’ – must be false. For this reason, the number represented by ‘$\sqrt{2}$’ must **not** be a rational number. Therefore, according to Definition 11.1, ‘$\sqrt{2}$’ is an irrational number.

———————————

This theorem confirms our suspicion that ‘$\sqrt{2}$’ is not a rational number. So now we are immediately faced with the question: ‘Is ‘$\sqrt{2}$’ a number at all?’ We are already aware that ‘$\sqrt{2}$’ can be approximated indefinitely by a sequence of finite decimal expansions and that it is the number (although not rational) that measures the length of a right-angled triangle of two equal sides of '1' unit. There-fore, the infinite decimal for '$\sqrt{2}$' represents the coordinate of a point on the number line (refer to Application 11.1 below) and so certainly deserves to be considered a number. We will be exploring the number system that '$\sqrt{2}$' belongs to in Chapter Twelve when we extend the Rational Number System.

There are an unlimited number of irrational numbers. As well as our example '$\sqrt{2}$', there are an infinite number of irrational numbers of the form '$\sqrt{n}$', where 'n' is any positive integer that is not a perfect square, as we can't calculate the square root of a negative integer.

This completes our introduction to irrational numbers (even though, as yet, we have not been able to define addition and multiplication of these numbers). In Chapter Twelve our goal will be to extend the Rational Number System by creating a number system that enables us to add and multiply both rational and irrational numbers. We call this new set of numbers and its associated operations the 'Real Number System'.

Applications of Irrational Numbers as Infinite Decimals

In this section we again consider some simple applications of the theory developed in this chapter that highlight the key properties of irrational numbers.

Application 11.1

Demonstrate by construction that '$\sqrt{2}$' exists on the number line.

The standard construction used to demonstrate the existence of the number '$\sqrt{2}$' follows directly from its use as the length of the hypotenuse of a right-angle triangle with sides of length '1' (refer to Figure 11.4).

In Figure 11.8 we have illustrated the fact that the hypotenuse is of length '$\sqrt{2}$'. First, we construct a circle that is centred on the origin with a radius equal to the length of the hypotenuse of the right-angled triangle. This circle intersects the number line at the point labelled 'P'. The coordinate 'P' is the real distance of the point 'P' from the origin which is the radius '$\sqrt{2}$'.

This actual distance '$\sqrt{2}$' can't be expressed as the ratio of two integers and hence is not rational (according to Theorem 11.1). This application completes our

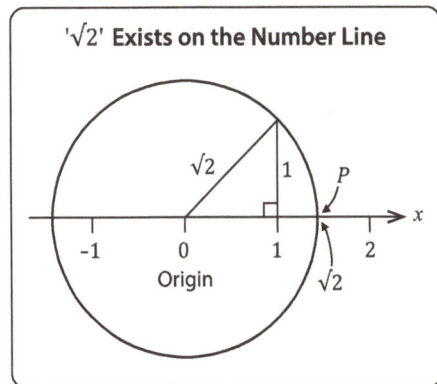

Figure 11.8

illustration of the infinite decimal '$\sqrt{2}$' as a number that represents the coordinate of a point on the number line.

Application 11.2

Find the first six terms of the non-increasing sequence and non-decreasing sequence that best approximate '$\sqrt{2}$'.

Let 'S_n' be the largest finite decimal to 'n' places such that '$(S_n)^2 < 2$'. According to the definition of the truncation of '$\sqrt{2}$', the term 'S_n' is given by: '$S_n = T(\sqrt{2}, n)$'. In the Approach Section of this chapter we established that for '$\sqrt{2}$' we have:

$$S_1 = T(\sqrt{2}, 1) = 1.4$$
$$S_2 = T(\sqrt{2}, 2) = 1.41$$
$$S_3 = T(\sqrt{2}, 3) = 1.414$$
$$S_4 = T(\sqrt{2}, 4) = 1.4142$$
$$S_5 = T(\sqrt{2}, 5) = 1.41421$$
$$S_6 = T(\sqrt{2}, 6) = 1.414213$$

Note that '$S_n \leq S_{n+1}$' for the terms 'S_1' through to 'S_6' so that this sequence of six terms is non-decreasing.

Let 'R_n' be the smallest finite decimal to 'n' places such that '$(R_n)^2 > 2$'. According to the definition of the truncation of '$\sqrt{2}$', we have '$R_n = T(\sqrt{2}, n) + 10^{-n}$'. In the Approach Section of this chapter we established that for '$\sqrt{2}$' we have:

$$R_1 = T(\sqrt{2}, 1) + 10^{-1} = 1.5$$
$$R_2 = T(\sqrt{2}, 2) + 10^{-2} = 1.42$$
$$R_3 = T(\sqrt{2}, 3) + 10^{-3} = 1.415$$
$$R_4 = T(\sqrt{2}, 4) + 10^{-4} = 1.4143$$
$$R_5 = T(\sqrt{2}, 5) + 10^{-5} = 1.41422$$
$$R_6 = T(\sqrt{2}, 6) + 10^{-6} = 1.414214$$

Note that '$R_n \geq R_{n+1}$' for the terms 'R_1' through to 'R_6' so that this sequence of six terms is non-increasing.

This completes our illustration of the application of the truncation function to aid our understanding of non-increasing and non-decreasing sequences that are approaching the infinite decimal '$\sqrt{2}$'.

In the next application, we give additional evidence that '$\sqrt{2}$' is a number.

Application 11.3

Show that the non-decreasing sequence and non-increasing sequence that best approximate '$\sqrt{2}$' with rational numbers, converge to each other.

Given 'S_n' is the largest finite decimal to 'n' places such that '$(S_n)^2 < 2$', we have '$S_n = T(\sqrt{2}, n)$'. The approximations 'S_n', where '$S_n \leq S_{n+1}$', are non-decreasing by the construction of 'S_n'. Likewise, as 'R_n' is the smallest finite decimal to 'n' places such that '$(R_n)^2 > 2$', we have '$R_n = T(\sqrt{2}, n) + 10^{-n}$'. The approximations '$R_n$', where '$R_n \geq R_{n+1}$' are non-increasing by the construction of 'R_n'. The two sequences '$(S_1, S_2, \ldots, S_n, \ldots)$' and '$(R_1, R_2, \ldots, R_n, \ldots)$' converge to each other as shown by:

$$|R_n - S_n| = 10^{-n} \qquad \text{... by subtracting } n^{\text{th}} \text{ terms}$$

In this way, the best upper and lower approximations to '$\sqrt{2}$' converge to each other and hence the two sequences are converging to the same number.

Our physical theories of the world would not be possible without reference to irrational numbers. Most variables in these theories are constructed from the physical quantities of mass, length and time which are represented by both rational and irrational numbers. However, when measuring physical quantities, we always represent this data as rational numbers with a specific number of decimal places to indicate the accuracy of these measurements.

Summary of Irrational Numbers as Infinite Decimals

In this chapter we proved that '$\sqrt{2}$' is an irrational number. Furthermore, we demonstrated that the Rational Number System has limitations in so far as not being able to assign numbers as coordinates to every point along a number line.

At this point, we have covered the objectives of this chapter and hence can summarise our findings up to now. The most important definitions that we have introduced in this chapter are:

1. Defining an **irrational number** as any number that is not a rational number. That is, a number that cannot be expressed as a fraction '$\frac{a}{b}$' where 'a' and 'b' are integers and '$b \neq 0$'.

2. Defining the (positive) **infinite decimal number** 'x' to be a number formed from two sets of digits as: '$a_n \ldots a_1 a_0.d_1 \ldots d_p \ldots$' where there is a rule for

determining the pth digit of 'x' after the decimal point for all positive integers 'p'.

3. Defining the **truncation operator** '$T(x, p)$' as the rational number generated by truncating the positive infinite decimal 'x' after the 'p^{th}' place so that:

$$T(x, p) = a_n \dots a_1 a_0.d_1 \dots d_p$$

The Theory Section of this chapter presented the most famous proof in Mathematics; namely, that the number '$\sqrt{2}$' is not a rational number. This result tells us that the Rational Number System is not adequate for assigning coordinates to all points along the number line. For example, '$\sqrt{2}$' is the physical length of the hypotenuse of a right-angle triangle with sides of length '1' and so is a length along the number line, yet it does not have a rational number as a coordinate.

Another consequence of '$\sqrt{2}$' not being rational, is that, while we can construct a sequence of rational numbers $(1, 1.4, 1.414, 1.4142, \dots)$ which has a limit of '$\sqrt{2}$', this limit is not a rational number. Our very first Axiom of the Rational Number System tells us that this system is closed under a finite number of addition and multiplication operations; however, when using the limit process on these operations (such as addition), the system is not closed under an 'infinite' number of these operations and so the system is '**not complete**'.

In this way, on the path to mastering the Rational Number System, we have discovered a fundamental inadequacy of this system which is its ability to assign coordinates to all points along the number line. In Chapter Twelve we will explore how we can remove this inadequacy in order to accommodate these new irrational numbers by extending the Rational Number System to form the '**Real Number System**'.

CHAPTER 12

THE REAL NUMBER SYSTEM USING INFINITE DECIMALS

Overview of the Real Number System using Infinite Decimals

In this chapter we complete our journey and develop mastery of all the elementary number systems. Up to this point in the book we have explored the following elementary number systems:

- The **Natural Number System** (Chapter Two)

- The **Integer Number System** (Chapter Three)

- The **Rational Number System** (Chapter Four – Rational Number System using Fractions, Chapter Five – Rational Number System using Indices, Chapter Ten – Rational Numbers using Repeating Decimals).

Our goal in this chapter will be to develop the fourth (and final) elementary number system called the '**Real Number System**'. This new number system will be constructed as an extension of the Rational Number System by defining general infinite decimals which incorporate both the rational and irrational numbers. These infinite decimal numbers are called **real numbers** and their essential properties are that they are:

- a number field (the best example of a number field is the Rational Numbers with their addition and multiplication operations satisfying the Axioms of the Rational Number System using Fractions as outlined in Table 4.4 in Chapter Four)

- totally ordered (the best example of a total order are the Rational Number System using Order with the symbol '<', which satisfies the Axioms of Order for the Rational Number System as outlined in Table 8.3 in Chapter Eight)

- complete (as defined at the end of Chapter Eleven in the Summary of Irrational Numbers as Infinite Decimals where we declare that a number system must be closed (i.e. contains limits) under any infinite number of additions when using the limit process on a sequence that is bounded by finite numbers).

The Real Number System will allow us to assign every infinite decimal number (as a coordinate) to a point along the number line and to every point along the number line we will be able to assign an infinite decimal number as a coordinate. This direct correlation between points and real numbers permits us to model physical quantities using real number variables.

Our modelling process (which commenced with us modelling equally-spaced points along a number line by using natural numbers) will make it possible for us to model **all** the other points between these equally-spaced points using real numbers in the form of infinite decimals.

Pictorial Representation of the Real Number System using Infinite Decimals

As we did for the elementary number systems in previous chapters, we will provide a pictorial representation of real numbers. In Figure 12.1 you will see that we have labelled several points along the number line using real numbers as their coordinates.

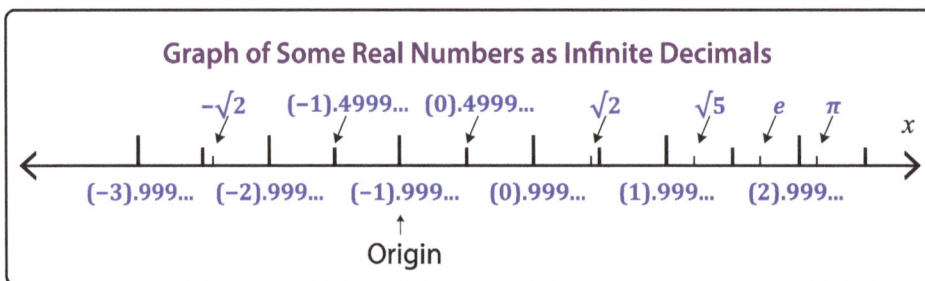

Figure 12.1

It is worth noting that the point labelled with the coordinate '$\sqrt{2}$', for example, is the point of intersection of a small vertical line with the main number line. For every point of intersection between a vertical line and the horizontal number line (labelled 'x' in Figure 12.1) we will be able to assign a real number coordinate.

Background and Context of the Real Number System using Infinite Decimals

By the 16th century, modern decimal notation was widely used and mathematicians were starting to recognise that all rational numbers could be represented by decimal notation. Likewise, they observed that irrational numbers could be approximated to any accuracy by using decimals with a limited number of decimal places but with no repeating pattern to these decimal places.

The word 'real' – used to describe the 'Real Number System' – was introduced by Rene` Descartes in the 17th century to distinguish these numbers from the other newly emerging numbers called 'complex' or 'imaginary' numbers belonging to the 'Complex Number System' (a non-elementary number system). The complex number system is the next natural extension of the Real Number System.

So far, the major milestones in our journey exploring the elementary number systems have consisted of:

1. Formalising counting with **Natural Numbers** (symbolised by '$\mathbb{N}$') by modelling the elementary properties of the simple counting process

2. Extending the Natural Number System by modelling the property that every natural number can be assigned an additive inverse and then naming the extended number system the **Integer Number System** (symbolised by '$\mathbb{Z}$')

3. Extending the Integer Number System by modelling the property that every non-zero integer number can be assigned a multiplicative inverse and then naming the extended system the **Rational Number System** (symbolised by '$\mathbb{Q}$').

In this chapter we will reach our fourth (and final) milestone by extending the Rational Number System by modelling the property that every point along the number line (including points associated with irrational numbers) can be represented by using infinite decimals and every infinite decimal can be assigned to a point along the number line. We call this extension of the Rational Number System, the **Real Number System** (symbolised by '$\mathbb{R}$').

The development of other branches of Mathematics during the 18th Century (such as Calculus) made extensive use of the Real Number System without completely defining it. In 1871 Georg Cantor provided the first rigorous definition of the Real Number System and this has been widely accepted and expanded on in the intervening years.

Historically, several methods have been developed and used to describe the Real Number System. The remainder of this chapter is devoted to outlining the meaning of the Real Number System using infinite decimals. The approached used to develop the 'Real Numbers' as 'Infinite Decimals' follows closely that given by T. Gowers in his supplementary notes to his lecture notes at Cambridge University.

Approach to the Real Number System using Infinite Decimals

We will begin our investigation of the Real Number System by showing how we can model the points along the number line that aren't described by rational numbers. Initially, we select an arbitrary point along the real number line and then outline the process for generating the infinite-decimal coordinate (from finite-decimal coordinates) associated with this point. This concept is illustrated in Figure 12.2 below.

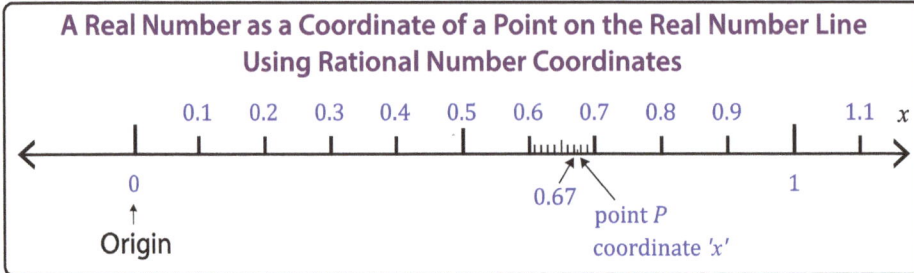

Figure 12.2

First, we assume that the point P has coordinate 'x'. From Figure 12.2 we observe that the point P is between the coordinates '0' and '1'. Hence, the decimal representation for P begins with '$T(x, 0) = 0$' (i.e. a zero before the decimal point) where '$T(x, n)$' is the truncation function of 'x' after the nth place (formally outlined in the Definition Section below).

Next, we observe the point P lies between the coordinates '0.6' and '0.7'. Hence, the decimal representation for P can be refined to '$T(x, 1) = 0.6$'. We then go on to observe that the location for point P can be further refined again to be between the coordinates '0.67' and '0.68'. Consequently, the decimal representation for P can be better approximated by '$T(x, 2) = 0.67$'. We can continue subdividing the interval from '0.67' and '0.68' into ten equal intervals and get a better approximation for the value of the coordinate of P.

In this chapter we will assume that we can continue on indefinitely to determine which coordinates a point lies between. In this way, we can determine the next digit that occurs in the improving approximation of the coordinate of 'x' which is the infinite decimal coordinate of the point P.

Language of the Real Number System using Infinite Decimals

In this section, our aim is to assign meaning to the alphabetic symbols of the language of the Real Number System. We will then use this language to express the definitions of the Real Number System that are essential for its development. Our starting alphabet we developed in the previous chapter is outlined in Table 12.1 on the following page.

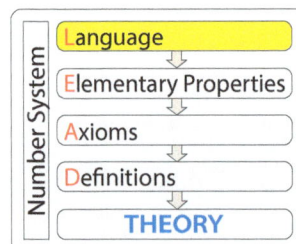

12.1 The Alphabet of the Rational Number System and Irrational Numbers	
Symbols	**Meaning from:**
'0', '1', '+', '×', '(', ')'	The Axioms of the Natural Number System
'=', '≠'	The Axioms of Logic
'2', '3', '4', '5', '6', '7', '8', '9'	Definition 2.1
'a', 'b', 'c', ..., 'x', 'y', 'z', ...	The variables that can be rational or irrational numbers
','	Definition 2.3
'-1'	The Axioms of the Integer Number System
'$-$'	Definitions 3.4 and 3.5
'$\frac{1}{\square}$'	The Axioms of the Rational Number System as Fractions
'$\frac{a}{b}$'	Definition 4.2
'÷'	Definition 4.4
'$\square^{-1}$'	The Axioms of Rational Number System using Index Form
'$(a \times b^{-1})^m$'	Definition 5.6
'.'	Definition 6.4
'$a_n \ldots a_1 a_0.d_1 \ldots d_m$'	Definition 6.5
'$\pm$'	Definition 7.8
'<'	The Axioms of the Rational Number System using Order
'≤', '>', '≥'	Definition 8.2
'$\|x - y\|$'	Definition 8.7
'{', '}'	Definition 9.1
'$(,)$'	Definition 9.2 (overloading the parentheses symbols)
'∞'	Definition 9.3
'$\lim_{n \to \infty} a_n$'	Definition 9.10 (refer to the note following this definition)
'$a_n \ldots a_1 a_0.d_1 \ldots d_m \ldots$'	Definition 10.1
'%'	Definition 10.3

12.1 The **Alphabet** of the Rational Number System and Irrational Numbers	
'$\sqrt{}$'	'$\sqrt{}$' is called the 'square root' symbol
'$T(x, n)$'	Definition 11.3

Table 12.1

Given that we are able to use a sequence of finite decimal numbers (i.e. rational numbers) to approximate 'real numbers' with any degree of accuracy, we would expect these 'real numbers' to conform to the properties of all the basic Axioms of the Rational Number System. This expectation is justified by the way we assign meaning to the symbols of the Real Number System as extensions of the symbols of the Rational Number System.

In this way, we will show that the elementary properties of the Real Number System can be derived from the Axioms of the Rational Number System which, in turn, confirms our intuition about these 'real numbers'.

The Elementary Property of the Real Number System using Infinite Decimals

Up to this point, we have developed some of the language for the Real Number System and have established the concept of an irrational number (Chapter Eleven). There are **no more elementary** properties associated with this new number system; however, the Real Number System has one **non-elementary** property.

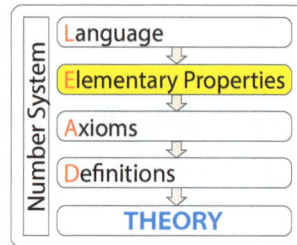

Completeness Property (non-elementary)

For the Rational Number System, we observed a sequence of rational numbers $(1.4, 1.41, 1.414, 1.4143, ...)$ where each number is the smallest decimal number (to that number of decimal places) whose square is less than '2'. This sequence approaches the number '$\sqrt{2}$' which we have shown to be an **irrational** number. Consequently, we can say that there exists an infinite sequence of rational numbers that does **not** have its limit **in** the Rational Number System (and as such does not belong to the Rational Number System).

The non-elementary property of the Real Number System that we will be demonstrating in this chapter is that the limit of a sequence of real numbers will always be a real number. This concept is called the **completeness** property. From the example described above it is apparent that the Rational Number System does not have the completeness property.

The Axioms of the Real Number System using Infinite Decimals

Although there are **no** additional number-system axioms, the Axioms of the Rational Number System will provide the axioms for use in the Real Number System.

Therefore, we can express the set of elementary properties for the rational numbers within the Real Number System as axioms, whereas for the irrational numbers these same properties are provable as theorems. These theorems of the Real Number System which cover the irrational numbers will be proven in the Theory Section below.

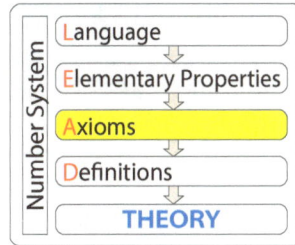

The Axioms of Logic

Before we can reason correctly from the axioms that apply to the rational number subsystem of the Real Number System, we again need to use the 'Axioms of Logic' as they apply to this rational number subsystem. You will recall from Table 2.5 in Chapter Two, that the 'Axioms of Logic' give us the properties of the equal sign '=', that is, how the equal sign is related to rational numbers as well as addition and multiplication of these rational numbers.

By using this equal sign associated with the Rational Number System, we will be able to define the equal sign for the Real Number System. We will then prove that the Axioms of Logic can be extended to also apply to the Real Number System. This will allow us to achieve our goal of extending the Rational Number System (which is an **ordered field**) to the Real Number System which we will demonstrate is a '**complete ordered field**'.

Definitions of the Real Number System using Infinite Decimals

Up until now we have intuitively referred to 'infinite decimals' as an extension of our introduction to infinite decimals in Chapter Eleven. However, our approach to developing the Real Number System is dependent on assigning a specific format to 'infinite decimals'. You will recall from Chapter Eleven that we were able to define positive infinite decimals, however, we were unable to perform operations with these new numbers.

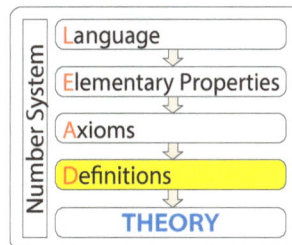

In order to define the equal '=' sign for 'infinite decimals' and be able to develop a definition of addition and multiplication for these 'infinite decimals', it is necessary to place several restrictions on the format that these 'infinite decimals' can take.

These format restrictions include:

1. Every rational number must be represented by a **unique** 'infinite decimal'.

2. There must be a rule for **uniquely** determining each digit in the string of digits of an 'infinite decimal'.

3. The decimal string of every 'infinite decimal' must be **non-decreasing**, that is, as you move along the decimal string of a number from left-to-right, each new number formed by adding the next decimal place cannot make that number less than the previous number.

4. The format of an 'infinite decimal' must be **compatible** with the addition and multiplication operators for these 'infinite decimals'.

The definition of an 'infinite decimal' that meets these restrictions is given as our first definition.

Definition 12.1   Infinite Decimals

An **infinite decimal** 'x' is a string of symbols of the form:

$$(d_0).d_1d_2 \dots d_m \dots$$

where the symbols in this string are interpreted as follows:

- There exists a method of uniquely determining the integer 'd_0' and each digit, '$d_1, d_2, \dots, d_m, \dots$' after the decimal point.

- The subscript 'm' is a non-negative integer.

- The parentheses symbols in '(d_0)' indicate that this is an infinite decimal.

- The symbol 'd_0' can be assigned any integer: positive, negative or zero.

- The symbols '$d_1, d_2, \dots, d_m, \dots$' are assigned values from the set of digits: {0, 1, 2, 3, 4, 5, 6, 7, 8, 9}.

- If '$d_1 = 0$', '$d_2 = 0$', '$d_3 = 0$', etc. (i.e. all digits after the decimal point are zero), then this decimal can be written as the infinite decimal '$(d_0 - 1).999\dots$'.

- In the case of a last non-zero digit following the decimal point '$(d_0).d_1 d_2 \dots d_m000\dots$' (i.e. '$d_m > 0$'), then '$d_m$' is changed to '$d_m - 1$' and the string of '$000\dots$' is changed to '$999\dots$'. (See Note 3 below).

––––––––––––––––––

Notes:

1. For the infinite decimal 'x', given by '$(d_0).d_1d_2 \dots d_m \dots$', the first '$n$' terms in the sequence of partial sums '$S_n(x)$' for this infinite decimal can be written as:

$$S_1(x) = d_0 + 0.d_1$$

$$\vdots$$

$$S_n(x) = d_0 + 0.d_1 d_2 \ldots d_n$$

$$\therefore \quad S_n(x) = d_0 + d_1 \times 10^{-1} + d_2 \times 10^{-2} + \ldots + d_n \times 10^{-n}$$

Consequently, the nth term after the decimal point of the infinite decimal is: $S(x, n) - S(x, n-1) = d_n \times 10^{-n}$.

2. From this definition of infinite decimals, it is apparent that the integer to the left of the decimal point, namely 'd_0', can be positive, zero or negative. Likewise, since we are always adding on the non-negative numbers '$d_1 \times 10^{-1}$', '$d_2 \times 10^{-2}$', etc. to the right of the decimal point, this sequence of partial sums '$S_n(x)$' is non-decreasing. In this way, we can declare that infinite decimals are non-decreasing by definition.

3. The final bullet point of this definition covers how to convert any finite decimal number into a unique infinite decimal format since any rational number such as '123' can be written as '123.000...' which can be converted to the infinite decimal format: '(122).999...'. Similarly, a finite decimal number such as '123.3458' can be written as '123.3458000...' which can be converted to the infinite decimal as '(123).3457999...'.

4. To arrive at a unique representation of infinite decimals, this definition forces us to replace '1.000...', '0.000...' and '−1.000...' with the infinite decimals '(0).999...', '(−1).999...' and '(−2).999...' respectively, since these are the appropriate infinite decimal representations of the rational numbers '1', '0' and '−1'. In Chapter Eleven we observed that when we multiply the non-decreasing sequence of approximations to '$\sqrt{2}$' by themselves we obtain the non-decreasing sequence:

$$\left(1.96, \ 1.9881, \ 1.999396, \ 1.99996164, \ \ldots\right)$$

which is approaching '2'. In this way, the last bullet point in the above definition aims to ensure consistency when adding and multiplying infinite decimals.

5. In the case of an infinite decimal, e.g. '(−3).999...', this number will be negative even though its infinite decimal fraction part is positive (i.e. '0.999...' is a positive number). This infinite decimal is 'equivalent to' '−2'.

Several examples in which this definition is applied to obtain the infinite decimal equivalent of some typical rational numbers are as follows:

1. '0' is written as: '(−1).999...'

2. '1' is written as: '(0).999...'

3. '1.5' is written as: '(1).4999...'

4. '−1' is written as: '(−2).999...'

5. '−1.5' is written as: '(−2).4999...'

6. '−1.23456' is written as: '(−2).76543999...'

7. '−2.4999...' is written as: '(−3).4999...'.

We give a simple example to show how a rational number can be transformed to another rational number so it will be ready to be expressed as an infinite decimal number. We illustrate the truth of Item 6 from the above list when we transform the rational number '−1.23456' into a different format and then express it as an infinite decimal.

$$-1.23456 = -1 - 0.23456 \qquad \text{... by defn of negative decimals}$$

$$= -1 - 0.23456 + 0 \quad \text{... by adding '0'}$$

$$= -1 - 0.23456 - 1 + 1 \text{... by subst. '}0 = -1 + 1\text{'}$$

$$= -2 + 1 - 0.23456 \quad \text{... by simplifying terms}$$

$$= -2 + 0.76544 \qquad \text{... by simplifying terms}$$

$$\therefore \quad -1.23456 = -2 + 0.76543999... \text{ ... by last bullet Definition 12.1}$$

The key observation is that we have been able to transform a decimal number with a negative component to the right of the decimal point into two numbers where the number with only a decimal fraction component is positive. By converting the original number '−1.23456' to '−2 + 0.76544', we can write the equivalent infinite decimal immediately as: '(−2).76543999...'.

At this point it is worth emphasising the fact that, up until this chapter, you will always have written decimal numbers in a format that allows the decimal fraction part of that decimal number to turn out positive or negative depending on the number. In this chapter we now restrict this familiar format of decimal numbers so that the decimal fraction part will always be positive. This change in the way we write traditional decimal numbers is demonstrating that this traditional format is not the 'best' way to work within the Real Number System.

The first operation we wish to perform on these infinite decimals is an extension of the truncation operation for the Rational Number System that we first encountered in Chapter Eleven in the form of Definition 11.3. This extended truncation operation is denoted by '$T(x, n)$' and now operates on infinite decimals 'x' (such as when we wish to define equality of two infinite decimals).

Definition 12.2 Truncation Operator on Infinite Decimals

In the case of an infinite decimal 'x', '$T(x, n)$' represents the finite decimal obtained by **truncating** 'x' after the nth decimal place where 'n' can be assigned any non-negative integer.

———————————

Notes:

1. If the infinite decimal 'x' is given by '$(d_0).d_1d_2 ... d_n ...$' where 'd_n' is the digit in the nth decimal place, then the truncations of 'x' are:

$$T(x, 0) = d_0$$

$$T(x, 1) = d_0 + 0.d_1$$

$$T(x, 2) = d_0 + 0.d_1d_2$$

$$\vdots$$

$$T(x, n) = d_0 + 0.d_1d_2 ... d_n$$

$$T(x, n + 1) = d_0 + 0.d_1d_2 ... d_n\, d_{n+1}$$

$$\vdots$$

Consequently, the truncations '$T(x, 0)$', '$T(x, 1)$', '$T(x, 2)$', etc. are finite decimals (and hence rational numbers). From the definition of partial sums, '$S_n(x)$' in the Note following Definition 12.1, we see that, in the case of infinite decimals, '$S_n(x) = T(x, n)$'.

2. By defining an infinite decimal as having a unique sequence of digits after the decimal point, we have removed any ambiguity in the value associated with the truncation process. The uniqueness of the truncation process will be an essential criteria when defining the equality symbol '$=$' for infinite decimals in Definition 12.3 below.

We will now illustrate the use of this truncation process for the familiar example of '$\sqrt{2}$'.

Example 12.1

In Chapter Eleven, we observed that the infinite decimal 'x' representing the irrational number '$\sqrt{2}$' has its first four terms as truncations of 'x' (and hence rational numbers) as follows:

$$T(x, 0) = T(\sqrt{2}, 0) = 1$$

$$T(x, 1) = T(\sqrt{2}, 1) = 1.4$$

$$T(x, 2) = T(\sqrt{2}, 2) = 1.41$$

$$T(x, 3) = T(\sqrt{2}, 3) = 1.414$$

———————————

In Definition 12.3 below, we will introduce the equality symbol for infinite decimals which will be essential for proving some of the properties of the Real Number System.

Definition 12.3   Equality Symbol '=' for Infinite Decimals

If 'x' and 'y' are real numbers, then we say 'x' is equal to 'y' (denoted by '$x = y$') if (and only if) the rational number equations '$T(x, n) = T(y, n)$' are true for all non-negative integer values of 'n'.

———————————

This definition relies on the unique expression of infinite decimals using rational numbers as covered in Definition 12.1. Another definition that relies on the unique expression of infinite decimals using rational numbers is the 'less than' symbol '$<$' used to define the order of infinite decimals. Defining the order of two infinite decimals 'x' and 'y' will prove essential for Definition 12.9 in which we establish the 'limit' of a sequence of infinite decimals.

Definition 12.4   Order of Infinite Decimals

If 'x' and 'y' are infinite decimals, then we say 'x' is less than 'y' (denoted by '$x < y$') if (and only if) there exists a non-negative integer 'n' such that the rational number inequality '$T(x, n) < T(y, n)$' is true.

———————————

Notes:

1. In this case 'n' is the truncation variable which is assigned a non-negative integer and provides the decimal place to the right of the decimal point along the string of digits representing the infinite decimals 'x' and 'y'.

2. It is also worth noting that the order relation for the rational numbers which result from truncations of the infinite decimals 'x' and 'y' can be used to extend the order relation to infinite decimals.

Now that we have a definition of the 'less than' order symbol for infinite decimals, we are able to define the other 'order' relation symbols (see Definition 12.5 below).

Definition 12.5   Order Relation Symbols for Infinite Decimals

The other order relationships for the infinite decimals 'x' and 'y' are defined as:

'$x \leq y$' if (and only if) either '$x < y$' or '$x = y$'

'$x > y$' if (and only if) '$y < x$'

'$x \geq y$' if (and only if) either '$y < x$' or '$x = y$'.

––––––––––––––––

The first appropriate use of this definition of order is to define positive and negative infinite decimals as outlined in Definition 12.6 below.

Definition 12.6   Positive and Negative Infinite Decimals

A **positive infinite decimal** is an infinite decimal 'x' such that '$x > (-1).999...$'. A **negative infinite decimal** is an infinite decimal 'x' such that '$x < (-1).999...$'.

––––––––––––––––

Note:

In order to be consistent with infinite decimals, we have used the infinite decimal equivalent '$(-1).999...$' of the rational number '0' in the above definition.

In Definition 12.1, it is straightforward to see that we have defined each infinite decimal to be a non-decreasing series. For example, the note above outlines how the infinite decimal '$(-1).999...$' starts at '-1' and we can then progressively add on the positive rational numbers '0.9', '0.09', '0.009' etc. to ensure this infinite decimal is non-decreasing.

The next step will be to define non-decreasing sequences of infinite decimals as outlined in Definition 12.7 below.

Definition 12.7   Non-decreasing Sequences of Infinite Decimals

Let $(x_0, x_1, x_2, ..., x_m, ...)$ be an infinite sequence of infinite decimals. We say this sequence is **non-decreasing** if (and only if) these infinite decimals satisfy the following order relations:

$$x_0 \leq x_1 \leq x_2 \leq ... \leq x_m \leq ...$$

The sequence of truncations of '$\sqrt{2}$' provides the simplest example of a sequence of non-decreasing finite decimals (according to Chapter Eleven) which approach '$\sqrt{2}$'. To create a non-decreasing sequence of infinite decimals from the truncations of an infinite decimal, we write the notation of this new sequence as $(z_0, z_1, z_2, ..., z_m, ...)$. In the case of the infinite decimal for '$\sqrt{2}$', we truncate '$\sqrt{2}$' to a rational number and then change this rational number to an infinite decimal using Definition 12.1 as follows:

$$z_0 = (0).999...\qquad \text{where: } T(\sqrt{2}, 0) = 1$$
$$z_1 = (1).3999...\qquad \text{where: } T(\sqrt{2}, 1) = 1.4$$
$$z_2 = (1).40999...\qquad \text{where: } T(\sqrt{2}, 2) = 1.41$$
$$z_3 = (1).413999...\qquad \text{where: } T(\sqrt{2}, 3) = 1.414$$
$$z_4 = (1).4141999...\qquad \text{where: } T(\sqrt{2}, 4) = 1.4142$$
$$z_5 = (1).41420999...\qquad \text{where: } T(\sqrt{2}, 5) = 1.41421$$
$$z_6 = (1).414212999...\qquad \text{where: } T(\sqrt{2}, 6) = 1.414213$$

The above list provides the first seven infinite decimals in a non-decreasing sequence of infinite decimals approaching the infinite decimal as represented by the symbol '$\sqrt{2}$'.

A key property of a sequence of infinite decimals involves determining whether this sequence is 'bounded'. A '**bound**' for a sequence of numbers is a number that is greater than or equal to every number in that sequence. If a sequence has a bound, there are an infinite number of bounds of that sequence and the **smallest bound is the 'limit' of the sequence**. This concept in relation to infinite decimals is covered in Definition 12.8 below.

Definition 12.8 **Bounded Non-decreasing Sequences of Infinite Decimals**
Let $(x_0, x_1, x_2, ..., x_m, ...)$ be an infinite non-decreasing sequence of infinite decimals. This sequence is **bounded** by the infinite decimal 'y' if (and only if) 'y' satisfies the order relation for all assignments to the sequence variable 'm', '$x_m \leq y$' for all non-negative integer values assigned to 'm'.

***

Example:

1. Since '$(1).999... > \sqrt{2}$', then '$(1).999...$' is also a bound for the sequence of infinite decimals formed from the truncations of '$\sqrt{2}$'. There is clearly an infinite number of bounds for this sequence, including: '$(1).4999...$', '$(1).41999...$', '$(1).414999...$', '$(1).4142999...$', etc.

2. It is straightforward to see that if 'y' is an infinite decimal and 'z_m' is the **infinite decimal representation** of '$T(y, m)$' for each 'm', then the terms of the sequence $(z_0, z_1, z_2, ..., z_m, ...)$ obey the following inequalities:

$$z_0 \leq y, z_1 \leq y, z_2 \leq y, ...$$

3. According to Definition 12.8, the integer to the left of the decimal point in the infinite decimals in a bounded sequence of infinite decimals, can

only increase to a finite integer. Otherwise, it is an **unbounded** infinite non-decreasing sequence and the number to the left of the decimal point must increase indefinitely. In this way, for the infinite decimal 'y', the infinite decimal '$(T(y, 0)).999...$' is a bound for 'y' because '$(T(y, 0)).999... \geq y$'.

An obvious property of the non-decreasing sequence of infinite decimals bounded by '$\sqrt{2}$' outlined above is that, for each decimal place, the truncations of this sequence are achieving their final value as we progress along the sequence. For example, when we truncate the terms of the infinite sequence 'approaching' '$\sqrt{2}$', for the first seven terms of the rational number sequence at the fourth place, we obtain the following results:

$$T(z_0, 4) = 0.9999$$

$$T(z_1, 4) = 1.3999$$

$$T(z_2, 4) = 1.4099$$

$$T(z_3, 4) = 1.4139$$

$$T(z_4, 4) = 1.4141$$

$$T(z_5, 4) = 1.4142$$

$$T(z_6, 4) = 1.4142$$

$$T(z_7, 4) = 1.4142$$

From these truncations we see that the sequence of terms '$T(z_m, 4)$' has achieved its final value since '$T(z_5, 4) = T(z_6, 4) = ... = 1.4142$'.

In Theorem 12.1 (as discussed in the Theory Section below) we will prove that if the sequence $(x_0, x_1, x_2, ... , x_m, ...)$ is a bounded non-decreasing sequence of infinite decimals, then the truncations of this sequence $(T(x_0, n), T(x_1, n), T(x_2, n), ...)$, i.e. truncated after the nth decimal place, eventually becomes a constant decimal rational number for each value of 'n'.

We use this key result to define the '**limit**' of a bounded non-decreasing sequence of infinite decimals (see Definition 12.9 below).

Definition 12.9   Limit of an Infinite Decimal Sequence

Let $(x_0, x_1, x_2, ... , x_m, ...)$ be a bounded non-decreasing sequence of infinite decimals. We can define the infinite decimal 'y' as the **limit** of the sequence of infinite decimals $(x_0, x_1, x_2, ... , x_m, ...)$ where for each non-negative truncation integer 'n' there exists a subscript integer 'N_n' such that for all cases of '$m > N_n$' the truncations '$T(x_m, n)$' are constant.

Hence, the **limit** 'y' is given by its truncations as follows:

$$T(y, n) = T(x_{N_n + 1}, n) = T(x_{N_n + 2}, n) = T(x_{N_n + 3}, n) = \ldots$$

In this way, the infinite decimals in the sequence $(x_0, x_1, x_2, \ldots, x_m, \ldots)$ have become constant at the nth decimal place after passing the $N_n{}^{th}$ infinite decimal forward along the sequence $(x_0, x_1, x_2, \ldots, x_m, \ldots)$.

Note:

1. Although in Definition 10.1 of Chapter Ten we used the definition of a limit to define repeating decimals, here we are using bounded non-decreasing sequences of infinite decimals and their truncations at the nth place to define the limit of infinite decimals. The ordering of infinite decimals is the key to enabling this different definition of a limit to be used.

2. Given an infinite decimal 'y', then let 'z_n' be the infinite decimal formed from the truncation at the nth decimal place of 'y' (i.e. '$T(y, n)$' according to Definition 12.1). Consequently, the infinite decimal 'y' is both the limit of the sequence of infinite decimals $(z_0, z_1, z_2, \ldots, z_n, \ldots)$ and also a bound for this sequence.

By its very construction, this definition of a limit encompasses the 'completeness' property for infinite decimals. This property is defined in Definition 12.10 below.

Definition 12.10   Completeness of the Set of Infinite Decimals

The set of infinite decimals is called **complete** because for every bounded non-decreasing sequence of infinite decimals $(x_0, x_1, x_2, \ldots, x_m, \ldots)$, there exists an infinite decimal 'y' such that 'y' is the limit of the sequence $(x_0, x_1, x_2, \ldots, x_m, \ldots)$.

Note:

In Theorem 12.1 in the Theory Section below, we will cover a construction of 'y' from the properties of infinite decimals and the bounded condition on sequences of infinite decimals. In this theorem we also prove that infinite decimals are complete by definition.

Our immediate objective is to now demonstrate that this definition of the limit of a sequence of infinite decimals can be used to obtain definitions for the operations of '$+$' and '$\times$' of infinite decimals. This definition is modelled on the equivalent addition and multiplication definitions for repeating decimals we studied in Definition 10.2 of Chapter Ten. In this case, we relied heavily on the two key theorems from Chapter Nine – namely Theorems 9.4 (Sum of Two Infinite Sequences) and 9.5 (Product of Two Infinite Sequences).

We now use this model of addition of repeating decimals and the definition of the limit of a bounded non-decreasing sequence of infinite decimals to define addition for the set of infinite decimals (see Definition 12.11 below).

Definition 12.11   Addition of Infinite Decimals

We define **addition** of the two bounded, non-decreasing infinite decimals 'x' and 'y' (which is denoted as '$x + y$') from the sum of the truncations of 'x' and 'y' – namely, from the sequence:

$$\left(T(x, 0) + T(y, 0)\, ,\, T(x, 1) + T(y, 1),\, ... \,,\, T(x, m) + T(y, m),\, ... \right)$$

We have to turn this sequence of truncation into infinite decimals in order to use the definition of a limit given in Definition 12.9 above. In this way, we let this new sequence of infinite decimals be called $\left(z_0, z_1, ... , z_m, ... \right)$ and be defined from the sequence of truncations of 'x' and 'y' above, by the following:

'z_0' is the infinite decimal equivalent of '$T(x, 0) + T(y, 0)$'

'z_1' is the infinite decimal equivalent of '$T(x, 1) + T(y, 1)$'

'z_2' is the infinite decimal equivalent of '$T(x, 2) + T(y, 2)$'

$$\vdots$$

'z_m' is the infinite decimal equivalent of '$T(x, m) + T(y, m)$'

$$\vdots$$

We define the **addition** '$x + y$' to be the limit of the bounded non-decreasing sequence of infinite decimals $\left(z_0, z_1, ... , z_m, ... \right)$.

For this definition to be valid, we must demonstrate that the sequence $\left(z_0, z_1, ... , z_m, ... \right)$ is bounded. However, we know that as '$T(x, 0) + 1 > T(x, m)$' for all positive values of 'm' (given the decimal fraction to the right of the decimal point can at most be '0.999...9') and similarly '$T(y, 0) + 1 > T(y, m)$' for all positive values of 'm', it follows that '$T(x, 0) + T(y, 0) + 2 > T(x, m) + T(y, m)$' and this also holds true for their infinite decimal equivalents. Therefore, '$T(x, m) + T(y, m)$' is bounded for all values of 'm'.

Note:

In the special case where 'x' and 'y' are repeating decimals, the definition above gives the same sum '$x + y$' of the limit of the sequences of partial sums (according to Definition 10.2 and Theorem 9.4).

One of the first applications of the addition operator for infinite decimals outlined in the Theory Section below demonstrates that the sum of an infinite decimal and its additive inverse (see Definition 12.12) gives the 'zero' infinite decimal – namely, '$(-1).999...$'.

Definition 12.12 Additive Inverse of an Infinite Decimal

Let 'x' be an infinite decimal so that 'x' is given by '$(d_0).d_1d_2 ... d_m ...$' for any integer 'd_0' and digits '$d_1, d_2, ... , d_m, ...$'. We define the **additive inverse** of 'x', denoted by '$-x$' as the infinite decimal:

$$-x = (e_0).e_1e_2 ... e_m ...$$

where: '$e_0 = -d_0 - 1$' and '$e_m = 9 - d_m$' for '$m = 1, 2, 3, ...$'. If 'x' takes the form '$x = (d_0).d_1d_2 ... d_m999...$', then we have to apply the last bullet point of Definition 12.1 which gives us the conversion of a finite decimal into a unique infinite decimal.

Notes:

1. To illustrate the last sentence in the above algorithm, we find the additive inverse of the infinite decimal '$x = (0).999...$' (i.e. the equivalent of the rational number '1'). From this algorithm, we obtain '$-x$' as the repeating decimal '$-1+ 0.000...$' which then has to be converted to its infinite decimal format as '$(-2).999...$'.

2. From the previous note, it is straightforward to see that this result can be generalised to '$-x + x = (-1).999...$'. According to Theorem 12.2(3), we will observe that the additive inverse does indeed have this property.

3. This process of generating the additive inverse of an infinite decimal 'x' indicates that the negative sign '$-$' is a unary operator for infinite decimals.

4. As we did with the rational numbers, we can define 'subtraction' of real numbers by using this additive inverse. In this way, we can define the subtraction of the infinite decimal 'x' from the infinite decimal 'y' (denoted as '$y - x$') as '$y + -x$'.

Example 12.2

Write out the additive inverse of '$\sqrt{2}$' (i.e. '$-\sqrt{2}$' to fifty decimal places) given:

$\sqrt{2} = (1).41421356237309504880168872420969807856967187537694...$

From Definition 12.12 we see that the additive inverse of '$\sqrt{2}$' (when written as the infinite decimal '$-\sqrt{2}$' to fifty decimal places) is:

$$-\sqrt{2} =$$

$$(-2).58578643762690495119831127579030192143032812462305...$$

Therefore, when we add these two infinite decimals we expect to obtain:

$$\sqrt{2} + -\sqrt{2} = (-1).999...9999... \qquad ... \text{ to '50' decimal places.}$$

The next step is to use the model of multiplication of repeating decimals using Theorem 9.5 and the definition of the limit of a bounded non-decreasing sequence of infinite decimals to define multiplication for the set of infinite decimals.

The definition of the additive inverse above signals that using negative numbers in multiplications of infinite decimals will not be as straightforward as it was for addition of infinite decimals. We can circumvent this difficulty by defining multiplication first for positive infinite decimals and then using the unary operator '$-$' to handle the other three cases.

Definition 12.13   Multiplication of Infinite Decimals

We define **multiplication** of two positive infinite decimals 'x' and 'y' (which is denoted as '$x \times y$') to be the limit of the product of the truncations of 'x' and 'y', namely, the limit of the sequence:

$$\Big(T(x, 0) \times T(y, 0), \; T(x, 1) \times T(y, 1), \; ... , \; T(x, m) \times T(y, m), \; ... \Big)$$

As we did for Definition 12.11 above, we have to turn this sequence of truncations into infinite decimals in order to use the definition of a limit given in Definition 12.9 above. In this way, we let this sequence of infinite decimals be called $\Big(z_0, z_1, ... , z_m, ... \Big)$ and be defined from the sequence of truncations of 'x' and 'y' above, by the following:

'z_0' is the infinite decimal equivalent of '$T(x, 0) \times T(y, 0)$'

'z_1' is the infinite decimal equivalent of '$T(x, 1) \times T(y, 1)$'

'z_2' is the infinite decimal equivalent of '$T(x, 2) \times T(y, 2)$'

$\vdots$

'z_m' is the infinite decimal equivalent of '$T(x, m) \times T(y, m)$'

$\vdots$

We define the **multiplication** '$x \times y$' to be the limit of the bounded non-decreasing sequence of infinite decimals $(z_0, z_1, \dots, z_m, \dots)$. Using exactly the same reasoning as in Definition 12.11, we have '$(T(x, 0) + 1)(T(y, 0) + 1) > T(x, m) \times T(y, m)$' and this also holds true for their infinite decimal equivalents. Therefore, '$T(x, m) \times T(y, m)$' is bounded for all values of 'm' and '$x \times y$' is truly the limit.

For negative infinite decimals, we define the remaining three cases as follows:

1. $-x \times y = -(x \times y)$

2. $x \times -y = -(x \times y)$

3. $-x \times -y = x \times y$

_____

Notes:

1. In the special case where 'x' and 'y' are infinite decimals with a repeating pattern, the definition above models the definition of the product '$x \times y$' as the product of the limit of the sequences of partial sums (according to Definition 10.2 and Theorem 9.5).

2. From this definition we would expect the result '$(-2).999\dots \times x = -x$' to be a true infinite decimal equation.

Similarly to the addition operator, now that we can use the multiplication operator, our first application will be to define the multiplicative inverse of an infinite decimal.

Definition 12.14 Multiplicative Inverse of an Infinite Decimal

If 'x' is a positive infinite decimal and we denote its **multiplicative inverse** by the symbols 'x^{-1}', then '$x \times x^{-1} = (0).999\dots$'. We use the following algorithm to calculate 'x^{-1}' using the infinite decimal sequence $(z_0, z_1, z_2, \dots, z_n)$ to approach 'x^{-1}' using the following steps:

1. Let an infinite decimal take the form '$z_0 = (e_0).999\dots$' where 'e_0' represents a non-negative integer. If a value of 'e_0' exists such that: '$x \times z_0 = (0).999\dots$', then we have found the multiplicative inverse of 'x'. Otherwise, we can determine the smallest non-negative integer value of 'e_0' such that '$x \times z_0 > (0).999\dots$' and set '$T(x^{-1}, 0) = T(z_0, 0) = e_0$' and then proceed to Step 2.

2. Let the infinite decimal 'z_1' take the form '$z_1 = (e_0).e_1 999\dots$' where 'e_0' represents the non-negative integer from Step 1. If the digit 'e_1' is such that: '$x \times z_1 = (0).999\dots$', then we have found the multiplicative inverse of 'x'. Otherwise, we can determine the smallest digit value of 'e_1' such that

'$x \times z_1 > (0).999...$' and set '$T(x^{-1}, 1) = T(z_1, 1) = e_0.e_1$' and then proceed to Step n.

$$\vdots$$

n.   Let 'z_{n-1}' take the form '$z_{n-1} = (e_0).e_1e_2 ... e_{n-2}e_{n-1}999...$' where '$e_0$' represents the non-negative integer from Step 1 and '$e_1, e_2, ..., e_{n-2}$' represents the digits from Step 2 through to Step $n-1$. If the digit 'e_{n-1}' is such that: '$x \times z_{n-1} = (0).999...$', then we have found the multiplicative inverse of 'x'. Otherwise, we can determine the smallest digit value of 'e_{n-1}' such that '$x \times z_{n-1} > (0).999...$' and set '$T(x^{-1}, n-1) = T(z_{n-1}, n-1) = e_0.e_1e_2 ... e_{n-2}e_{n-1}$' and then proceed to calculate the next decimal place of the multiplicative inverse by this method.

Using this algorithm we can find the **multiplicative inverse** 'x^{-1}' to any number of decimal places 'n' since '$T(x^{-1}, n)$' is formed from the truncations of 'z_n', where:

$$T(x^{-1}, n) = T(z_n, n)$$

If 'x' and 'x^{-1}' are negative infinite decimals (i.e. '$x < (-1).999...$'), then we apply this definition to '$-x$' and use our algorithm to find its multiplicative inverse '$(-x)^{-1}$' and then, finally, define the multiplicative inverse of '$x < (-1).999...$' as '$-((-x)^{-1})$'.

Notes:

1.   We use the infinite decimals '(z_n)', which are equivalent to finite decimals (i.e. they end with the string '999...'), to help us find the nth digit in the infinite decimal form of the multiplicative inverse.

2.   In Theorem 12.4(3) below we demonstrate that 'x^{-1}' (as we defined above) deserves its name as the multiplicative inverse because it satisfies the equation '$x \times x^{-1} = (0).999...$' for '$x \neq (-1).999...$'.

3.   As with the rational numbers, we can define division of the real numbers by using their multiplicative inverses, so that division of the infinite decimal 'x' by the infinite decimal 'y', can be denoted as '$\frac{x}{y}$', but is defined by '$x \times y^{-1}$'.

We will now use this algorithm to find the multiplicative inverse of the rational number '2.1235762' by finding the multiplicative inverse of its infinite decimal equivalent given by '$(2).1235761999...$'. We can find '$T(x^{-1}, 7)$' for this multiplicative inverse, 'x^{-1}', by using the multiplicative-inverse algorithm (Definition 12.14) in this more familiar context to help us to understand this process.

Example 12.3

Use Definition 12.14 to find '$T(x^{-1}, 7)$' for the multiplicative inverse of 'x' where 'x' is assigned the value '(2).1235761999...'.

We start this process with the integer digit to the left of the decimal point and observe:

Step 1: $(2).1235761999... \times (0).9999... > (0).999...$

Since the '0' in the number '(0).9999...' on the left-hand side of this inequality is the smallest non-negative integer which satisfies this inequality, it follows that:

'$z_0 = (0).9999...$' so that: '$T(x^{-1}, 0) = T(z_0, 0) = 0$'

Step 2: $(2).1235761999... \times (0).3999... < (0).999...$

But: $(2).1235761999... \times (0).4999... > (0).999...$

Hence: '$z_1 = (0).4999...$' so that: '$T(x^{-1}, 1) = T(z_1, 1) = 0.4$'

Step 3: $(2).1235761999... \times (0).46999... < (0).999...$

But: $(2).1235761999... \times (0).47999... > (0).999...$

Hence: '$z_2 = (0).46999...$' so that: '$T(x^{-1}, 2) = T(z_2, 2) = 0.47$'

Step 4: $(2).1235761999... \times (0).470999... > (0).999...$

Since the next number after '(0).46999...' (and preserving this format) is '(0).470999...', then this is the smallest number that satisfies the condition '$x \times z_n > (0).999...$'.

Hence: '$z_3 = (0).470999...$' so that: '$T(x^{-1}, 3) = T(z_3, 3) = 0.470$'

Step 5: $(2).1235761999... \times (0).4708999... < (0).999...$

But: $(2).1235761999... \times (0).4709999... > (0).999...$

Hence: '$z_4 = (0).4709999...$' so that: '$T(x^{-1}, 4) = T(z_4, 4) = 0.4709$'

Step 6: $(2).1235761999... \times (0).47090999... > (0).999...$

As we did in Step 3, we can conclude:

Hence: '$z_5 = (0).47090999...$' so that: 'T$(x^{-1}, 5) = T(z_5, 5) = 0.47090$'

Step 7: $(2).1235761999... \times (0).470902999... < (0).999...$

But: $(2).1235761999... \times (0).470903999... > (0).999...$

Hence: '$z_6 = (0).470903999...$' so that: '$T(x^{-1}, 6) = T(z_6, 6) = 0.470903$'

Step 8: $(2).1235761999... \times (0).4709036999... < (0).999...$

But: $(2).1235761999... \times (0).4709037999... > (0).999...$

Hence: '$z_7 = (0).4709037999...$' so that: '$T(x^{-1}, 7) = T(z_7, 7) = 0.4709037$'

Consequently, the integer before, and the first seven digits after, the decimal point of the multiplicative inverse of '$(2).1235761999...$' are given by:

$$T(x^{-1}, 7) = 0.4709037$$

We now use this multiplicative-inverse algorithm again to find the multiplicative inverse of the irrational infinite decimal '$\sqrt{2}$', i.e. '$(\sqrt{2})^{-1}$', to six decimal places.

Example 12.4

Use Definition 12.14 to find '$T(x^{-1}, n)$' for the multiplicative inverse of '$\sqrt{2}$'. We calculate the value of '$T((\sqrt{2})^{-1}, 6)$' by starting with '$T((\sqrt{2})^{-1}, 0)$' as follows:

Step 1: $\sqrt{2} \times (0).9999... = (1).41421356237309504880... > (0).999...$
Since the '0' in the number '$(0).9999...$' on the left-hand side of this inequality is the smallest non-zero integer which satisfies this inequality.
Hence: '$z_0 =(0).9999...$' so that: '$T((\sqrt{2})^{-1}, 0) = T(z_0, 0) = 0$'.

Step 2: $\sqrt{2} \times (0).6999... = (0).98994949366116653416... < (0).999...$
But: $\sqrt{2} \times (0).7999... = (1).13137084989847603904... > (0).999...$
Hence: '$z_1 =(0).7999...$' so that: '$T((\sqrt{2})^{-1}, 1) = T(z_1, 1) = 0.7$'.

Step 3: $\sqrt{2} \times (0).70999... = (1).00409162928489748464... > (0).999...$
Since the next number after '$(0).6999...$' (and preserving this format to two decimal places) is '$(0).70999...$', then this is the smallest number that satisfies the condition '$x \times z_2 > (0).999...$'.
Hence: '$z_2 =(0).70999...$' so that: '$T(x^{-1}, 2) = T(z_2, 2) = 0.70$'.

Step 4: $\sqrt{2} \times (0).706999... = (0).99984898859777819950... < (0).999...$
But: $\sqrt{2} \times (0).707999... = (1).00126320216015129455... > (0).999...$
Hence: '$z_3 = (0).707999...$' so that: '$T(x^{-1}, 3) = T(z_3, 3) = 0.707$'.

Step 5: $\sqrt{2} \times (0).7070999... = (0).99999040995401550900... < (0).999...$
But: $\sqrt{2} \times (0).7071999... = (1).00013183131025281851... > (0).999...$
Hence: '$z_4 = (0).7071999...$' so that: '$T(x^{-1}, 4) = T(z_4, 4) = 0.7071$'.

Step 6: $\sqrt{2} \times (0).70710999... = (01).00000455208963923995... < (0).999...$

This one-step calculation is the same as for Step 3.

Hence: '$z_5 = (0).70710999...$' so that: '$T(x^{-1}, 5) = T(z_5, 5) = 0.70710$'.

Step 7: $\sqrt{2} \times (0).707105999... = (0).99999885923538974757... < (0).999...$

But: $\sqrt{2} \times (0).707106999... = (1).00000030944895212067... > (0).999...$

Hence: '$z_6 = (0).707106999...$' so that: '$T(x^{-1}, 6) = T(z_6, 6) = 0.707106$'.

Consequently, we have identified the first six decimal places after the decimal point for the multiplicative inverse of '$\sqrt{2}$'. You will observe that we can check this result as we would expect our new infinite decimal arithmetic to give '$(\sqrt{2})^{-1} = \frac{\sqrt{2}}{2}$'. Fortunately, it is straightforward to divide '$\sqrt{2}$' by '2' and obtain this same result by an alternative method.

––––––––––––––––

Having now defined addition and multiplication for infinite decimals, as well as the inverses for these two operations, we are able to demonstrate that infinite decimals also have the properties of being commutative and associative under these operations and that the two operations ('$+$' and '$\times$') are related by the distributive property. Refer to the Theory Section below to see how these properties hold as theorems for the irrational numbers.

It will also be shown that the Axioms of Logic for the Rational Number System determine the properties of logic for the Real Number System using Infinite Decimals.

The number system of the infinite decimals which includes the operations of addition and multiplication defined above and which has the same properties as the Rational Numbers (proved below) is called the '**Real Number System using Infinite Decimals**'. We will formalise this concept in Definition 12.15 below.

Definition 12.15   Real Number System using Infinite Decimals

The set of infinite decimals with its associated operations of '$+$' and '$\times$' and the corresponding properties of the Rational Number System using recurring decimals, as well as the definitions of the inverses of the operations '$+$' and '$\times$', and the order symbol '$<$' is called the **Real Number System using Infinite Decimals**. The set of Real Numbers is denoted by the symbol '$\mathbb{R}$'.

––––––––––––––––

Our alphabet for the Real Number System using Infinite Decimals is outlined in Table 12.2 on the following page.

12.2 The Symbols of the Alphabet of the Real Number System using Infinite Decimals

Symbols	Meaning from:
'0', '1', '+', '×', '(', ')'	The Axioms of the Natural Number System
'=', '≠'	The Axioms of Logic
'2', '3', '4', '5', '6', '7', '8', '9'	Definition 2.1
'a', 'b', 'c', ..., 'x', 'y', 'z', ...	Real number variables
'.'	Definition 2.3
'-1'	The Axioms of the Integer Number System
'$-$'	Definitions 3.4 and 3.5
'$\frac{1}{\square}$'	The Axioms of the Rational Number System as Fractions
'$\frac{a}{b}$'	Definition 4.2 (refers to fraction symbol)
'÷'	Definition 4.4 (refers to division symbol)
'$\square^{-1}$'	The Axioms of Rational Number System using Index Form
'$(a \times b^{-1})^m$'	Definition 5.6
'.'	Definition 6.4
'$a_n \dots a_1 a_0 . d_1 \dots d_m$'	Definition 6.5
'$\pm$'	Definition 7.8
'$\lvert x - y \rvert$'	Definition 8.7 (refers to absolute value symbol)
'{', '}'	Definition 9.1 (refers to set symbols)
'$(\, , \,)$'	Definition 9.2 (overloading the parentheses symbols as either sequence delineators or infinite decimal numbers)
'∞'	Definition 9.3 (refers to symbol for infinity)
'$\lim\limits_{n \to \infty} a_n$'	Definition 9.10 (refer to the note following this definition)
'$a_n \dots a_1 a_0 . d_1 \dots d_m \dots$'	Definition 10.1 (refers to finite decimal)
'%'	Definition 10.3 (refers to percentage symbol)
'$\sqrt{}$'	'$\sqrt{}$' is called the 'square root' symbol
'$T(x, p)$'	Definition 11.3 (refers to the truncation operator)
'$(d_0).d_1 d_2 \dots d_m \dots$'	Definition 12.1 (refers to infinite decimals)

12.2 The Symbols of the **Alphabet** of the Real Number System using Infinite Decimals	
'y' as limit of (x_m)	Definition 12.9 (refers to limit for infinite decimals)
'x^{-1}'	Definition 12.14 (refers to infinite decimals)

Table 12.2

Table 12.2 incorporates the notation of the Real Number System using Infinite Decimals which permits us to generalise the work we have covered on repeating decimals and to represent real numbers using infinite decimals.

Intuitively, we would expect certain properties of the Real Number System to be inherited from the Rational Number System. Since irrational numbers can be approximated as closely as we like by rational numbers, we would expect many properties of the rational numbers to be imposed on the irrational numbers when using the limit process. However, the property of being rational is not one of these properties.

In this chapter, we have finally started exploring the last of the elementary number systems – the Real Number System using Infinite Decimals. The properties of the Real Number System captured in the Axioms of the Rational Number System and the Axioms of Logic can be proved in theorems for all the irrational numbers. We will use these axioms, our definitions and theorems (below) to achieve our objective of developing mastery in all the elementary number systems.

Theory of the Real Number System using Infinite Decimals

In this section we derive the basic properties of the Real Number System and demonstrate ways in which these properties make the Real Number System an extension of the Rational Number System.

Conceptually we will prove that the infinite decimals of the Real Number System are the appropriate numbers for assigning coordinates to all points along the number line.

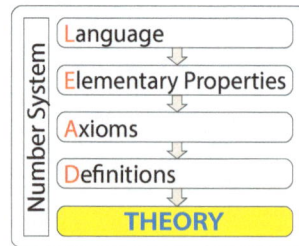

Number System:
- Language
- Elementary Properties
- Axioms
- Definitions
- **THEORY**

Theorem 12.1   Bounded Sequences of Infinite Decimals

For an infinite sequence $(x_0, x_1, x_2, \ldots, x_m, \ldots)$ of bounded non-decreasing infinite decimals, the truncations of this sequence $(T(x_0, n), T(x_1, n), T(x_2, n), \ldots)$, i.e. truncated after the nth decimal place, eventually becomes a constant decimal rational number for each value of 'n'. For each assignment to 'n', we are creating the truncations of the infinite decimal 'y' given by '$T(y, n)$'.

Proof

Given $\left(x_0, x_1, x_2, \ldots, x_m, \ldots\right)$ is a non-decreasing sequence, then so is the sequence:

$$\left(T(x_0, n),\ T(x_1, n),\ T(x_2, n),\ \ldots,\ T(x_m, n),\ \ldots\right)$$

Since the sequence $\left(x_0, x_1, x_2, \ldots, x_m, \ldots\right)$ is bounded, the truncated sequence given by $\left(T(x_0, n),\ T(x_1, n),\ T(x_2, n),\ \ldots\right)$ is also bounded and can only increment a finite number of times before it becomes a constant finite decimal at the nth place. This fixed constant finite decimal is defined as '$T(y, n)$' for the nth termination of some infinite decimal 'y'.

Similarly, if the same sequence is now truncated at the $(n + 1)^{th}$ place after the decimal point – i.e. $\left(T(x_0, n+1),\ T(x_1, n+1),\ T(x_2, n+1),\ \ldots\right)$ – this sequence will become constant after another finite number of changes. Since each change to the first 'n' places after the decimal point has already been taken into account as a result of all changes to all terms of the sequence truncated at the 'n^{th}' place, the first 'n' places of these two sequences are the same and the fixed finite decimal at the $(n + 1)^{th}$ place after the decimal point can be labelled '$T(y, n + 1)$'.

Every decimal place of some infinite decimal 'y' can be determined by the above process. Consequently, we can declare $(T(y, 0), T(y, 1), T(y, 2), \ldots)$ are the truncations that define the infinite decimal 'y'.

Hence the theorem is proved and for every bounded non-decreasing sequence of infinite decimals, these infinite decimals eventually becomes constant at a decimal place. That is, for each fixed decimal place, i.e. value of 'n', at some point along the sequence, i.e. for some value of 'm', these decimal places no longer change'.

––––––––––––

The importance of this theorem is that it gives a process for determining the nth decimal place and consequently the $(n+1)^{th}$ decimal place can also be calculated. Therefore, all decimal places can be calculated and the corresponding infinite decimal is denoted 'y' and it is **the** limit of this infinite, bounded, non-decreasing sequence of infinite decimals.

Definition 12.11 allows us to prove that addition of infinite decimals satisfies the properties given by the first three axioms for the addition operator of the Rational Number System. These properties for infinite decimals are covered in Theorem 12.2 on the following page.

Theorem 12.2 Addition is Closed, Commutative, Associative and has an Inverse

If we consider 'x', 'y' and 'z' to be infinite decimals, then these infinite decimals satisfy the following equations:

1. The number '$x + y$' is an infinite decimal   ... Closure Property '$+$'
2. $x + y = y + x$ ... Commutative Property '$+$'
3. $(x + y) + z = x + (y + z)$ ... Associative Property '$+$'
4. $x + (-x) = (-1).999...$ ... Inverse Property '$+$'

Proof

Equation 1:

First, the Closure Property for addition follows on from the definition of the 'addition' operator for infinite decimals in Definition 12.11 and the definition of a 'limit' in Definition 12.9.

Equation 2:

From the definition of addition of infinite decimals (Definition 12.11):

The sum of the infinite decimals 'x' and 'y' (denoted by '$x + y$') is the limit of the sequence of infinite decimals formed from the sequence of rational numbers $\big(T(x, 0) + T(y, 0), T(x, 1) + T(y, 1), T(x, 2) + T(y, 3), ...\big)$.

However, the truncations '$T(x, n)$' and '$T(y, n)$' are rational numbers which satisfy the Commutative Axiom '$T(x, n) + T(y, n) = T(y, n) + T(x, n)$'. Therefore, we can rearrange this addition sequence as follows: $\big(T(y, 0) + T(x, 0), T(y, 1) + T(x, 1), T(y, 2) + T(x, 2), ...\big)$. In this way, the sequence of infinite decimals formed from this last sequence has as a limit '$y + x$' according to Definition 12.11 again. Using the definition of the equal sign '$=$' for infinite decimals (Definition 12.3) we have:

$$x + y = y + x$$

Equation 3:

The proof that infinite decimals are associative under the operation of addition follows similar reasoning to the proof in Equation 2 above.

Equation 4:

In this last equation you will recall that '$(-1).999...$' is the representation of '0' as an infinite decimal and is used in the statement of the Additive Inverse Property outlined above. The first step is to prove that the general infinite decimal

'$x = (d_0).d_1d_2 \ldots d_m \ldots$' and its additive inverse '$-x$' given by '$-x = (-d_0- 1).e_1e_2 \ldots e_m \ldots$', where the digits are '$e_m = 9 - d_m$' are indeed additive inverses such that '$x + -x = (-1).999\ldots$'.

We know that 'x' is the limit of the sequence of infinite decimals formed from their finite decimal equivalents which are:

$$\big(T(x, 0),\ T(x, 1),\ T(x, 2),\ \ldots,\ T(x, m),\ \ldots\big)$$
$$=\big(d_0,\ d_0 + 0.d_1,\ d_0 + 0.d_1d_2,\ \ldots,\ d_0 + 0.d_1d_2 \ldots d_m,\ \ldots\big)$$

Also, '$-x$' is the limit of the following sequences of infinite decimals formed from their finite decimal equivalents which are:

$$\big(T(-x, 0),\ T(-x, 1),\ T(-x, 2),\ \ldots,\ T(-x, m),\ \ldots\big)$$
$$=\big(-d_0- 1,\ -d_0- 1 + 0.e_1,\ -d_0- 1 + 0.e_1e_2,\ \ldots,\ -d_0- 1 + 0.e_1e_2\ldots e_m,\ \ldots\big)$$

Now using the definition of '$x+y$' (i.e. Definition 12.11) and observing that '$e_m + d_m = 9$' and extending these sums to infinite decimals gives the first four individual terms of the sequence whose limit is '$x + -x$' as follows:

$$T(x, 0) + T(-x, 0) = d_0 + (-d_0- 1) = -1$$

If we write '-1' as an infinite decimal, we get: $(-2).999\ldots$

$$T(x, 1) + T(-x, 1) = (d_0 + 0.d_1) + (-d_0- 1 + 0.e_1) = -1 + 0.d_1 + 0.e_1$$
$$= -1 + 0.9$$

Hence, writing '$-1 + 0.9$' as an infinite decimal gives: '$(-1).8999\ldots$'.

$$T(x, 2) + T(-x, 2) = d_0 + 0.d_1d_2 + -d_0- 1 + 0.e_1e_2$$
$$= -1 + 0.d_1d_2 + 0.e_1 e_2$$
$$= -1 + 0.99$$

Hence, writing '$-1 + 0.99$' as an infinite decimal gives: '$(-1).98999\ldots$'.

$$T(x, 3) + T(-x, 3) = d_0 + 0.d_1d_2d_3 + -d_0 - 1 + 0.e_1e_2e_3$$
$$= -1 + 0.d_1d_2d_3 + 0.e_1e_2e_3$$
$$= -1 + 0.999$$

Hence, writing '$-1 + 0.999$' as an infinite decimal gives: '$(-1).998999\ldots$'.

In these equations we can observe that the terms of the sequence have the digit '8' moving to the right so that the limit of this bounded non-decreasing sequence is settling to the infinite decimal '$(-1).999...$' which is the infinite decimal representation of '0'. Consequently, for every general infinite decimal 'x', its additive inverse is '$-x$' (according to Definition 12.12).

In this way, the addition operator for infinite decimals has the Closure, Commutative, Associative and Inverse properties. This explanation concludes the proof of this theorem.

Having now demonstrated that the addition operator for infinite decimals has the same properties as the addition operator for rational numbers, we can show that the multiplication operator for infinite decimals also has the same properties as the equivalent operator for rational numbers. In order to make the proof of this theorem more readable we will first prove that the algorithm in Definition 12.14 leads to some simple inequalities.

Theorem 12.3   Rational Number Inequalities for 'x' and 'x^{-1}'

Let 'x' be a positive infinite decimal and '$T(x, n)$' and '$T(x^{-1}, n)$' be the truncations of 'x' and 'x^{-1}' at the nth place, respectively. Also, let 'q_n' be the infinite decimal equivalent of '$T(x^{-1}, n)$' for 'n' which is a non-negative integer. Then the following inequalities hold true:

1. $x \times q_n < (0).999...$
2. $T(x, n) \times T(x^{-1}, n) < 1$
3. $(T(x, n) + \frac{1}{10^n}) \times (T(x^{-1}, n) + \frac{1}{10^n}) > 1$

Proof
Equation 1:

From Definition 12.14 we already know that 'z_n' is an infinite decimal in the form: '$z_n = (e_0).e_1 e_2 ... e_n 999...$' for an integer '$e_0$' and digits '$e_1, e_2, ... , e_n$' and '$T(x^{-1}, n) = T(z_n, n) = (e_0).e_1 e_2 ... e_n$'. Therefore, the equivalent infinite decimal 'q_n' corresponding to '$T(x^{-1}, n)$' so that '$q_n = (e_0).e_1 e_2 ... e'_n 999...$' where '$e_n$' is not equal to '0' and '$e'_n = e_n - 1$'. However, if '$e_n = 0$' the previous decimal place needs to be adjusted to '$e_{n-1} - 1$' provided 'e_{n-1}' is also not '0'.

Consequently, 'q_n' is the next smallest infinite decimal less than 'z_n' and 'z_n' is the smallest infinite decimal such that '$x \times z_n > (0).999...$', and hence '$q_n$' is the largest decimal of this form such that '$x \times q_n < (0).999...$'. The case of '$x < 0$' can be dealt with using Definition 12.13 for multiplication of negative infinite decimals. This explanation concludes the proof of Equation 1.

Equation 2:

This equation follows on easily from Equation 1 by us first observing that since 'x' is a non-decreasing decimal and so cannot end in a string of '000...' we can conclude that the infinite decimal equivalent of '$T(x, n)$' must be less than 'x'.

Secondly, if we now use this inequality and replace the infinite decimals in Equation 1 with their rational number equivalents and use '$T(x, n)$' instead of 'x', we obtain the rational number inequality '$T(x, n) \times T(x^{-1}, n) < 1$'. As with Equation 1, the case of '$x < 0$' can be dealt with using Definition 12.13 for multiplication of negative infinite decimals. This explanation concludes the proof of Equation 2.

Equation 3:

This final equation now follows on easily from Equations 1 and 2. First, we observe that if we truncate 'x' at the nth place we could be 'chopping off' a string of the form '999...'. Consequently, the infinite decimal equivalent of '$T(x, n) + \frac{1}{10^n}$' must be greater than or equal to 'x'.

Secondly, in the proof of Equation 1 we see that 'z_n' is the infinite decimal equivalent of '$T(x^{-1}, n) + \frac{1}{10^n}$'. Therefore, from the equation '$x \times z_n > (0).999...$' the rational number equivalent equation follows, namely:

$$\text{'}(T(x, n) + \frac{1}{10^n}) \times (T(x^{-1}, n) + \frac{1}{10^n}) > 1\text{'}.$$

Similar to the previous equations, the case of '$x < 0$' can be dealt with using Definition 12.13 for multiplication of negative infinite decimals. This explanation concludes the proof of Equation 3 and the theorem as a whole.

––––––––––––––

The properties of the infinite decimal multiplication operator and the way in which these properties are derived from their rational number equivalent operator are shown in Theorem 12.4 below.

Theorem 12.4   Multiplication is Closed, Commutative, Associative and has an Inverse

If 'x', 'y' and 'z' are infinite decimals, then these decimals satisfy the following equations:

1.   The number '$x \times y$' is an infinite decimal ... Closure Property '$\times$'

2. $x \times y = y \times x$ ... Commutative Property '$\times$'

3. $(x \times y) \times z = x \times (y \times z)$ ... Associative Property '$\times$'

4. $x \times x^{-1} = (0).999...$ ... Inverse Property '$\times$'

Proof

Equation 1:

The Closure Property for '×' follows from the definition of the 'multiplication' operator for infinite decimals in Definition 12.13 and the definition of a 'limit' given in Definition 12.9.

Equation 2:

This equation follows on from Definition 12.13 and involves the same process we used to prove Equation 2 of Theorem 12.2.

Equation 3:

Here we follow the same process as we did in Equation 3 of Theorem 12.2 to prove that infinite decimals are associative under the operation of multiplication.

Equation 4:

In this last equation you will recall that '(0).999...' is the representation of '1' as an infinite decimal. Lastly, we will prove that every infinite decimal '$x \neq 0$' with its multiplicative inverse 'x^{-1}' (according to Definition 12.14) is consistent with the multiplicative inverse for rational numbers. In this way, if the infinite decimal 'x' has multiplicative inverse 'x^{-1}', then we would expect the product of the truncations of these infinite decimals will become progressively closer to '1'.

We assume 'x' is a positive infinite decimal. According to Theorem 12.3:

$$T(x, n) \times T(x^{-1}, n) \; < \; 1$$

and

$$(T(x, n) + \tfrac{1}{10^n}) \times (T(x^{-1}, n) + \tfrac{1}{10^n}) \; > \; 1$$

The first of these inequalities can be rearranged as:

$$1 - T(x, n) \times T(x^{-1}, n) > 0$$

If we multiply the second of these first two inequalities above we obtain:

$$(T(x, n) + \tfrac{1}{10^n}) \times (T(x^{-1}, n) + \tfrac{1}{10^n})$$

$$= \; T(x, n) \times T(x^{-1}, n) + \tfrac{1}{10^n} \times T(x^{-1}, n) + T(x, n) \times$$

$$\tfrac{1}{10^n} + \tfrac{1}{10^n} \times \tfrac{1}{10^n}$$

$$= \; T(x, n) \times T(x^{-1}, n) + (T(x, n) + T(x^{-1}, n) + \tfrac{1}{10^n}) \tfrac{1}{10^n} > 1$$

$$\therefore (T(x, n) + T(x^{-1}, n) + \tfrac{1}{10^n}) \tfrac{1}{10^n} > 1 - T(x, n) \times T(x^{-1}, n) > 0$$

Since the infinite decimals 'x' and 'x^{-1}' are bounded, there are two inequalities: '$T(x, 0) + 1 > T(x, n)$' and '$T(x^{-1}, 0) + 1 > T(x^{-1}, n)$' for all values of '$n$'. Likewise, the inequality '$1 > \frac{1}{10^n}$' holds for all values of '$n > 0$'. Therefore, by using these three inequalities, we can simplify the previous long inequality to:

$$(T(x, 0) + T(x^{-1}, 0) + 3)\frac{1}{10^n} > 1 - T(x, n) \times T(x^{-1}, n) > 0$$

Let 'm' be a fixed integer such that '$10^m > (T(x, 0) + T(x^{-1}, 0) + 3)$'. We can then substitute into the above inequality to obtain:

$$\frac{1}{10^{(n-m)}} > 1 - T(x, n) \times T(x^{-1}, n) > 0 \qquad \text{... for fixed '}m\text{'}$$

Rearranging this inequality gives:

$$0 < 1 - T(x, n) \times T(x^{-1}, n) < \frac{1}{10^{(n-m)}}$$

Since 'n' can be made arbitrarily larger than 'm' (which is fixed), this implies the term '$T(x, n) \times T(x^{-1}, n)$' can be made as close to the value '1' as we choose. This, in turn, implies that 'x^{-1}' is the inverse of 'x'.

In this way, the multiplication operator for infinite decimals satisfies the Closure, Commutative, Associated and Inverse properties and the theorem is proved.

Having now established the Closure, Commutative, Associative and Identity Theorems for infinite decimals, the next theorem to prove is that infinite decimals also have the Distributive property.

Theorem 12.5   Distributive Property for Infinite Decimals

If 'x', 'y' and 'z' are infinite decimal variables, then these decimals satisfy the following equation:

$$x \times (y + z) = x \times y + x \times z$$

This equation is called the Distributive property for infinite decimals.

Proof

In the case where 'x', 'y' and 'z' are positive infinite decimals, on the right-hand side of the above equality the infinite decimal is given by '$x \times y + x \times z$'. Let the terms of the infinite sequence '$w_0, w_1, w_2, ...$' be given by 'w_n', where 'w_n' is the infinite decimal equivalent of the finite decimals '$T(x, n)T(y, n) + T(x, n)T(z, n)$'.

From our definitions of addition and multiplication of infinite decimals, the limit of the infinite decimal sequence 'w_n' is the infinite decimal given by '$x \times y + x \times z$'. However, from the properties of finite decimals we have:

$$T(x, n)T(y, n) + T(x, n)T(z, n) = T(x, n)(T(y, n) + T(z, n))$$

In this way, the terms of the infinite sequence '$w_0, w_1, w_2, ...$' also have the infinite decimal term '$x \times (y + z)$' as a limit and hence by the definition of the '$=$' sign, the distributive property holds for infinite decimals. So far in this theorem we have assumed that 'x', 'y' and 'z' are positive infinite decimals. However, cases where 'x', 'y' and 'z' are not positive can be dealt with in a straightforward manner by using Definition 12.13 for the multiplication of negative infinite decimals.

Consequently, we have proved that the infinite decimals have the Distributive property under the operations of addition and multiplication.

The last set of properties of infinite decimals we will endeavour to prove also follows from their relationship to rational numbers. These are the set of logical properties known as the Axioms of Logic for Rational Numbers.

Theorem 12.6   Logical Properties for Infinite Decimals

If 'x', 'y' and 'z' are infinite decimals, then these decimals have the following properties:

1. If '$x = y$' then '$y = x$' ... Symmetry property
2. If '$x = y$' and '$y = z$' then '$x = z$' ... Transitivity property
3. If '$x = y$' and '$z = w$' then '$x + z = y + w$' ... Equality of Addition
4. If '$x = y$' and '$z = w$' then '$x \times z = y \times w$' ... Equality of Products

Proof

We will show that each of these properties follows in a straightforward way from the equivalent properties of the Rational Number Subsystem of the Real Number System.

Property 1:

According to Definition 12.3, if '$x = y$', then the rational number arithmetic equations '$T(x, n) = T(y, n)$' are true for all non-negative integer values of 'n'. By the symmetry of rational numbers it follows that '$T(y, n) = T(x, n)$' for non-negative values of 'n'. Therefore, by definition it follows that '$y = x$'.

Property 2:

According to Definition 12.3 and the Transitive Axiom of Rational Numbers, if '$x = y$' and '$y = z$', then the rational number arithmetic equations '$T(x,n) = T(y,n)$' and '$T(y, n) = T(z, n)$' are true for all non-negative integer values of 'n'. By the Transitive Axiom of Rational Numbers it follows that '$T(x, n) = T(z, n)$' are true for all non-negative integer values of 'n'. Hence, by Definition 12.3 the equation '$x = z$' must be true and the infinite decimals also have the Transitivity property.

Properties 3 and 4:

Properties 3 and 4 follow the same approach as the previous properties and rely on the rational number equations: '$T(x, n) + T(z, n) = T(y, n) + T(w, n)$' and the equations: '$T(x, n) \times T(z, n) = T(y, n) \times T(w, n)$' being true when the rational number equations: '$T(x, n) = T(y, n)$' and '$T(z, n) = T(w, n)$' are true.

In this way, it is apparent that the logical properties of the infinite decimals are true directly from the logical properties of the equivalent Rational Number System Axioms of Logic.

The proof of these logical properties of infinite decimals concludes the Theory Section for this chapter. We have now demonstrated that the Real Number System has the properties that allow us to call it a complete ordered field. Our approach to building the Real Number System – which includes rational numbers as a subsystem and the irrational numbers as a subset – clearly illustrated that the Real Number System is a natural extension of our model of the Rational Number System.

At this point we have completed the main goal of this chapter and one of our major objectives for this book; namely that we can model the 'complete' number line – which we can now call the Real Number Line – by assigning real numbers as coordinates to every point along this number line.

Applications of Real Numbers using Infinite Decimals

One of the primary uses of the Real Number System is in labeling points along the number line with real numbers. This direct correlation between points and real numbers allows us to represent physical variables in the world of science, economics, etc. For example, most variables in our physical theories are constructed from the physical quantities of mass, length, time and electric charge which are represented by real numbers. In this section we will explore a few applications of this theory to the physical world around us from the vast number of possible applications of the Real Number System.

To make our lives simpler, mathematicians have agreed to use special symbols for the most common irrational numbers. We have already encountered these symbols in labelling the irrationals generated by taking the square root of certain integers, for example, '$\sqrt{2}$', '$\sqrt{3}$', '$\sqrt{5}$', '$\sqrt{6}$', etc. The other two most common irrational numbers are pi (which is denoted by 'π') and Euler's number (which is denoted by 'e'). Although the constant 'π' was known to Ancient Greeks, the constant 'e' wasn't discovered until the 17th century. Jacob Bernoulli discovered the constant 'e' when he was searching for a solution to a problem in compound interest. Mathematically 'e' is the ubiquitous constant like pi and is defined by the following formula:

$$e = \lim_{n \to \infty} (1 + \tfrac{1}{n})^n = 2.71821...$$

As with the representation of numbers in the Rational Number System, it is common practice to write terms and equations in the Real Number System with a mixture of number forms. For example, we can write the algebraic equation '$x + 2\sqrt{3} = \sqrt{3}$' and manipulate its terms as though the numbers are rational numbers and there are no other considerations. It is only when we want to find the infinite decimal expression of 'x' that we have to substitute the infinite decimal that '$\sqrt{3}$' represents. We will take advantage of this convenience of notation in the following applications.

Application 12.1

Write the following rational numbers in infinite decimal format: (a) '$2\frac{1}{2}$', (b) '$-3\frac{3}{4}$', (c) '$\frac{10}{3}$', (d) '$-\frac{7}{3}$', (e) '$5\frac{21}{99}$', (f) '-1.478', (g) '17.99', (h) '-0.1', (i) '$\frac{\frac{1}{99}}{13}$', (j) '$\frac{3\frac{1}{2}}{5\frac{1}{4}}$'.

Using Definition 12.1, we can write each rational number in infinite decimal format as follows:

(a) '$2\frac{1}{2} = 2.5$' as an infinite decimal is '$(2).4999...$'.

(b) '$-3\frac{3}{4} = -3 - \frac{3}{4} = -3 - 1 + (1 - \frac{3}{4}) = -4 + \frac{1}{4} = -4 + 0.25$' as an infinite decimal is '$(-4).24999...$'.

(c) '$\frac{10}{3} = 3 + \frac{1}{3} = 3 + 0.333...$' as an infinite decimal is '$(3).333...$'.

(d) '$-\frac{7}{3} = -2 - \frac{1}{3} = -2 - 1 + (1 - \frac{1}{3}) = -3 + \frac{2}{3}$' as an infinite decimal is '$(-3).666...$'.

(e) '$5\frac{21}{99} = 5 + 0.212121...$' as an infinite decimal is '$(5).212121...$'.

(f) '$-1.478 = -1 - 1 + (1 - 0.478) = -2 + 0.522$' as an infinite decimal is '$(-2).521999...$'.

(g) '$17.99 = 17 + 0.99$' as an infinite decimal is '$(17).98999...$'.

(h) '$-0.1 = -1 + 1 - 0.1 = -1 + 0.9$' as an infinite decimal is '$(-1).8999...$'.

(i) '$\dfrac{1}{\frac{99}{13}} = \dfrac{13}{99} = 0.131313...$' as an infinite decimal is '$(0).131313...$'.

(j) '$\dfrac{3\frac{1}{2}}{5\frac{1}{4}} = \dfrac{7}{2} \times \dfrac{4}{21} = \dfrac{28}{42} = \dfrac{2}{3} = 0.666...$' as an infinite decimal is '$(0).666...$'.

Application 12.2

Given 'x' is the infinite decimal '$x = (3).2148569...$', then we can:

(a) Write out the first five partial sums of 'x' using the truncation process.

(b) Demonstrate that this is a non-decreasing sequence of partial sums.

(c) Express the digit in the seventh place after the decimal point in terms of truncations.

The solutions to these three questions are:

(a) The first five partial sums of 'x' are: '$T(x, 0) = 3$', '$T(x, 1) = 3.2$', '$T(x, 2) = 3.21$', '$T(x, 3) = 3.214$' and '$T(x, 4) = 3.2148$'.

(b) For the partial sums to be non-decreasing we require: '$T(x, 0) \leq T(x, 1) \leq T(x, 2) \leq T(x, 3) \leq T(x, 4)$'. Substituting for these truncations gives '$3 \leq 3.2 \leq 3.21 \leq 3.214 \leq 3.2148$'. From the order relation for rational number we know these multiple inequality relations are all true statements.

(c) The digit in the seventh place after the decimal point is given by '$T(x, 7) - T(x, 6) = 9$'.

Application 12.3

Find the additive inverses of the infinite decimals representing the rational numbers:

(a) '$2\frac{1}{2}$', (b) '$-3\frac{3}{4}$', (c) '$\frac{10}{3}$', (d) '$-\frac{7}{3}$', (e) '$5\frac{21}{99}$' from Application 6.1 above.

The derivation of these additive inverses is given in the following statements:

(a) Given that '$2\frac{1}{2} = 2.5$' is the infinite decimal '(2).4999...', its additive inverse is obtained by applying the inverse algorithm in Definition 12.11 which gives: '$-2.5 = -2 + -1 + 1 + -0.5 = -3 + 0.5 = (-3).5000...$'. This can be written as an infinite decimals as '$(-3).4999...$'. To check this result, when we add these two numbers together we obtain '$(-1).999...$' which is '0' as an infinite decimal. This is the expected result.

(b) Given that '$-3\frac{3}{4}$' is the infinite decimal '$(-4).24999...$', we can generate its additive inverse as '$(-(-4+1)).75000...$' which as an infinite decimal is '$(3).74999...$'. Once again, checking this result by adding gives '$(-1).999...$'.

(c) Given that '$\frac{10}{3}$' is the infinite decimal '$(3).333...$', we can generate its additive inverse as '$(-(3+1)).666...$' which is the infinite decimal '$(-4).666...$'. A quick check of this result by adding gives '$(-1).999...$' which is the outcome we expected.

(d) Given that '$-\frac{7}{3}$' is the infinite decimal '$(-3).666...$' we can generate its inverse as '$(-(-3+1)).333...$' which is the infinite decimal '$(2).333...$'. Once again, a quick check gives the expected result of '$(-1).999...$'.

(e) Given that '$5\frac{21}{99}$' is the infinite decimal '$(5).212121...$' we can generate its additive inverse as '$(-(5+1)).787878...$' which is the infinite decimal '$(-6).787878...$'. Another check gives the addition of these two number as '$(-1).999...$' which is the expected result.

In Application 12.4 we will perform the equivalent exercise for multiplication, that is, we will find the multiplicative inverses of the first infinite decimals in Application 12.1. As these examples are rational numbers we will be able to find their multiplicative inverses as rational numbers and check that their infinite decimal equivalent satisfies Definition 12.13.

Application 12.4

Find the multiplicative inverses of the given infinite decimals and verify that they are multiplicative inverses to the fourth decimal place. The given numbers are: (a) '$2\frac{1}{2}$', (b) '$-3\frac{3}{4}$', (c) '3', (d) '$\sqrt{2} = 1.414213...$'.

(a) Given that '$2\frac{1}{2} = 2.5$' is the infinite decimal '$(2).4999...$' its multiplicative inverse is '$\frac{2}{5} = 0.4$' as a rational number which has the infinite decimal equivalent of '(0).3999...'. Hence, our goal is to check that '$T((0).3999..., 4)$' as an infinite decimal '$z_4 = (0).3998999...$' is the largest decimal such that:

$(2).4999... \times z_4 < (0).999...$ ... by objective

$(2).4999... \times (0).3998999... = 0.99974999...$ ... by Definition 12.14

∴ $(2).4999... \times z_4 < 0.999...$ ... by simplifying terms

We observe that adding '1' to the fourth place of 'z_4' gives '(0).3999...' and the inequality no longer holds as '$(2).4999... \times (0).3999... = (0).99$ $9...\not< 1$'. Therefore, '(0).3998999...' is the largest infinite decimal at the fourth place which satisfies the condition of being a multiplicative inverse as required by Definition 12.14.

Note:

Adding '1' to the nth decimal place of the truncation of the infinite decimal '(0).3999...', i.e. to '$T((0).3999..., n)$', equals '0.4' and we get the equivalent rational equation, '$\frac{5}{2} \times 0.4 = 1$', hence '$\frac{5}{2} \times 0.4 \not< 1$'.

(b) Given that '$-3\frac{3}{4}$', which is the infinite decimal '$(-4).24999...$', we can apply Definition 12.14 and find the multiplicative inverse of '$3\frac{3}{4}$' or '(3).74999...' as an infinite decimal. The multiplicative inverse of '$\frac{15}{4}$' is the fraction '$\frac{4}{15} = 0.2666...$' as a rational number which has the infinite decimal equivalent of '(0).2666...'. Hence, our goal is to check that '(0).2666...' is the multiplicative inverse of '(3).74999...' at the fourth decimal place and that it satisfies Definition 12.14. That is, the infinite decimal equivalent of '$T((0).2666..., 4)$' (namely '(0).2665999...') is the largest infinite decimal to four decimal places which satisfies the following:

$(3).74999... \times (0).2665999... < 0.999...$ ... by given question

But: $(3).74999... \times (0).2665999... = (0).99974999... < (0).999...$

                                       ... by multiplying

The next largest product of infinite decimal given by incrementing the fourth place is given by:

$(3).74999... \times (0).2666999... = (1).000124999...$

                                       ... by multiplying

Therefore, the infinite decimal equivalent of '(0).2665999...' is the largest infinite decimal at the fourth place which satisfies the condition of being a multiplicative inverse of '$-3\frac{3}{4}$' as required by Definition 12.13. We take the negative of '(0).2665999...' and get '(−1).7333999...' as the multiplicative inverse of '(−4).24999...' to the fourth decimal place.

(c) Given that '3' is the infinite decimal '(2).999...' we can apply Definition 12.14 to find the multiplicative inverse as an infinite decimal. The multiplicative inverse of '3' is '$\frac{1}{3}$' or '(0).333...' as an infinite decimal. So our goal is to check that '(0).333...' is the multiplicative inverse of '(2).999...' at the fourth decimal place by satisfying the conditions of Definition 12.14. That is, the infinite decimal equivalent of '$T((0).333... , 4)$', is given by the infinite decimal '(0).3332999...' and is the largest decimal at the fourth decimal place, satisfying the following:

$$(2).999... \times (0).3332999... < 0.999...$$

... by given question

But: $(2).999... \times (0).3332999... = (0).9998999...$

... by multiplication

And: $(2).999... \times (0).3333999... = (1).0001999...$

... by multiplication

Once again, '0.3332999...' is the largest decimal at the fourth place which satisfies the condition of being a multiplicative inverse of '3' as required by Definition 12.14.

(d) Given that '$\sqrt{2}$' is the infinite decimal '(1).4142135...', this implies that '$\sqrt{2}$' lies between '1.4142135' and '1.4142136'. From the calculator we see that the multiplicative inverse is '$(\sqrt{2})^{-1} = (0).7071...$'. We apply Definition 12.14 to determine whether this number satisfies the condition of being a multiplicative inverse to the first four decimal places. We will prove this result by using the equivalent infinite decimal of '$T(\sqrt{2}, 7) = 1.4142135$' – namely, '1.4142134999...' – and multiplying by both of the following infinite decimals: '(0).7070999...' and '(0).7071999...'.

$$(1).4142134999... \times (0).7070999... = (0).99999036584999...$$

$$(1).4142134999... \times (0).7071999... = (1).0001317871999...$$

Therefore, '0.7071' is the largest decimal at the fourth place whose infinite decimal equivalent (i.e. '(0).7070999...') when multiplied by '$\sqrt{2}$' is less than '1' and so qualifies as the multiplicative inverse of '$\sqrt{2}$' to four decimal places.

These applications reinforce most of the key concepts about real numbers using infinite decimals. This explanation completes the Application Section of this chapter.

Summary of Real Numbers using Infinite Decimals

In this chapter we have demonstrated the complete connection between the location of a point on the number line and its representation as an infinite decimal. This simple property allows us to extend the Rational Number System using Infinite Decimals.

At this point, we have covered the objectives of this chapter and hence can summarise our findings up to now. The most important definitions we have introduced in this chapter are:

1. An **infinite decimal** 'x' in decimal format is a string of symbols that take the form '$(d_0).d_1 d_2 ... d_n ...$' where:

    -   'd_0' is any integer and '(d_0)' indicates it is an infinite decimal.
    -   the symbols '$d_1, d_2, ..., d_n, ...$' are assigned values from the set of digits.
    -   If '$(d_0).d_1 d_2 ... d_n 000...$' ends in an infinite string of '0', then it can be written as '$(d_0).d_1 d_2 ... d'_n 999...$' where '$d'_n = d_n - 1$'.

2. The infinite decimals 'x' and 'y' are **ordered** by '$<$' so that '$x < y$', if (and only if) there exists a non-negative integer 'n' such that '$T(x, n) < T(y, n)$'.

3. The infinite non-decreasing sequence of infinite decimals, $(x_0, x_1, x_2, x_3, ...)$ is **bounded** by the infinite decimal 'y' if (and only if) 'y' satisfies the order relation '$x_n \leq y$' for all infinite decimals 'x_n' in the sequence.

4. The infinite decimal 'y' is the **limit** of the bounded sequence of infinite decimals $(x_1, x_2, x_3, ...)$ based on the infinite decimal equivalent of the truncations of this sequence given by $(T(x_0, n), T(x_1, n), T(x_2, n), T(x_3, n), ...)$. We say that '$y$' is the limit of $(x_1, x_2, x_3, ...)$ if (and only if) the finite decimal '$T(y, n)$' is the constant decimal that the infinite decimal equivalent of the sequence of truncations must become after a sufficient number of terms of the sequence are considered for each value of 'n'.

5. The set of infinite decimals is called **complete** if for every bounded sequence of infinite decimals, $(x_0, x_1, x_2, x_3, ...)$ there exists an infinite decimal 'y' such that 'y' is the limit of the sequence $(x_0, x_1, x_2, x_3, ...)$.

6. We define **addition** of two infinite decimals 'x' and 'y' (denoted by '$x + y$') as the limit of the bounded non-decreasing sequence $(z_1, z_2, z_3, \ldots, z_n, \ldots)$ where 'z_n' is the infinite decimal equivalent of the rational number '$T(x, n)$ $+ T(y, n)$'.

7. We define the **additive inverse** of an infinite decimal 'x', denoted by '$-x$', as the non-decreasing infinite decimal '$-x = (e_0).e_1e_2 \ldots e_m \ldots$', where the digits '$e_m$' are chosen so that '$x + -x = (-1).999\ldots$'.

8. We define **multiplication** of two positive infinite decimals 'x' and 'y' (denoted by '$x \times y$') as the limit of the bounded non-decreasing sequence $(z_1, z_2, z_3, \ldots, z_n, \ldots)$ where 'z_n' is the infinite decimal equivalent of the rational number '$T(x, n) \times T(y, n)$'. For negative infinite decimals, we define the remaining three cases as follows:

    (i) $-x \times y = -(x \times y)$

    (ii) $x \times -y = -(x \times y)$

    (iii) $-x \times -y = x \times y$.

9. We define the **multiplicative inverse** of an infinite decimal 'x', denoted by 'x^{-1}', where '$x \neq (-1).999\ldots$', as the infinite decimal '$x^{-1} = (e_0).e_1e_2 \ldots e_m \ldots$', where the digits '$e_m$' can be chosen so that '$x \times x^{-1} = (0).999\ldots$'.

10. We define the **Real Number System using Infinite Decimals** as the set of infinite decimals with its associated operations of '$+$' and '$\times$' and the corresponding properties of the Rational Number System, as well as the definitions of the inverses of the operations '$+$' and '$\times$', and the order symbol '$<$'. The set of **Real Numbers** is denoted by the symbol '$\mathbb{R}$'.

CONCLUSION

Summary of the MSSM Program

The MSSM (Mastering Secondary School Mathematics) Program (on which this textbook is based) involves the common-sense approach of starting with the simplest and most familiar area of Mathematics – namely counting. As we have observed throughout this book, we begin this process of mastering Mathematics by giving meaning to the alphabet of 'counting' symbols using the LEAD sequence of steps (**L**anguage to **E**lementary Properties to **A**xioms to **D**efinitions) to develop the Theorems of Natural Number Arithmetic.

To gain a useful visual representation of these Natural Numbers, we marked out equally-spaced intervals with vertical markers along a straight line. Starting with the number '0' at the left-most marker, we then proceeded to add the number '1', '2', '3', etc above these markers by proceeding from left to right, respectively. In other words, we discovered a simple way of assigning numbers to equally-spaced points along the line (called the number line).

Subsequently, we extended the most elementary number system – the Natural Number System – to the next simplest number system which is the Integer Number System. We achieved this by adding one more symbol and one more elementary property to give this symbol meaning in relation to the symbols of the Natural Number System. In this way, we built on our knowledge developed with our work on the Natural Number System and effectively transferred it to the Integer Number System. Similarly, we were also able to produce a visual representation of the Integer Numbers by extending the Natural Number Line. This process involved plotting the negative integers to the left of the '0' symbol in the same symmetrical way in which we had previously laid out the numbers to the right of '0' in the Natural Number System.

Our next step was to extend the Integer Number System to produce the Rational Number System by adding numbers of the form '$\frac{a}{b}$' (called fractions) where 'a' and 'b' are integers and 'b' is not equal to '0'. Once again, the Rational Number System is derived from the knowledge we developed during our investigations of the Integer Number System but now we can add an unlimited number of new symbols for more points along the number line.

Having studied these three number systems, we are able to label all of those points along the number line which have rational numbers as coordinates. However, in 600 B.C. the Ancient Greeks proved that there were points along this number line which couldn't be labelled by a rational number. These points (one of which was labelled by '$\sqrt{2}$') are said to have coordinates that are irrational numbers.

The Ancient Greeks were unable to formalise a number system that could incorporate irrational numbers, so they focused their attention on a geometrical approach to describing numbers and physical properties of the world around them. It took another 2,000 years before mathematicians invented the modern definition of a 'limit' and mastered the concept of 'infinity'. This, in turn, laid the foundations for treating irrational numbers as the 'limit of certain sequences' or as the 'limit of infinite decimal expansions'. By incorporating these irrational numbers into the Rational Number System, we arrive at the last of the elementary number systems – the Real Number System.

With the Real Number System at our disposal, we now have a number system that lets us truly model the physical world around us. At this stage we can now declare that to every physical point along a straight line we can assign a unique real number and to every real number we can identify a unique physical point along a line. By implication, we can further declare that using this Real Number System we can model distance, time, mass, temperature and most other physical quantities.

Although it doesn't do it justice, if you had to summarise the outcome of this textbook, you could say that it encompasses the 'Mathematics of a Line'.

The MSSM Program – Unlocking the 'Secrets' of Other Areas of Secondary School Mathematics

Exercises for consolidating the concepts and results of this book are provided in the form of tutorials that ensure the correct learning outcomes of the MSSM Program are being achieved. These tutorials also ensure the connection between the mathematical learnings and exercises at school are being related to, and reinforced by, the MSSM Program. The overall purpose is to facilitate students to deconstruct their mathematical learning and build the best foundation for mastering Mathematics. These tutorials are available from WCMPT Pty Ltd – a Brisbane-based company which specialises in providing world-class Mathematics and Physics tutoring through the MSSM Program.

The value of the MSSM Program presented in this book is highlighted in follow-on extension topics provided by WCMPT Pty Ltd as summarised in the table on the following page.

No.	Topic	Summary of Topic
1.	Euclidean Geometry	Mathematics in the plane (or flat surface)
2.	Coordinate Geometry	Mathematics of coordinates of points in the plane
3.	Trigonometry	Mathematics of angles in the plane
4.	Algebra	Mathematics of algebraic terms and equations
5.	Kinematics	Mathematics of moving points in the plane (Laws of Motion)
6.	Newtonian Mechanics	Mathematics of Newton's Laws of Motion
7.	Pre-calculus	Prerequisites for Calculus
8.	Software packages	Modern scientific calculators (mathPad and Wolfram) Email Word processing Spreadsheet Presentation Graphics Notes.
9.	Meaningful Access to Online Content	Foundation for Learning from Online Content

1. Euclidean Geometry

Euclidean Geometry is the mathematics of the plane and 3-dimenional space described in terms of points, lines, circles and angles.

Using the LEAD Approach, the MSSM Program outlined in this textbook provides a seamless transition to understanding Elementary Geometry as described in the most famous and invaluable of all mathematics books: *The Elements* by Euclid (300B.C.).

2. Coordinate Geometry

Coordinate Geometry is an alternate way of expressing the geometrical relation-ship between points, lines and angles in a plane or 3-dimensionsal space using a coordinate system made up of number lines (usually placed at right angles for simplicity).

The MSSM Program provides a simple but comprehensive Introduction to Coordinate Geometry based on the LEAD Approach to learning Mathematics. This approach is based on work done in the first extension topic – Euclidean Geometry.

3. Trigonometry

The branch of Mathematics called Trigonometry focuses on the relationships between lengths and angles of triangles in the plane. This topic incorporates a specialised combination of results from Euclidean Geometry and Coordinate Geometry which were originally developed to study the solar system and the universe beyond, but is now key to the study of the Sciences and Engineering.

The MSSM Program provides a simple but comprehensive Introduction to Trigonometry which builds on the topics covered so far.

4. Algebra

We identify Algebra with the basic Algebra learnt in primary and secondary school for solving simple algebraic equations. A working knowledge of Algebra is essential for understanding the sciences, engineering, economics and medicine. In Chapter Seven of this book, we have laid out the solid foundations of Algebra.

The MSSM Program provides the simple and natural extension to Algebra which is required to develop the student's knowledge to achieve mastery of Secondary School Mathematics Algebra.

5. Kinematics

The branch of Mathematics called Kinematics studies the motions of points (or 'particles', 'bodies' or 'objects') in the plane or in 3-dimensional space by ignoring their 'masses' and the 'forces' causing their motion. This is the major logical step on the path to learning Elementary Physics which has as its foundation the principles of Newtonian Mechanics.

The MSSM Program provides a simple but comprehensive Introduction to Kinematics which builds on all previous topics covered.

6. Newtonian Mechanics

Newtonian mechanics (also known as Classical Mechanics) is the study of the motion of bodies, thus making it one of the largest subjects in science, engineering and technology. Newtonian Mechanics describes the motion of physical objects from projectiles to mechanical and terrestrial systems, as well as astronomical objects, such as asteroids, planets and stars.

Newtonian mechanics also requires the concept of a system (i.e. mechanical system, terrestrial system, solar system, galaxy, etc) and describes the motion of objects such as projectiles and mechanical systems when part of a larger system. This lays the foundations for the invaluable skills of systemic thinking and root-cause analysis.

The axiomatic approach by Newton is consistent with the LEAD Approach adopted by the MSSM Program. WCMPT provides a 27-chapter Introduction to Newtonian Mechanics which is suitable for secondary school mathematics and physics students.

7. Pre-calculus

A Pre-calculus course involves studying those areas of Mathematics which are prerequisites to comprehending a first course in Calculus. As Calculus is the most important and central part of Mathematics (due to the enormous number of applications it relates to in the physical world around us), it is clear that developing mastery in pre-calculus is critical to a student deriving the full benefits of an education in Mathematics.

The MSSM Program offers the key insights required to understand the Pre-calculus topics which are essential to understanding the full significance and methods of Calculus.

8. Software Packages

In researching and writing this textbook, a standard list of mathematics-related application software was used to develop and present the MSSM Program (as listed in the Table above under the 'Summary of Topics' column).

As applications are being developed, improved and providing better value for money, the list of applications mentioned above may no longer be relevant for all schools and students. WCMPT will adapt to the most common applications available to students at any particular time period.

The first application in this list deserves special mention as it is a modern version of a scientific calculator (but it is significantly more powerful). The combination of MathPad and Wolfram Alpha applications will allow a student to solve nearly every mathematical equation they encounter in Secondary School Mathematics, once it has been correctly formulated and manipulated into a standard equation.

The MSSM Program insists that every solution to a problem be validated by some method and that modern applications are extremely valuable tools in making this step easy and instructive. A subset of skills required for developing Mastery in

Secondary School Mathematics will involve the mastery of the type of applications listed in the table above.

9. Meaningful Access to Online Content

There is a very large volume of content available in online mathematics sites. The challenge for students to overcome is the frustration of conflicting terminology, non-standard notation, relevance and trying to learn by 'doing'. This process of 'learning by doing' involves attempting to gain an understanding of Mathematics by carrying out a 'recipe' or executing a sequence of steps (i.e. an algorithm) to obtain the right answer. This clearly doesn't work as a stand-alone strategy.

The LEAD Approach to learning was specifically designed to create a more efficient style of learning Mathematics. This approach provides the foundation of the basic concepts in Secondary School Mathematics and hence creates a significant amount of the mathematical context required to gain meaningful access to invaluable online content.

Objectives of the MSSM Program

A key objective of the MSSM Program is to bridge the gap between this approach and the one adopted by a syllabus-based mathematics program so that students can use their existing texts to fully develop their problem-solving capabilities. If students are able to convert conceptual reasoning into real-world problem solving they will be rewarded with a positive feedback loop. This, in turn, will build students' basic skills, excitement, motivation and a confident attitude towards life-long learning.

What are corporate businesses and government organisations looking for in graduate students these days? The business world wants creative conceptual thinkers who can view the world systemically, strive for high measurable performance levels and focus their intuition on available opportunities – and then follow a process of implementing solutions which deliver real-world benefits.

Some philosophies of learning implicitly have the view: 'You either have it or you don't'. In contrast, the MSSM Program espouses the philosophy: 'You develop it and thrive, or you don't and just try to survive!'

REFERENCES

During the process of researching and writing this textbook, I referred to the following books (which I also recommend as being useful resources for every secondary-school student's personal Mathematics library):

Dunham, William (1990), *Journey Through Genius: The Great Theorems of Mathematics*, Wiley.

Gowers, Timothy (2002), *Mathematics: A Very Short Introduction*, Oxford University Press.

Heath, Sir Thomas L. (translator), *Euclid: The Thirteen Books of The Elements* (Vols. 1 & 2), Dover.

Newton, Sir Isaac (1995 edition), *The Principia: The Mathematical Principles of Natural Philosophy*, Prometheus.

ABOUT THE AUTHOR

After completing high school, Paul McNamara won a university scholarship to study a science degree and pursue his interest in Pure and Applied Mathematics (as well as Physics). By the third year of his degree, he was gaining good results in the majority of his subjects and so was promoted to the Honours program in Physics.

After leaving university, Paul pursued a career in the emerging field of computing. He became a computer technician – applying basic logic and physics to manufacturing and repairing of computer hardware. After working for two years in manufacturing, Paul realised that in order to create a successful career in computing he required an electrical engineering degree (or equivalent). So he returned to university as a mature-age student. During this period, Paul worked as a university tutor and was also accepted into the Masters of Pure Mathematics program.

At this stage, Paul decided to follow his passion for understanding the Theory of Relativity and switched faculties to enrol in a Master of Physics degree. Within six months, the Physics faculty decided to upgrade his course to a PhD program which led to Paul spending five years tutoring in Physics and gaining his PhD degree (with a thesis in General Relativity). Following this seven year detour, Paul then returned to his previous career in computing.

After spending another seven years working full-time in the computer and business fields, Paul made the decision to branch out and pursue consultant/contractor projects. Since then, he has worked as a project manager, a program manager and a consultant.

Nevertheless, throughout his whole career, Paul's passion for Mathematics and Physics has never wavered; his hobby has been to continue to improve his understanding of their foundations.

Paul currently lives in Brisbane with his wife, and has two children. He spends his leisure time catching up with family and friends, and enjoying his sporting passion – tennis.